Thomas Winking

2006

D0622191

Euler Through Time:
A New Look
at Old Themes

Euler Through Time:
A New Look at Old Themes

V. S. Varadarajan

AMS
AMERICAN MATHEMATICAL SOCIETY
www.ams.org

2000 *Mathematics Subject Classification.* Primary 01A70; Secondary 01A50, 11-03, 40-03.

For additional information and updates on this book, visit
www.ams.org/bookpages/euler

Library of Congress Cataloging-in-Publication Data
Varadarajan, V. S.
  Euler through time : a new look at old themes / V. S. Varadarajan.
    p. cm.
  Includes bibliographical references.
  ISBN 0-8218-3580-7 (acid-free paper)
  1. Mathematics–History–18th century. 2. Number theory–History–18th century. 3. Sequences
(Mathematics)–History–18th century. 4. Mathematics–History–19th century. 5. Number theory–
History–19th century. 6. Sequences (Mathematics)–History–19th century. 7. Euler, Leonhard,
1707–1783. I. Title.

QA24.V37   2006
510—dc22

                                                                              2005057177

    Copying and reprinting. Individual readers of this publication, and nonprofit libraries
acting for them, are permitted to make fair use of the material, such as to copy a chapter for use
in teaching or research. Permission is granted to quote brief passages from this publication in
reviews, provided the customary acknowledgment of the source is given.
    Republication, systematic copying, or multiple reproduction of any material in this publication
is permitted only under license from the American Mathematical Society. Requests for such
permission should be addressed to the Acquisitions Department, American Mathematical Society,
201 Charles Street, Providence, Rhode Island 02904-2294, USA. Requests can also be made by
e-mail to reprint-permission@ams.org.

© 2006 by the American Mathematical Society. All rights reserved.
The American Mathematical Society retains all rights
except those granted to the United States Government.
Printed in the United States of America.

∞ The paper used in this book is acid-free and falls within the guidelines
established to ensure permanence and durability.
Visit the AMS home page at http://www.ams.org/

10 9 8 7 6 5 4 3 2 1   11 10 09 08 07 06

Contents

Preface

The origins of this book go back to a course on the history of mathematics that I gave at UCLA in the winter quarter of 2001. In most universities such a course follows a standard curriculum that starts with the Babylonians and works its way to more recent topics. I decided to do it differently and focus attention on the work of a single great figure. I chose Euler, attracted both by his universality and the great relevance of his work to what should be, if not what is, the undergraduate program in mathematics today. I discussed mostly Euler's work on infinite series and products, and was guided by the chapter on Euler in A. Weil's beautiful book *Number Theory : An Approach through History from Hammurapi to Legendre* that had been published in 1984. The result, as far as I was concerned, was predictable. I fell under the magical spell of Euler's personality and mathematics. It was a huge personal discovery for me to learn how alive Euler's themes still are. I started writing this book after the course, having been encouraged by my friends that it would be a good thing to try to do.

Before I get to Euler I really want to emphasize an important point. I believe that writing historically about mathematics should not be limited to questions like who did what when and to whom he/she wrote about it. I am of the opinion that the history of mathematics and mathematicians should go beyond these concerns, however legitimate they are. I feel that a complete mathematical history should pay great attention to the historical evolution of ideas and how they mesh with what we know and are interested in today. To modify a famous remark, the history of mathematics is too important to be left entirely to historians. No one has done this type of historical writing more brilliantly than A. Weil, as one can tell by his historical memoirs that are scattered throughout his *Oeuvres Scientifiques.* I have tried to follow his example; how far I have succeeded can be judged only by the readers.

No single person or book can describe completely the many-sided genius of Euler or his sunny and equable temperament that informs it. I have thus tried to limit my focus. I have been concerned only with the task of telling what the themes of Euler were and how they can be connected to current interests. Moreover, I have limited myself mostly to his work on infinite series and products and its repercussions in modern times, namely the theory of zeta values, and divergent series and integrals (Chapters 3 and 5). In Chapter 2 I have given a brief overview of some other parts of his work, for instance in elliptic integrals and number theory. His work on elliptic integrals is the forerunner of the modern theory of elliptic curves and abelian varieties, and his work on number theory raised questions which could only be understood fully after the development of class field theory. In Chapter 6 I have sketched a brief account of the theory of Euler products which he started but which really started to unfold with the work of Dirichlet and which, in the course of a long history, finally reached its climactic developments with the work of the number theorists of the late 19th century (class field theory) and the currently very active Langlands program. Parts of this chapter may be regarded as a very brief introduction to the Langlands program.

When writing any book, the question is always this: for whom is the book being written? To a large extent I was guided by a desire to reach the beginning graduate students and the advanced undergraduates, and to communicate to them the marvelous fact that things like class field theory, Borel summation, elliptic curves, and so on, did not come out fully grown from the primeval ocean (of the ancient Hindus for example) but grew out of small beginnings, and many of these go back to Euler. I feel that this historically motivated method of teaching is the best suited to convey the organic structure of mathematics. This of course is not the preferred way of teaching nowadays, where the students learn coherent sheaves and adeles before learning the Euler-Fermat theorem of primes which are sums of two squares. It actually happened in a high-powered course on D-modules I was attending a few years back that the first differential equation that was written, a few weeks after some very heavy stuff on D-modules, was Euler's, and the lecturer got it wrong because he forgot that the invariant operator on $\mathbf{C}^\times$ is not d/dz but $z(d/dz)$! I am convinced that the only way to produce young mathematicians who are not imprisoned by the small number of ideas they learn in a conventional graduate education is to emphasize the unity of mathematics from the beginning, and for this, the historical method, called the biogenetic method by Shafaraevitch, is the only possible one.

Thus this book is not a conventional historical essay on Euler—of these there are many wonderful examples—but rather a discussion of some of the Eulerian themes and how they fit into the modern perspective. More than anyone else, I know what the shortcomings of my attempt are. For example, although I have tried to keep the exposition as elementary as possible, here and there are places where this has been impossible to maintain, and I have had to assume familiarity on the part of the reader of more advanced material. But I have tried to organize everything in such a manner that a beginning graduate student, as well as a mathematician who does not always have a specialized knowledge of the topics treated, will find things understandable as well as enjoyable.

It only remains for me to give thanks to the people who helped me in this effort: to Don Babbitt and Sergei Gel'fand, who kept encouraging me throughout this enterprise; to Marina Eskin, who attended my course on Euler with an enthusiasm that was infectious; to Pierre Deligne, who was extraordinarily generous, as he always is, in sharing with me his insights on many aspects of Euler's work during his visit to UCLA in the spring of 2005; to Anita Colby, science librarian at UCLA, who always gave her help freely, instantly, and with a smile, for her help in bringing many items of Euleriana to my attention; to Richard Tsai for helping me with calculations using MATHEMATICA; to Aaron Pearl for help in photographing pages from Euler's papers; to Boyan Kostadinov, my student, for his help in translating articles by and about Euler from Russian as well as in proofreading; and to my wife, Veda, whose understanding and support have been the great steadying influences in my life.

V. S. Varadarajan
Pacific Palisades

CHAPTER 1

Leonhard Euler (1707-1783)

1.1. Introduction

No one can dispute the statement that Euler was the greatest mathematician and natural philosopher of the 18th century and one of the greatest of all time. He worked on all branches of mathematics, both pure and applied, known in his time. To the end of his life he carried in his mind the entire corpus of mathematics and physics of his epoch. He achieved universality in the most effortless manner, and many of the themes he worked on are still active today. He created new branches of mathematics, like combinatorial topology, graph theory, and the calculus of variations. He was the founder of modern differential and integral calculus as we know them today, and his books introduced algebra and calculus and their applications to enormous numbers of students. It could be said without exaggeration that he did to analysis what Euclid did to geometry, except that Euler himself created a huge part of what went into his books. With his lifelong interest and beautiful contributions, he started the process of establishing number theory as a major discipline in mathematics, a process that was completed by Gauss and the publication of his monumental *Disquisitiones Arithmeticae*. Ever since, number theory has attracted the attention and interest of the greatest mathematicians. As Laplace said, all the mathematicians of his time were his students.

Euler is easily the most prolific mathematician of all time. The range and volume of his output is simply staggering. He published over 850 papers, almost all of substantial length, and more than 25 books and treatises. In 1907 the Swiss Academy of Sciences established the **Euler Commission** with the charge of publishing the complete body of work consisting of all of his papers, manuscripts, and correspondence. This project, known as *Opera Omnia* [1], began in 1911 and is still in progress. His scientific publications, not counting his correspondence, run to over 70 volumes, each between approximately 300 and 600 pages. Thousands of pages of handwritten manuscripts are still not in print. He was in constant communication with all the great scientists of his day, and his correspondence covers several thousand pages, taking up the entire Series IV of *Opera Omnia*. The first volume of this series is completely devoted just to a catalogue and brief summaries of letters written by him and to him. *Opera Omnia*, of which I shall give more details later, includes his books on differential and integral calculus, calculus of variations, his great treatises on algebra and on analysis, his unfinished treatise on number theory, and his justly famous *Letters to a German Princess on Different Subjects in Natural Philosophy*, which is one of the most popular books on science ever written, translated into eight languages and reprinted countless times [2]. When publication is completed, *Opera Omnia* will be a gigantic landmark in the history of science. Its pages will contain a substantial part of all mathematical research that was carried out in the middle of the 18th century. The 18th century can thus be truly said to be the age of Euler. Furthermore, large parts of the mathematics of the 19th century flowed out of his work. What is even more remarkable is that some of his themes have generated new interest and attention even today. In spite of the fact that he was almost totally blind in the last fifteen years of his life, he wrote over 400 memoirs during that time, about half of his entire output, many of which required colossal calculations that he carried out entirely in his head. The memoirs he submitted to the St. Petersburg Academy were still being published decades after his death; indeed, one of them was not published till 1862, almost 80 years after his death.

Here is a list, which is at best partial, of topics Euler worked on in his lifetime, many of which were founded by him and in almost all of which his work was pioneering:

Differential and integral calculus
Logarithmic, exponential, and trigonometric functions
Differential equations, ordinary and partial
Elliptic functions and integrals
Hypergeometric integrals
Classical geometry
Number theory
Algebra
Continued fractions
Zeta and other (Euler) products
Infinite series and products
Divergent series
Mechanics of particles
Mechanics of solid bodies
Calculus of variations
Optics (theory and practice)
Hydrostatics
Hydrodynamics
Astronomy
Lunar and planetary motion
Topology
Graph theory

Euler was awarded many honors in his life. He was a member of both the St. Petersburg and Berlin Academies of Sciences. He was elected a member of the Royal Society of London in 1749 and the Académie des Sciences of Paris in 1755. His teacher, Johann Bernoulli, not an easy man to get along with, called him *the incomparable Leonhard Euler* and *mathematicorum princeps*. Several lunar features have been named after him, and the list of mathematical and other scientific discoveries named after him is almost endless: Euler line of a triangle, Euler angles of a rotation, Euler-Lagrange equations of the calculus of variations, Eulerian integrals, Euler characteristic, Euler equations of motion of solid bodies, Euler equations of fluid mechanics, Euler function $\varphi(n)$, Euler-Maclaurin sum formula, Euler's constant, Euler products, and so on. The ten franc bill in Switzerland has his picture on it. Some of the problems he worked on are still open, and his work links with an astonishing amount of contemporary research in both pure and applied mathematics. His books *Introductio in Analysin Infinitorum* and *Mechanica*, which went through many editions, brought calculus and mechanics to the entire scientific world. There was no longer any necessity of reading the obscure papers of Leibniz or the works of Newton couched in an opaque geometrical language that was unsuitable for most problems.

To survey Euler's life and work in detail in a single book is impossible. However there are many excellent accounts which survey parts of his life and work, and I have made free use of these. Without even remotely attempting any sort of completeness, I have listed some of these in [3]. One of the most interesting is the account of A. Weil [3a]: it deals just with Euler's work on number theory, which occupies a bare(!) 4 volumes of the 70 plus volumes of his *Opera Omnia*, but still is over 120 pages long. In addition there are also the introductions to the various volumes of *Opera Omnia* written by experts which deal in depth with his work contained in those volumes. All of this suggests that a complete *scientific biography* of Euler, treating all of his work with historical accuracy and placing it in a modern perspective, will be so complex that its length will exceed any reasonable bound.

A major part of Series I of *Opera Omnia* is concerned with analysis. To Euler this very often meant working with infinite series and transformations of series and integrals. He was, without any doubt, the greatest master of infinite series and products of his, and indeed of any, time. Perhaps only Jacobi and Ramanujan from the modern era can evoke comparable wonder and admiration as formalists. Before Euler, infinite series and products made their appearance only in isolated works (Leibniz, Gregory, Wallis, etc.) and always in an auxiliary manner. Euler was the first to treat them systematically and in great depth, not only for their applications, but also for their own intrinsic interest. In this area he discovered some of the most beautiful formulae ever to be found in mathematics. His ideas on summing divergent series, which he used brilliantly, for instance, in discovering the functional equation of the zeta function at the integer points, led directly to the modern theory of summation of divergent series that was at the center of research in analysis in the 19th century and in the early part of the 20th (see [4], Chs. 1, 2). In recent years summation of divergent series has again become powerful in a wide range of problems, ranging from quantum field theory to dynamical systems.

With its emphasis on concepts and structures, modern mathematics has mostly shied away from formulae. Nevertheless, many of the greatest peaks of mathematics are described by formulae of one sort or another. Here is a sample of some of them that Euler discovered.

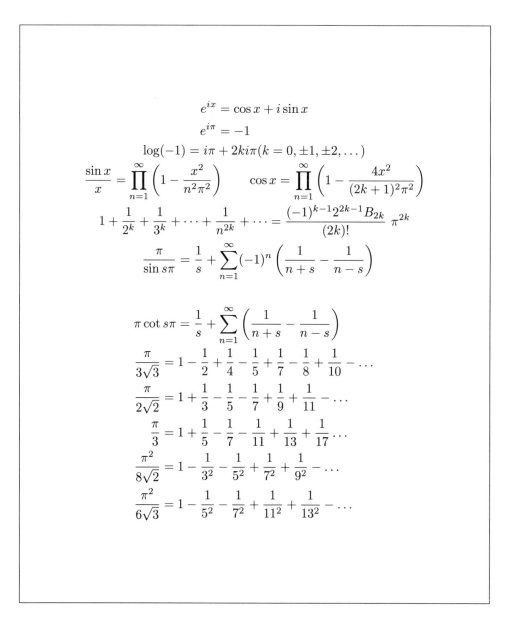

$$e^{ix} = \cos x + i \sin x$$

$$e^{i\pi} = -1$$

$$\log(-1) = i\pi + 2ki\pi (k = 0, \pm 1, \pm 2, \dots)$$

$$\frac{\sin x}{x} = \prod_{n=1}^{\infty} \left(1 - \frac{x^2}{n^2\pi^2}\right) \qquad \cos x = \prod_{n=1}^{\infty} \left(1 - \frac{4x^2}{(2k+1)^2\pi^2}\right)$$

$$1 + \frac{1}{2^k} + \frac{1}{3^k} + \cdots + \frac{1}{n^{2k}} + \cdots = \frac{(-1)^{k-1} 2^{2k-1} B_{2k}}{(2k)!} \pi^{2k}$$

$$\frac{\pi}{\sin s\pi} = \frac{1}{s} + \sum_{n=1}^{\infty} (-1)^n \left(\frac{1}{n+s} - \frac{1}{n-s}\right)$$

$$\pi \cot s\pi = \frac{1}{s} + \sum_{n=1}^{\infty} \left(\frac{1}{n+s} - \frac{1}{n-s}\right)$$

$$\frac{\pi}{3\sqrt{3}} = 1 - \frac{1}{2} + \frac{1}{4} - \frac{1}{5} + \frac{1}{7} - \frac{1}{8} + \frac{1}{10} - \cdots$$

$$\frac{\pi}{2\sqrt{2}} = 1 + \frac{1}{3} - \frac{1}{5} - \frac{1}{7} + \frac{1}{9} + \frac{1}{11} - \cdots$$

$$\frac{\pi}{3} = 1 + \frac{1}{5} - \frac{1}{7} - \frac{1}{11} + \frac{1}{13} + \frac{1}{17} \cdots$$

$$\frac{\pi^2}{8\sqrt{2}} = 1 - \frac{1}{3^2} - \frac{1}{5^2} + \frac{1}{7^2} + \frac{1}{9^2} - \cdots$$

$$\frac{\pi^2}{6\sqrt{3}} = 1 - \frac{1}{5^2} - \frac{1}{7^2} + \frac{1}{11^2} + \frac{1}{13^2} - \cdots$$

$$\sum_{n=1}^{\infty} \frac{1}{n^s} = \prod_{p \text{ prime}} \left(1 - \frac{1}{p^s}\right)^{-1}$$

$$\frac{1 - 2^{m-1} + 3^{m-1} - \text{etc.}}{1 - 2^{-m} + 3^{-m} - \text{etc.}} = -\frac{1.2.3. \ldots (m-1)(2^m - 1)}{(2^{m-1} - 1)\pi^m \cos \frac{m\pi}{2}}$$

$$\sum_{n>m>0} \frac{1}{n^2 m} = \sum_{n>0} \frac{1}{n^3}$$

$$\prod_{n=1}^{\infty} (1 - x^n) = 1 + \sum_{n=1}^{\infty} (-1)^n \left(x^{\frac{3n^2 - n}{2}} + x^{\frac{3n^2 + n}{2}}\right)$$

$$1 - 1!x + 2!x^2 - 3!x^3 + \ldots = \frac{1}{1+} \frac{x}{1+} \frac{x}{1+} \frac{2x}{1+} \frac{2x}{1+} \frac{3x}{1+} \frac{3x}{1+} \text{ etc.}$$

$$1 - 1! + 2! - 3! + \ldots = 0.596347362123 \ldots$$

$$\sum_{k=0}^{m} f(k) = \int_0^m f(x)dx + \frac{1}{2}(f(0) + f(m))$$

$$+ \sum_{k \geq 1} \frac{B_{2k}}{(2k)!} \left(f^{(2k-1)}(m) - f^{(2k-1)}(0)\right)$$

$$\frac{\partial}{\partial y} F(x, y, y') = \frac{d}{dx}\left(\frac{\partial}{\partial y'} F(x, y, y')\right)$$

1.2. Early life

Leonhard Euler was born in Basel, Switzerland, on April 15, 1707. His father, Paul Euler, was a parish priest, and soon after Leonhard's birth, he settled in a small village nearby where Leonhard grew up. Paul Euler wanted his son to become a priest, and so Leonhard's early education emphasized theology and related subjects. But it soon became clear that Euler's interests and abilities were in mathematics. Paul Euler had attended the lectures in Basel of the great Jacob Bernoulli. Jacob Bernoulli held the mathematics chair at the University of Basel and was one of the most distinguished mathematicians of Europe in his time. The University of Basel was founded in 1460 and was a leading center of learning and research at the beginning of the 18th century. Because of his experience in attending the lectures of Jacob Bernoulli, Paul Euler was able to instruct his young son in mathematics in his early years. With this encouragement Euler studied for himself difficult works in algebra when he was in his early teens. But these studies did not fully satisfy Leonhard, who was discovering his passion for mathematics. By the time Leonhard was ready to begin higher studies, Jacob Bernoulli had died, and his younger brother, Johann Bernoulli, had succeeded him at the University of Basel to the chair of mathematics. Johann Bernoulli was regarded as the foremost mathematician in Europe of his day. Paul Euler had roomed with him when they were both students at Basel. Leonhard was introduced to Johann Bernoulli and his sons, Nicolaus and Daniel. Daniel would go on to become a famous mathematician and scientist in his own right as well as a lifelong friend of Euler. Johann Bernoulli recognized Euler's genius very early and pushed him to study the masters, helping him in this endeavor. Euler recalls in his autobiographical writings how he was able to meet regularly with Johann Bernoulli and discuss with him the difficulties he had encountered during his mathematical studies and how he worked very hard so as not to bother his mentor with unnecessary questions [3b]:

> *I soon found the opportunity to gain introduction to the famous professor Johann Bernoulli, whose good pleasure it was to advance me further in the mathematical sciences. True, because of his business he flatly refused to give me private lessons, but he gave me much wiser advice, namely to get some more difficult mathematical books and work through them with all industry, and wherever I should find some check or difficulties, he gave me free access to him every Saturday afternoon and was so kind as to elucidate all difficulties, which happened with such greatly desired advantage that whenever he had obviated one check for me, because of that ten others disappeared right away, which is certainly the way to make a happy advance in the mathematical sciences.*

This period of regular contact with a great mathematician like Johann Bernoulli was decisive in the development of Euler as a mathematician. Johann Bernoulli, together with his brother, Jacob, had studied the works of Leibniz on calculus carefully and had made many contributions to problems involving the properties of curves which are the solutions of many types of extremal problems, such as the isochrone and the brachistochrone. These problems must have captured Euler's imagination, because he went on later to formulate these questions in great generality and derived what we now call the Euler-Lagrange equations in the calculus of variations. Johann Bernoulli was a combative and irascible person, as evidenced by the many instances of quarrels with his elder brother and even his own sons on questions of priority, but Euler was his favorite disciple and remained so throughout his life. Perhaps this says more about Euler's personality than that of Johann Bernoulli.

Euler completed his university studies in 1726 and began his independent investigations immediately thereafter. He participated in a competition organized by the Paris Academy of Sciences on the most efficient way to arrange masts on a ship, and although he had never been on a ship, his entry received the second prize and a favorable mention and was published. The prizes of the Paris Academy were substantial in terms of money, and they were generally offered for the solution of some important problem. They were therefore influential in fostering research and for the most part were administered fairly. Euler was to win it twelve times in his career. He also participated in a competition for a professorship in physics

at Basel and wrote a monograph on sound, *Dissertatio physica de sono*, supporting his application. In this monograph he summarized the existing knowledge on
acoustics, formulated mathematically with complete clarity the basic questions, and
answered some of them. Although he did not succeed in getting the position, this
work would remain a classic for years to come. He came in second, possibly because
he was thought to be too young for that position. By this time his friends Daniel
and Nicolaus Bernoulli had already gone to St. Petersburg and were members of
the newly established Academy of Sciences there. This academy had been founded
in 1725 by Catherine I, the widow of Peter the Great, following the plans of the
late great czar. Through the recommendation of the Bernoulli brothers Euler was
offered a position in the Academy in 1726. The position was in physiology, and Euler joined the University at Basel to study the subject and look for mathematical
applications in it as a preparation for his move to St. Petersburg. He left Basel to
go to St. Petersburg and joined the Academy in 1727, two years after the Academy
was established. But by then Nicolaus Bernoulli had died, and Euler was invited
to be an adjunct in mathematics, a position that was more suitable for his talents
and interests. He became a professor of physics in 1730, and when Daniel Bernoulli
left for Switzerland in 1733, Euler succeeded him to the chair in mathematics at
the Academy. He stayed in St. Petersburg, for 14 years altogether, before he went
to Berlin in 1741 and joined the Academy there. He worked 25 years in Berlin before returning to St. Petersburg, where he worked till his death in 1787. He never
returned to Switzerland, although he retained his Swiss citizenship till the end.

Before we begin a sketch of Euler's academic life it may be useful to set down
a time line that details the main events of his career.

1707   Born in Basel, Switzerland, April 15.

1725   Peter the Great and his widow Catherine establish the St. Petersburg
        Academy of Sciences in St. Petersburg, Russia.

1727   Euler moves to St. Petersburg and becomes an adjunct in mathematics.

1733   Euler takes over the chair in mathematics after Daniel Bernoulli returns
        to Basel. Gets married and buys a house.

1735   Solves the problem of finding the sum of $\sum_{n \geq 1} \frac{1}{n^2}$ and acquires an inter-
        national reputation.

1738   Euler loses the vision in his right eye after a serious illness.

1741   Political turmoil in Russia after death of the czarina and the regency.
        Euler leaves Russia to join the Academy of Sciences in Berlin, Prussia.

1762   Catherine (the Great) II becomes the czarina in Russia and starts the
        efforts to get Euler back.

1766   Euler returns to St. Petersburg. His eyesight begins to deteriorate.

1771   Euler loses the vision in his left eye also.

1783   Dies in St. Petersburg on September 18.

It may probably be helpful to have a feeling for the chronology of Russian imperial succession to follow the events involving Euler. Peter I (the Great) died in 1725, and his widow, Catherine I, followed him on the throne. The St. Petersburg Academy was established in the same year by her, following closely Peter's plans. She died in 1727, and Peter II followed her as the czar. But the rule of Peter II lasted only till 1730, when Anna Ivanovna, niece of Peter I, succeeded him. Her reign lasted till 1740, when Ivan VI succeeded to the throne. Being too young, his mother, Anna Leopoldovna, acted as the regent. But the regency was very short lived, and Anna and Ivan VI were overthrown by Elizabeth Petrovna, daughter of Peter I, in a coup. She ruled from 1741 to 1762. Moscow University was founded in 1755 during her reign. In 1762 Peter III became the czar but was murdered almost immediately, and his widow, Catherine II (the Great Catherine), gained power and ruled Russia as the czarina till 1796 with great distinction. Apart from renewing the glory of the St. Petersburg Academy by getting Euler to come back, she created the beginnings of the great art collections of the Hermitage museum, expanded Russian influence politically as well as militarily, and made Russia a great European power. One may legitimately compare her reign to that of Elizabeth I of England.

1.3. The first stay in St. Petersburg: 1727-1741

The years 1727-1741 at St. Petersburg were very productive for Euler scientifically and very comfortable personally. It was during his first stay in St. Petersburg that his transformation into a mathematician of the foremost rank took place. His colleagues at the Academy were first rate scientists and the conditions offered by the Academy were very generous, a fact which he acknowledged in a letter written in 1749 [3c]:

> ... I and all others who had the good fortune to be for some time with the Russian Imperial Academy cannot but acknowledge that we owe everything which we are and possess to the favorable conditions we had there. ...

In return these scientists were expected to publish their research and add prestige to the Academy and advise the czar on whatever question for which their help was sought. This freedom to work on problems that interested him apart from the time he was assisting the government was priceless for Euler, and his mathematical personality bloomed under these conditions. Euler himself was in charge of projects involving cartography (which produced severe strain on his eyes, as he remarked once in a letter), shipbuilding, and general questions of military science.

Initially Euler had difficulties in his career mainly because of the interference of the administrator for the Academy, one Schumacher, whose main interest seemed to lie in the suppression of talent wherever it might rear its inconvenient head ([3b], p. xv). When Schumacher contrived to place Euler on the same level as a few mediocre colleagues, Euler protested strongly to him (see [3d], p. 23):

> It seems to me that it is very disgraceful to me, that I, who up to now have had more salary than the others, shall now be set equal to them. ... I think that the number of those who have carried [mathematics] as far as I is pretty small in the whole of Europe, and none of them will come for 1000 rubles.

But the growing stature of Euler in the world of mathematics and physics eventually overcame all obstacles, and Euler rose rapidly in the ranks of the Academy. During the early years in St. Petersburg Daniel Bernoulli was a constant companion to Euler, both personally and scientifically. But gradually Daniel Bernoulli got tired of life in Russia and the constant intrigues of Schumacher and could not wait to go back to Switzerland. When an opportunity came he took it and left for Basel to become a professor of anatomy and botany there. However it must be said that he never again reached afterwards the level of scientific work that he had attained in Russia. This is understandable since in St. Petersburg there was the constant presence of and inspiring interaction with Euler. After Daniel Bernoulli's departure Euler was appointed to the chair of mathematics and paid handsomely, so that a measure of financial security was achieved. Euler got married in 1733 to Katherina Gsell, the daughter of a Swiss painter named Gsell who was working in St. Petersburg, and purchased a house on the banks of the Neva River close to the Academy. Euler's son, Johann Albrecht, who would become a high-level scientist and a member of the Academy and with whom Euler would collaborate much later in his life, was born in 1734.

In St. Petersburg Euler's genius flourished, and he worked on a variety of subjects. Papers streamed from his pen to fill the pages of the proceedings of the Academy. He produced fundamental memoirs on number theory, infinite series, calculus of variations, mechanics, as well as applications of these topics to various questions in music, cartography, shipbuilding, and so on. He prepared close to 100 memoirs for publication on a variety of topics. He acquired a European reputation as one of the greatest living mathematicians and mathematical physicists through the originality and prolific nature of his work, and the remarkable manner in which he could develop applications of mathematics to all sorts of situations. For a detailed account of Euler's first years in St. Petersburg, the reader is referred to [3d]. Here I shall discuss only the highlights of his work in mathematics during the stay in St. Petersburg.

Very early in his stay at St. Petersburg Euler became acquainted with Christian Goldbach. This soon developed into a lifelong friendship which played a great role in Euler's life, both personally and scientifically. Goldbach was "an energetic and intelligent Prussian for whom mathematics was a hobby, the entire realm of letters an occupation, and espionage a livelihood" (see [3b], p. xv). Goldbach was a catalyst for Euler, suggesting to him possible questions and acquainting him with what he had learned in his travels. They exchanged a huge number of letters, most of them on problems of number theory and analysis. The very first letter that Goldbach wrote to Euler in December 1729 contains a postscript asking whether Euler knew of Fermat's statement that all numbers of the form $2^{2^n} + 1$ were primes. Euler was initially cool to the suggestion of Goldbach but became very interested in number theory soon afterwards and disproved the assertion about Fermat primes. It was Goldbach who stimulated Euler to look deeply into Fermat's discoveries and supply the proofs of all of Fermat's statements. Eventually Euler would go far beyond Fermat, discovering towards the end of his life the law of quadratic reciprocity for which he could not supply a proof. It was proved by Gauss.

It was during this period that he solved one of the most famous open problems at that time, namely to find the sum of the series

$$1 + \frac{1}{2^2} + \frac{1}{3^2} + \frac{1}{4^2} + \cdots,$$

whose solution had eluded all the leading mathematicians of that era. Euler discovered that the sum is

$$\frac{\pi^2}{6}$$

and also that

$$1 + \frac{1}{2^4} + \frac{1}{3^4} + \cdots = \frac{\pi^4}{90}.$$

It is this work more than anything else that established him as the foremost mathematician of his time. Later on he would succeed in proving that

$$1 + \frac{1}{2^{2k}} + \frac{1}{3^{2k}} + \cdots = \frac{(-1)^{k-1}2^{2k-1}B_{2k}}{(2k)!}\pi^{2k}$$

where B_{2k} are rational numbers, named Bernoulli numbers by Euler, in view of the fact that they were originally introduced by Jacob Bernoulli in his *Ars Conjectandi*. It must be noted that the evaluation of the series

$$1 + \frac{1}{2^{2k+1}} + \frac{1}{3^{2k+1}} + \frac{1}{4^{2k+1}} + \cdots$$

is open to this day and appears to be out of reach at present.

Closely related to the evaluations of the series above is the question of computing them accurately to several decimal places. For low values of k this is indeed a problem because the series converge slowly:

$$\frac{1}{(n+1)^k} + \frac{1}{(n+2)^k} + \cdots \asymp \frac{1}{n^{k-1}}.$$

These questions of accurate numerical evaluation, which were very dear to Euler's heart, must have inspired him to make his remarkable discovery of what is now called the Euler-Maclaurin summation formula. Euler discovered it in 1734 and gave several applications; Maclaurin arrived at it apparently independently in 1738. With characteristic generosity Euler did not engage in any dispute about priority except to content himself to the statement that the result and its demonstration were publicly read before the Academy in 1734. The summation formula was a favorite of Euler. He used it in hundreds of problems and with many ingenious variations. The summation formula is

$$f(0) + f(1) + \cdots + f(m) = \int_0^m f(x)dx + \frac{1}{2}(f(0) + f(m))$$
$$+ \sum_{k \geq 1} \frac{B_{2k}}{(2k)!}\left(f^{(2k-1)}(m) - f^{(2k-1)}(0)\right).$$

The point is that the right side of the formula is a series which is typically divergent and so has to be treated as an asymptotic series. This requires great skill in deciding up to what stage one should carry the sum on the right, and Euler was a past master in this.

It was during these years that Euler went deeply into problems of finding curves satisfying extremal conditions, what we now call the calculus of variations. I have already remarked on the fact that Euler's interest in these problems goes back to his days of apprenticeship to Johann Bernoulli. But Euler was to take this subject from the realm of a few special problems to a far-reaching theory. This work would appear as a book in 1744, but much of the work must have been done during his St. Petersburg stay.

His two-volume work *Mechanica* also dates to this period. Euler was the first person to introduce analytical and algebraic methods in mechanics, thus breaking away from the tradition of Newton and his contemporaries, who insisted on geometric methods. In retrospect, Euler's methods revolutionized the treatment of problems by basing their solutions on differential equations. With hindsight we should not find this too surprising, since the mechanical trajectories are solutions of the Euler-Lagrange variational equations for a suitable functional (action) and so quite within the horizon of Euler at that time. It is fair to say that with the publication of *Mechanica*, Euler's reputation as a natural philosopher of the highest rank was solidified. He also published two volumes on *Scientia navalis* (hydrodynamics, shipbuilding and navigation).

Add to all of this his work on cartography, theory of music, and other duties executed in service of the government, and we get a picture of a man of prodigious energy and creativity, with his powers approaching a peak, working all the time. This enormous workload, especially the drudgery and close reading involved in his cartographic work, led to a severe illness, as a result of which he lost the sight in his right eye in 1738. He was reported to have said at that time that "this means less distraction"and continued to work at the same punishing rate.

1.4. The Berlin years: 1741-1766

The death of the czarina in 1740 and the resulting political turmoil and xenophobia in Russia created conditions that were very difficult, even dangerous, for foreigners in Russia. Euler got an offer from King Frederick II of Prussia to become the director of mathematics in the Prussian Academy of Sciences. He left St. Petersburg to go to Berlin in 1741 and joined the Academy as its top mathematician. He was to remain with the Academy for the next 25 years. He was compensated generously and was able to purchase a good house. But what was remarkable was that the St. Petersburg Academy still retained him as a member and continued to pay him a pension. In recompense, Euler sent almost half of his publications to the St. Petersburg Academy for publication, wrote several important books on its commission, advised the Academy on its manifold scientific activities, and served as its representative in the Western world. He was clearly relieved to be away from the danger and turmoil of Russia. In a letter to Johann Caspar Wettstein, Euler wrote (see [3d], p.159):

> *I can do just what I wish [in my research]. . . . The king calls me*
> *his professor, and I think I am the happiest man in the world . . . .*

In Berlin Euler reached the peak of his career. More than 100 memoirs were sent to the St. Petersburg Academy, and about 125 memoirs were published in the Berlin Academy on all possible topics in mathematics and physics. This was an astonishing burst of creativity, unparalleled before or since.

However, over the years the relations between him and King Frederick slowly deteriorated. The king was essentially an ignoramus so far as mathematics was concerned and was much taken with superficial cleverness. He admired French literati and figures of science (like Voltaire and d'Alembert) but without any deep understanding of their merits or their work. He certainly did not comprehend fully the monumental stature of Euler. When Euler first came to Berlin the king had appointed Maupertius as the president of the Academy. Maupertius of course was not at the level of Euler as a scientist, but Euler, as was typical of him, maintained cordial relations with Maupertius and was very influential in the affairs of the Academy. Maupertius died in 1759, and after that Euler functioned as the *de facto* president. The king never conferred on Euler the title of the president of the Academy. When it appeared around 1763 that Frederick was contemplating the appointment of d'Alembert as the president of the Berlin Academy (although d'Alembert eventually rejected the king's offer), Euler began to think about leaving Berlin. Eventually he wrote to the secretary of the St. Petersburg Academy about his wish to return. Meanwhile Catherine (the Great) II had become the czarina in Russia, and one of her most important priorities was to restore the St. Petersburg Academy to its former position of glory and importance. This meant going after Euler, and she ordered her ambassador in Prussia to make an offer to Euler on any terms that Euler would specify. At first the king refused Euler's requests to leave, but he could not prevail against the pressure of the formidable czarina. So he finally gave in and allowed Euler to leave with his family and several assistants, expressing his frustration and irritation in crude jokes to his companions. Euler returned to Russia in 1766 via Poland, where he was treated with great respect and warmth by the Polish king Stanislas, a former lover of Catherine II, thanks almost certainly to her suggestions. He returned to St. Petersburg virtually as a conquering hero, universally admired and respected.

While he was in Berlin Euler published several works. His great treatise on analysis, *Introductio in analysisin infinitorum*, in which he gave a majestic exposition of his theory of circular functions and their applications, was published in 1748. This book was a landmark in analysis and dominated the field for the next century or so. It is still beautiful to read. All later books on calculus cover only parts of the material of this book and without the compelling force of the *Introductio*. The monograph on the calculus of variations that was begun while he was in St. Petersburg was published in 1744. The treatise on mechanics of a solid body, *Theoria motus corporum solidorum seu rigidorum*, was published in 1765.

It was during these years that Euler obtained his famous infinite product expansion for $\sin x$, which allowed him to establish with complete rigor his earlier results on the zeta values. It was also during this period that he discovered the relation between the solutions of the Pell's equation $x^2 - Ny^2 = 1$ and the continued fraction for $\sqrt{N}$. The Euler equations for the motion of a rigid body are still among the basic examples of dynamical systems on a Lie group. His infinite product for $\zeta(s)$ as well as its functional equation dates back to this time. It is thus clear that this was his greatest period, when everything went wonderfully well.

1.5. The second St. Petersburg stay and the last years: 1766-1783

Euler's prodigious scientific activity continued unabated after his return to St. Petersburg even though another serious illness resulted in his losing almost entirely the sight in his other eye. He thus became almost totally blind after 1771. Unbelievably, almost half of his work, about 400 memoirs, was written during this second St. Petersburg stay, a substantial part of which was carried out during this period of near total blindness. His book on algebra was published during this period. He worked with assistants, many of whom were first-rate scientists in their own right, and one of whom was his own son, Johann Albrecht. He had a huge slate board fitted to his desk in his study on which he wrote in big letters so that he could dimly see what was being written. His memory was prodigious, and his ability to perform intricate calculations absolutely incredible. Still it must have been hard on him to rely on others in working out calculations. In a moving letter to Lagrange, answering a letter from him, he writes (see [3a], p. 168):

> I have had all your calculations read to me, concerning the equation $101 = p^2 - 13q^2$, and I am fully persuaded of their validity; but as I am unable to read or write, I must confess that my imagination could not follow the reasons for all the steps you have had to take, nor keep in mind the meaning of all your symbols. It is true that such investigations have formerly been a delight to me and that I have spent much time on them; but now I can only undertake what I can carry out in my head, and often I have to depend on some friend to do the calculations which I have planned....

It was during this period that he completed some of his work on lunar motion that included colossal calculations, many of which were carried out by him entirely in his head. Other books during this period include *Dioptrica* (1769-1771); the three volumes of *Institutiones calculi integralis* (1768-1770); and his treatise on algebra, *Vollständige Anleitung zur Algebra* (1770), first published in Russian. These books were dictated to his assistants.

Euler's wife, Katherina, died in 1773, and Euler married her sister, Abigail Gsell, three years later. Euler died on 18 September 1783 as a result of a cerebral hemorrhage. There was no indication of any problem till his very death, and he worked on problems literally till his last breath. His death was quick and painless. On the day of his death he was working with his assistants on the orbit of the recently discovered planet Uranus,* and on his slate board was found a calculation of how high a hot air balloon could rise, perhaps stimulated by the news of the first ascents of such balloons.[†] Eulogies were delivered by the Marquis de Condorcet and by Euler's great-grandson, P.-H. Fuss. For English translations of these see [5].

1.6. Opera Omnia

The first thing that strikes anyone is the sheer amount and scope of Euler's scientific work. The definitive catalogue of Gustav Eneström lists over 850 titles of memoirs with very few real repetitions.

The Swiss Academy of Sciences established the **Euler Commission** in 1907 with the charge to publish all of Euler's papers, correspondence, manuscripts, notes, and diaries. This of course required not only financial assistance but also the cooperation of literally hundreds of leading mathematicians and Eulerian scholars of great distinction around the world. Publication of the collected works of Euler

---

*Uranus was discovered by the English astronomer William Herschel in 1781. The determination of its orbit, however, proved troublesome and was not completed till 1846.

[†]Brothers Joseph and Etienne Montgolfier, paper manufacturers, were pioneers in the flights of hot air balloons and had succeeded in making several unmanned flights in 1783. Their first flight to carry living things (animals actually) took place on September 19, 1783. However, news of the earlier flights must have reached St. Petersburg during the last days of Euler.

began in 1911 but was stopped while still incomplete. It was resumed a few years ago and is now essentially complete, except for a few volumes of Series IV A and some of Series IV B; these are in preparation and scheduled to appear soon. Many of the volumes have substantial introductions written by modern experts in various fields. Each volume is between 300 and 600 pages approximately so that one is looking at a collection of over 30,000 pages in print. It is the result of a gigantic collective scientific effort unlike any the scientific world has ever seen. The collection is called *Opera Omnia* and is in four series. I have listed at the end the titles and the editors of all the volumes so that the reader can get an idea of the monumental nature of the effort that has gone into this project, as well as the prodigious scope of the man whose works are being published. There is nothing that is even remotely like this in the whole history of science.

1.7. The personality of Euler

The portrait of Euler that emerges from his publications and letters is that of a genial man of simple tastes and conventional religious faith. He was even wealthy, at least in the second half of his life, but ostentation was not a part of his lifestyle. His memory was prodigious, and contemporary accounts have emphasized this. He would delight relatives, friends, and acquaintances with a literal recitation of any song from Virgil's *Aeneis*, and he would remember minutes of Academy meetings years after they were held. He was not given to envy, and when someone made an advance on his work his happiness was genuine. For example, when he learnt of Lagrange's improvements on his work on elliptic integrals, he wrote to him that his admiration knew no bounds and then proceeded to improve upon Lagrange! (See [3a], p. 284.)

But what is most characteristic of his work is its clarity and openness. He never tries to hide the difficulties from the reader. This is in stark contrast to Newton, who was prone to hide his methods in obscure anagrams, and even from his successor, Gauss, who very often erased his steps to present a monolithic proof that was seldom illuminating. In Euler's writings there are no comments on how profound his results are, and in his papers one can follow his ideas step by step with the greatest of ease. Nor was he chary of giving credit to others; his willingness to share his summation formula with Maclaurin, his proper citations to Fagnano when he started his work on algebraic integrals, his open admiration for Lagrange when the latter improved on his work in calculus of variations are all instances of his serene outlook. One can only contrast this with Gauss's reaction to Bolyai's discovery of non-Euclidean geometry. Euler was secure in his knowledge of what he had achieved but never insisted that he should be the only one on top of the mountain.

Perhaps the most astonishing aspect of his scientific opus is its universality. He worked on everything that had any bearing on mathematics. For instance his early training under Johann Bernoulli did not include number theory; nevertheless, within a couple of years after reaching St. Petersburg he was deeply immersed in it, recreating the entire corpus of Fermat's work in that area and then moving well beyond him. His founding of graph theory as a separate discipline, his excursions in what we call combinatorial topology, his intuition that suggested to him the idea of exploring multizeta values are all examples of a mind that did not have any artificial boundaries. He had no preferences about which branch of mathematics was dear to him. To him, they were all filled with splendor, or *Herrlichkeit*, to use his own favorite word.

Hilbert and Poincaré were perhaps the last of the universalists of the modern era. Already von Neumann had remarked that it would be difficult even to have a general understanding of more than a third of the mathematics of his time. With the explosive growth of mathematics in the twentieth century we may never see again the great universalists. But who is to say what is and is not possible for the human mind?

It is impossible to read Euler and not fall under his spell. He is to mathematics what Shakespeare is to literature and Mozart to music: universal and *sui generis*.

Notes and references

[1] **Leonhardi Euleri Opera Omnia**, Edited by the Euler Commission of the Swiss Academy of Science in collaboration with numerous specialists, 1911-. Originally started by the publishing house of B. G. Teubner, Leipzig and Berlin. Birkhäuser, Boston and Basel, has continued publication and has made available all volumes of the *Opera Omnia*.

Series prima : *Opera mathematica.*

> This is Series I and contains all his papers on what we may call "pure mathematics". It has 29 volumes, 30 volume-parts.

Vollständige Anleitung zur Algebra.

> I-1. With supplements by Joseph Louis Lagrange, 651 pages, 1911. Heinrich Weber (Ed.).

Commentationes arithmeticae. Contributions to number theory.

> I-2 (611 pages, 1915), Ferdinand Rudio (Ed.).
>
> I-3 (543 pages, 1917), Ferdinand Rudio (Ed.).
>
> I-4 (431 pages, 1941), Rudolf Fueter (Ed.).
>
> I-5 (374 pages, 1944), Rudolf Fueter (Ed.).

Commentationes algebraicae ad theoriam aequationum pertinentes.

> I-6 (509 pages, 1921), Ferdinand Rudio, Adolf Krazer, and Paul Stäckel (Eds.).

Commentationes algebraicae ad theoriam combinationum et probabilitatum pertinentes

> I-7 (580 pages, 1923) Louis Gustave du Pasquier (Ed.).

Introductio in Analysin Infinitorum, parts 1, 2.

> I-8 (392 pages, 1922), Adolf Krazer and Ferdinand Rudio (Eds.).
>
> There are translations into English for both volumes:
>
> *Introduction to Analysis of the Infinite*, Book I (1988), Book II (1990), John D. Blanton, (Tr.), Springer-Verlag.
>
> I-9 (403 pages, 1945), Andreas Speiser (Ed.).

Institutiones calculi differentialis.

> I-10: (676 pages, 1913), Gerhard Kowalewski (Ed.).

Institutiones calculi integralis.

> This is in two parts. The first part, consisting of the first nine chapters, has been translated into English by John D. Blanton (Tr.) as *Foundations of Differential Calculus*, Springer-Verlag, 2000.
>
> I-11 (462 pages, 1913), Friedrich Engel and Ludwig Schlesinger (Eds.).
>
> I-12 (542 pages, 1914), Friedrich Engel and Ludwig Schlesinger (Eds.).
>
> I-13 (508 pages, 1914), Friedrich Engel and Ludwig Schlesinger (Eds.).

Commentationes analyticae ad theoriam serierum infinitarum pertinentes.

His papers on infinite series, products, zeta values, and related matters are in I-14, I-15, and I-16:

I-14 (617 pages, 1925), Carl Boehm and Georg Faber (Eds.).

I-15 (722 pages, 1922), Georg Faber (Ed.).

I-16/1 (355 pages, 1933), Carl Boehm (Ed.).

I-16/2 (332 pages, 1935), Carl Boehm (Ed.).

Commentationes analyticae ad theoriam integralium pertinentes.

Related matters involving integrals are in I-17, I-18, and I-19:

I-17 (457 pages, 1914), August Gutzmer (Ed.).

I-18 (475 pages, 1920), August Gutzmer and Alexander Liapounoff (Eds.).

I-19 (494 pages, 1932), Alexander Liapounoff, Adolf Krazer, and Georg Faber (Eds.).

Commentationes analyticae ad theoriam integralium ellipticorum pertinentes.

Papers on elliptic integrals are in I-20 and I-21:

I-20 (371 pages, 1912) Adolf Krazer (Ed.).

I-21 (380 pages, 1913), Adolf Krazer (Ed.).

Commentationes analyticae ad theoriam aequationum differentialium pertinentes.

Papers on differential equations are in I-22 and I-23:

I-22 (420 pages, 1936), Henri Dulac (Ed.).

I-23 (455 pages, 1938), Henri Dulac (Ed.).

Methodus inveniendi lineas curvas maximi minimive proprietate gaudentes sive solution problematis isopermietrici latissimo sensu accepti.

I-24 (308 pages, 1952), Constantin Carathéodory (Ed.).

Commentationes analyticae ad calculum variationum pertinentes.

I-25 (343 pages, 1952): Constantin Carathéodory (Ed.).

I-24 and I-25 contain his works on the calculus of variations.

Commentationes geometricae.

I-26 (362 pages, 1952), Andreas Speiser (Ed.).

I-27 (400 pages, 1954), Andreas Speiser (Ed.).

I-28 (381 pages, 1955), Andreas Speiser (Ed.).

I-29 (488 pages, 1956), Andreas Speiser (Ed.).

Series secunda: *Opera mechanica et astronomica.*

This is Series II and contains works devoted to mechanics and astronomy and has 31 volumes, 32 volume-parts.

Mechanica sive motus scientia analytice exposita.

II-1 (417 pages, 1912) and II-2 (460 pages, 1912), Paul Stäckel (Ed.).

Theoria motus corporum solidorum seu rigidorum ex primis nostrae cognitionis principiis stabilita et ad omnes motus qui in huiusmodi corpora cadere possunt accomodata.

II-3 (327 pages, 1948) and II-4 (359 pages, 1950), Charles Blanc (Ed.).

Commentationes mechanicae.

II-5 (326 pages, 1957), Joachim Otto Fleckenstein (Ed.).

Commentationes mechanicae ad theoriam motus punctorum pertinentes.

II-6 (302 pages, 1957), and II-7 (326 pages, 1958), Charles Blanc (Ed.).

Mechanica corporum solidorum.

II-8 (417 pages, 1965), and II-9 (441 pages, 1968), Charles Blanc (Ed.).

Commentationes mechanicae ad theoriam flexibilium et elasticorum pertinentes.

> II-10 (451 pages, 1947), Fritz Stüssi and Henri Favre (Eds.).
>
> II-11/1 (383 pages, 1957), Fritz Stüssi and Ernst Trost (Eds.).
>
> II-11/2 (435 pages, 1960), Clifford Ambrose Truesdell (Ed.).

Commentationes mechanicae ad theoriam fluidorum pertinentes.

> II-12 (288 pages, 1954), Clifford Ambrose Truesdell (Ed.).
>
> II-13 (375 pages, 1955), Clifford Ambrose Truesdell (Ed.).

Neue Grundsätze der Artillerie.

> II-14 (484 pages, 1922), Friedrich Robert Sherrer (Ed.).

Commentationes mechanicae ad theoriam machinarum pertinentes.

> II-15 (318 pages, 1957), Jakob Ackeret (Ed.).
>
> II-16 (327 pages, 1979), and II-17 (312 pages, 1982), Charles Blanc and Pierre de Haller (Eds.).

Scientia navalis.

> II-18 (427 pages, 1967), and II-19 (459 pages, 1972), Clifford Ambrose Truesdell (Ed.).

Commentationes mechanicae et astronomicae ad scientiam navalem pertinentes.

> II-20 (275 pages, 1974), and II-21 (241 pages, 1978), Walter Habicht (Ed.).

Theoria motuum lunae, nova methodo petractata.

> II-22 (412 pages, 1958), Leo Courvoisier (Ed.).

Sol et luna.

> II-23 (336 pages, 1969), J. O. Fleckenstein (Ed.).
>
> II-24 (326 pages, 1991), Charles Blanc (Ed.).

Commentationes astronomicae ad theoriam perturbationum pertinentes.

> II-25 (331 pages, 1960), Max Schürer (Ed.).
>
> II-26, and II-27 (In preparation).

Commentationes astronomicae ad theoriam motuum planetarum et cometarum pertinentes.

> II-28 (332 pages, 1959), Leo Courvoisier (Ed.).

Commentationes astronomicae ad praecessionem et nutationem pertinentes.

> II-29 (420 pages, 1961), Leo Courvoisier (Ed.).

Sphärische Astronomie und Parallaxe.

> II-30 (351 pages, 1964), Leo Courvoisier (Ed.).

Kosmische Physik (In preparation).

Series tertia. *Opera physica, Miscellanea.*

> This is Series III, dedicated to physics and other miscellaneous contributions.

Commentationes physicae ad physicam generalem et ad theoriam soni pertinentes.

> III-1 (591 pages, 1926), Eduard Bernoulli, Rudolf Bernoulli, Ferdinand Rudio, and Andreas Speiser (Eds.).

Rechenkunst. Accesserunt commentationes ad physicam generalem pertinentes et miscellanea.

> III-2 (431 pages, 1942), Edmund Hoppe, Karl Matter, and Johann Jakob Burckhardt (Eds.). *Dioptrica.*
>
> III-3 (510 pages, 1911), III-4 (543 pages, 1912), Emil Cherbuliez (Ed.).

Commentationes opticae.

> III-5 (395 pages, 1963), David Speiser (Ed.).
>
> III-6 (396 pages, 1963), and III-7 (247 pages, 1964), Andreas Speiser (Ed.).

III-8 (266 pages, 1969), Max Herzberger (Ed.).

III-9 (328 pages, 1973), Walter Habicht and Emil Alfred Fellmann (Eds.).

Magnetismus, Elektrizität, und Wärme.

III-10 (In preparation).

Lettres à une princesse d'Allemagne. 1st part, III-11 (312 pages, 1960), Andreas Speiser (Ed.).

Lettres à une princesse d'Allemagne Accesserunt: Rettung der göttlichen Offenbarung Eloge d'Euler par le Marquis de Condorcet. 2nd part.

III-12 (312 pages, 1960), Andreas Speiser (Ed.).

Series quarta. A: *Commercium epistolicum.*

This is Series IV A. It consists of 9 volumes of his correspondence.

Descriptio commercii epistolici.

Beschreibung, Zusammenfassung der Briefe und Verzeichnisse.

IV A-1 (684 pages, 1975), Adolf P. Juskevic, Vladimir I. Smirnov, and Walter Habicht (Eds.).

Commercium cum Johanne (I) Bernoulli et Nicolao (I) Bernoulli.

IV A-2 (747 pages, 1998), E. A. Fellman and G. K. Mikhailov (Eds.).

IV A-3, IV A-4 (In preparation).

Correspondance de Leonhard Euler avec A. C. Clairaut, J. d'Alembert et J. L. Lagrange.

IV A-5 (611 pages, 1980), Adolf P. Juskevic and Rene Taton (Eds.).

Correspondance de Leonhard Euler avec P.-L. M. de Maupertius et Frédéric II.

IV A-6 (456 pages, 1986), P. Costabel, F. Winter, A. T. Grigorjian, and A. P. Juskevic (Eds.).

IV A-7, IV A-8, IV A-9 (In preparation).

In addition there are the two volumes of the correspondence of Euler edited by P.-H. Fuss which detail quite vividly Euler's contacts with Goldbach, the Bernoullis, and others.

P. -H. Fuss, *Correspondance Mathematique et Physique*, Volumes I, II. The Sources of Science, Johnson Reprint Corporation, 1968.

Series quarta. B: *Manuscripta.*

This is Series IV B, in preparation. It will consist of about 7 volumes, including his unpublished manuscripts, notes, diaries, and so on.

[2] There is a recent French edition published under the direction of S. D. Chatterji, by Presses Polytechniques et Universitaires Romandes, 2003, Lausanne.

[3a] A. Weil, *Number Theory: An Approach through History from Hammurapi to Legendre*, Birkhäuser, 1984.

[3b] C. A. Truesdell, *Leonard Euler, supreme geometer*, introduction to the English translation *Elements of Algebra*, by Rev. John Hewlett, Springer- Verlag, 1972, of *Vollständige Anleitung zur Algebra*, I-1, *Omnia Opera.*

[3c] A. P. Youschkevitch, *Euler, Leonhard*, Dictionary of Scientific Biography, 4(1971), 467-484.

[3d] R. Calinger, *Leonhard Euler: The first St. Petersburg years*, Historia Mathematica 23(1996), 121-166.

[3e] W. Dunham, *Euler, The Master of Us All*, Mathematical Association of America, 1999.

[3f] Math. Magazine, 56 (November 1983).

[4] G. H. Hardy, *Divergent Series*, Oxford, 1973.

[5] J. S. D. Glaus, Translations into English of the eulogies of the Marquis de Condorcet and of P.-H. Fuss. See the website

http://www-groups.dcs.st-and.ac.uk/~history

CHAPTER 2

The Universal Mathematician

2.1. Introduction

The most striking aspect of Euler as a mathematician is his universality. The modern mathematician who came closest to being universal is Hilbert; however he operated quite differently from Euler. Hilbert's career was divided into more or less disjoint periods in which he was almost exclusively concerned with one area. But Euler was very different. He would again and again return to his themes, sometimes even after several decades. A look at the chronology of his works reveals that he did many different things during the same period. Thus to get even a bird's eye view of what Euler accomplished, it is better to break down his opus by the various areas and then look at his contributions to each of them: mechanics, physics, astronomy and planetary motion, optics (both theory and practice), combinatorial topology and graph theory, and so on. The sheer immensity of the task of giving an overview of Euler's contributions to various branches of mathematics becomes obvious when one starts reading the introductions to the various volumes of *Opera Omnia*. I certainly do not have the competence to even attempt such a daunting task. However, I shall try to sketch in an impressionistic manner his work in elliptic integrals, calculus of variations, and number theory, in which his work created new areas of research that are still flourishing in a remarkable manner.

2.2. Calculus

In the realm of calculus Euler's achievement was no less than a complete systematisation of the subject. He wrote the first thorough textbooks on the subject. His *Institutiones calculi differentialis* (*Opera Omnia*, I-10) on differential calculus and his *Institutiones calculi integralis* in three volumes (*Opera Omnia*, I-11, I-12, I-13) on integral calculus presented the subject of calculus in its definitive form. They have determined the manner in which calculus is learned and taught even today. His papers on analysis, spread over several volumes titled *Commentationes analyticae*, developed a complete theory of circular functions and their applications to various aspects of analysis (*Opera Omnia*, I-14 through I-20). He then summarized a large part of this in his book entitled *Introductio in analysin infinitorum* in two volumes (*Opera Omnia*, I-8, I-9). The *Introductio* is one of the most beautiful books on analysis. Later on, inspired by Fagnano's work, he studied elliptic integrals and discovered the famous addition theorem of the subject. His work in this area is contained in his two volumes entitled *Commentationes analyticae ad theoriam integralium ellipticorum pertinentes* (*Opera Omnia*, I-20, I-21). In the calculus of variations his initial approach was highly geometric, and in his book *Methodus inveniendi lineas curvas maximi minimive proprietate gaudentes sive solutio problematis isoperimetrici latissimo sensu accepti* (*Opera Omnia*, I-24), he laid the foundations of a systematic enlargement of the subject beyond the few special problems studied by Johann Bernoulli and others. But a few years after this work appeared, Lagrange came up with an algebraic method that was very powerful and went beyond what Euler had done. From then on Euler adopted Lagrange's methods, and his efforts made the calculus of variations, a name he himself coined in honor of Lagrange's great discovery of the concept of variational derivative of a functional, a distinct subject in its own right. His papers are in *Commentationes analyticae ad calculum variationum pertinentes* (*Opera Omnia*, I-25). Finally, during the course of his extensive work on infinite series, Euler realized that many of his transformations of infinite series involved divergent series. With wonderful insight he saw that it was important to have some sort of theory of divergent series in which one would be able to associate sums to divergent series. His great paper *De seriebus divergentibus*, communicated in 1755 and published in 1760 (*Opera Omnia*, I-14, 585–617), was the first step in the modern theory of divergent series that was created in the 19th and 20th centuries by Abel, Frobenius, Cesaro, Borel, Hardy, Littlewood, Wiener, and a host of other mathematicians. In modern times, the idea of giving sense to divergent integrals has proved to be extremely powerful in all parts of analysis, from partial differential equations to representation theory of Lie groups to quantum field theory (see Chapter 5 for the story).

Differential calculus. The creation and initial development of differential calculus was the great achievement in the century before Euler's. It was based on the discoveries of Newton and Leibniz, followed by those of many others, of which the Bernoulli brothers, Jacob and Johann, were very prominent. Euler, being a student of Johann Bernoulli and hence, first and foremost, an analyst, took differential calculus to its proper stage of development. In *Institutiones calculi differentialis*, published in 1755, he erected the complete theory of differentiation. He did this based on a careful study of finite differences and then, as we would do now, went to the stage when the difference in the independent variable became 0. He worked out all the formulae for the derivatives of the rational and algebraic functions as well as the transcendental functions. He introduced partial differentiation of functions of several variables and derived the formula (going back to Nicolaus Bernoulli)

$$\frac{\partial^2 F}{\partial x \partial y} = \frac{\partial^2 F}{\partial y \partial x}.$$

He showed by many examples how to find the derivatives of a function when it is only implicitly given. He derived the criterion for the differential expression $Pdx + Qdy$ to be a total differential dV, namely,

$$\frac{\partial P}{\partial y} = \frac{\partial Q}{\partial x}.$$

He discussed transformations of series by using successive differences which converted slowly diverging or slowly converging series into rapidly converging series, namely what later became known as *Euler summation*. He also treated in an abbreviated fashion the Euler-Maclaurin summation formula, which in modern terms is the generation of the translation action as the exponential of differentiation and therefore clearly belongs in a treatise on differential calculus (see Chapter 4). He illustrated the use of this formula by several examples, such as the calculation of the zeta values and Euler's constant. He also used many examples to calculate roots of equations by power series methods. In all of his examples he was always very conscious of the goal of numerical computation and developed his formulae always to that stage at which numerical calculation became possible. In this sense his calculus treatise is quite modern. Finally, he treated problems of maxima and minima and gave many illustrations of what we now know as L'Hopital's rule.

Integral Calculus. The three-volume treatise *Institutiones calculi integralis*, published in 1768-1770, was a far more ambitious and extensive undertaking and spanned a whole slew of areas: explicit evaluation of integrals of rational functions with or without trigonometric factors, the definition and study of what will be later called Eulerian integrals by Legendre, namely, the Gamma and Beta functions. These were discussed in several equivalent forms, leading to the evaluation of integrals

$$\int \frac{x^{m-1}}{1+x^n} dx, \quad \int \frac{x^{m-1} \pm x^{n-1}}{1+x^n} dx.$$

In particular he knew all the basic properties of these integrals. He considered the circular and elliptic integrals in the form

$$\int \frac{dx}{\sqrt{1-x^2}}, \quad \int \frac{dx}{\sqrt{A+Cx^2+Ex^4}},$$

and even the most general

$$\int \frac{dx}{\sqrt{A+2Bx+Cx^2+2Dx^3+Ex^4}}.$$

In addition he gave a systematic treatment of differential equations, both ordinary and partial, of first and second order. He discussed constant coefficient differential equations, both homogeneous and inhomogeneous, and their versions on the torus (Euler equations). He also treated the cases when the solutions of differential equations could be obtained as power series, thus paving the way for his successors like Frobenius and Fuchs. He knew the integral representations of the hypergeometric functions, the relation between the solutions of the Riccati equation and continued fractions. He studied the equation of a vibrating string and improved upon the work of D'Alembert. He was one of the influential participants in the controversy that began with Daniel Bernoulli's assertion that the solutions of the vibrating string equation could be expanded as a series of sines and cosines; this question of course would not be settled till Fourier and Dirichlet developed a more general concept of a real function than was hitherto understood and erected what may be regarded as the foundations of modern real analysis.

2.3. Elliptic integrals

Euler's work on elliptic integrals, which occupies almost all of volumes I-20, I-21 of *Opera Omnia*, marks the birth of the modern theory of differentials and their integrals on compact Riemann surfaces (or equivalently, smooth projective algebraic curves). It deals with the case of *elliptic* or genus 1 curves; its full extension to include all algebraic functions and their integrals was essentially the achievement of Abel and especially of his successor, Riemann.

Even though Euler's interest in elliptic integrals goes back to his early years with Johann Bernoulli, they were rekindled in a major way when, in December 1751, the two-volume work of C. G. Fagnano entitled *Produzioni Matematiche*, published in 1750, came to the Berlin Academy for Euler's review and attention. Among the works of Fagnano included in these volumes were articles published around the years 1714-1720 that contained the formula for the duplication of the arc length of the lemniscate. These papers ignited Euler's interest, and starting from January 1752, just a few weeks after he came across Fagnano's work, Euler wrote the first of a series of papers creating the new area of algebraic functions and their integrals and taking it to heights no one could have dreamt of. Among his remarkable discoveries are the famous addition and multiplication theorems for elliptic integrals. Of this achievement Weil writes [4], p. 246:

> With characteristic generosity Euler never ceased to acknowledge his indebtedness to Fagnano; but surely none but Euler would have seen in Fagnano's isolated results the germ of a new branch of analysis. . . .

The words *lemniscatus* or *lemniscus* in Latin refer to decorative ribbons, and *lemniscate* refers to the curve which is the locus of a point that moves in the plane in such a way that the *product* of its distances from two fixed points is a constant c. We consider the special case when the points are $(\pm a, 0)$ and $c = a^2$. Then its equation in Cartesian coordinates is

$$r^4 + 2a^2 r^2 - 4a^2 x^2 = 0, \qquad r^2 = x^2 + y^2.$$

The curve looks like a horizontal figure eight, thus justifying its name. We take $2a^2 = 1$ and get the equation

$$2x^2 = r^2 + r^4.$$

In polar coordinates this is

$$r^2 = \cos 2\theta = 1 - 2\sin^2 \theta$$

giving

$$2y^2 = r^2 - r^4$$

and hence the parametric representation

$$r \longmapsto \left(r\sqrt{\frac{1+r^2}{2}}, r\sqrt{\frac{1-r^2}{2}} \right) \qquad (0 \le r \le 1)$$

for the part of the lemniscate that we are interested in. Using this we get for the arc length integral the formula

$$s = s(r) = \int_0^r \frac{dt}{\sqrt{1-t^4}}.$$

Notice that this is very analogous to the arc length of the circle which is

$$\arcsin r = \int_0^r \frac{dt}{\sqrt{1-t^2}}.$$

However for the lemniscate the arc length integral cannot be expressed in terms of elementary functions. Nevertheless Fagnano discovered a most interesting formula that duplicates the arc length: namely, that

$$\int_0^r \frac{dt}{\sqrt{1-t^4}} = 2\int_0^u \frac{ds}{\sqrt{1-s^4}}$$

if

$$r = \frac{2u\sqrt{1-u^4}}{1+u^4}.$$

In order to motivate the huge leap from Fagnano to Euler, it may help to look at the analogous case of the circle. Here

$$\arcsin r = \int_0^r \frac{du}{\sqrt{1-u^2}},$$

and the formula

$$\sin 2x = 2\sin x \cos x$$

leads to the duplication formula

$$\arcsin r = 2\arcsin u \qquad \text{if } r = 2u\sqrt{1-u^2}.$$

In this case we can do more; the addition formula

$$\sin(x+y) = \sin x \cos y + \cos x \sin y,$$

which we can recast as

$$\arcsin u + \arcsin v = \arcsin r, \qquad r = u\sqrt{1-v^2} + v\sqrt{1-u^2},$$

leads to the addition theorem, i.e.,

$$\int_0^u \frac{dt}{\sqrt{1-t^2}} + \int_0^v \frac{dt}{\sqrt{1-t^2}} = \int_0^r \frac{dt}{\sqrt{1-t^2}} \qquad r = u\sqrt{1-v^2} + v\sqrt{1-u^2}.$$

Euler saw in Fagnano's work more than just a duplication formula. In the first place, he obtained by induction the formula for the transformation that multiplies the arc length by n: if $z \longmapsto u$ is the transformation that results in the multiplication of the lemniscatic arclength by n, then the transformation

$$z \longmapsto v = \frac{z\sqrt{\frac{1-u^2}{1+u^2}} + u\sqrt{\frac{1-z^2}{1+z^2}}}{1 - uz\sqrt{\frac{(1-u^2)(1-z^2)}{(1+u^2)(1+z^2)}}}$$

results in the multiplication of the lemniscatic arc length by $n+1$. He then proceeded to obtain his famous *addition theorem*, which included the duplication formula as a special case: if one puts

$$s(u) = \int_0^u \frac{dt}{\sqrt{1-t^4}},$$

then

$$(*) \qquad s(u) + s(v) = s(r), \qquad r = \frac{u\sqrt{1-v^4} + v\sqrt{1-u^4}}{1+u^2v^2}.$$

Taking $u = v$ we get the duplication formula. It is entirely characteristic that Euler formulated all this in the language of differential forms. Let

$$\omega(x) = \frac{dx}{\sqrt{1-x^4}}.$$

Then the addition theorem can be written as

$$\omega(r) = \omega(u) + \omega(v) \qquad r = \frac{u\sqrt{1-v^4} + v\sqrt{1-u^4}}{1 + u^2v^2}$$

and the multiplication theorem as the inductive definition of the transformation that gives

$$\omega(r) = n\omega(u).$$

It is not at all clear how Euler came to these remarkable results. I shall give Siegel's reconstruction of Euler's argument for the addition theorem [3], pp. 7-11. If one thinks of the duplication transformation as a special case of the addition theorem, one should try to obtain a symmetric algebraic expression in u and v that reduces to

$$\frac{2u\sqrt{1-u^4}}{1 + u^4}$$

when $u = v$. One of the simplest such is

$$\frac{u\sqrt{1-v^4} + v\sqrt{1-u^4}}{1 + u^2v^2}.$$

The real mystery is how Euler came to the above formula. To start with let us first try to prove the addition formula assuming the form given to it by Euler:

$$(*) \qquad s(u) + s(v) = s(r), \qquad r = \frac{u\sqrt{1-v^4} + v\sqrt{1-u^4}}{1 + u^2v^2}.$$

For the proof it is convenient to assume that r is fixed, so that v becomes a function of u which is equal to r when $u = 0$. Let

$$U = \sqrt{1-u^4}, \qquad V = \sqrt{1-v^4}.$$

Then

$$(1 + u^2v^2)r = uV + vU$$

and we get, differentiating with respect to u,

$$(2uv^2\,du + 2vu^2\,dv)(uV + vU) = (1 + u^2v^2)(V\,du + U\,dv + u\,dV + v\,dU).$$

We simplify this, keeping in mind the symmetry between u and v. We obtain

$$A\,du + B\,dv = 0$$

where

$$A = (1 + u^2v^2)(V + v(dU/du)) - 2uv^2(uV + vU)$$

and B is obtained from A by interchanging u and v. A simple calculation shows that A can be written as

$$A = \frac{1}{U}\left\{ UV(1 - u^2v^2) - 2uv(u^2 + v^2) \right\}.$$

The expression

$$C = UV(1 - u^2v^2) - 2uv(u^2 + v^2)$$

within braces is *symmetric* in u and v, and so we get

$$C\left(\frac{du}{U} + \frac{dv}{V} \right) = 0.$$

Now C reduces to $\sqrt{1-r^4}$ when $u = 0$; it is hence nonzero. Cancelling it, we obtain

$$\frac{du}{\sqrt{1-u^4}} + \frac{dv}{\sqrt{1-v^4}} = 0$$

from which we get, integrating from 0 to u,

$$\int_0^u \frac{du}{\sqrt{1-u^4}} + \int_r^v \frac{dv}{\sqrt{1-v^4}} = 0.$$

This implies at once that

$$\int_0^u \frac{dt}{\sqrt{1-t^4}} + \int_0^v \frac{dt}{\sqrt{1-t^4}} = \int_0^r \frac{dt}{\sqrt{1-t^4}} \quad \text{if } r = \frac{u\sqrt{1-v^4} + v\sqrt{1-u^4}}{1+u^2v^2}.$$

The integrations are all valid if u and v are positive and sufficiently small.

Once he had obtained the addition and multiplication theorems for the lemniscate, it was natural for Euler to ask if these results hold more generally. He did this when he obtained the addition theorem for the arbitrary elliptic integral where the expression $1-u^4$ under the radical sign is replaced by a generic quartic polynomial. Indeed, he realized that by a fractional linear transformation one can reduce a generic quartic to the form

$$P(u) = 1 + mu^2 + nu^4$$

so that it is enough to treat this case. (This is already very interesting because one needs to know the projective invariants of quartics, a subject that did not really get started till Clebsch, Gordan, and others began to study projective invariants of binary forms.) Then Euler obtained his general addition theorem in the form

$$\int_0^u \frac{dt}{\sqrt{P(t)}} + \int_0^v \frac{dt}{\sqrt{P(t)}} = \int_0^r \frac{dt}{\sqrt{P(t)}}$$

where

$$r = \frac{u\sqrt{P(v)} + v\sqrt{P(u)}}{1 - nu^2v^2}.$$

Concerning Euler's proof of the addition theorem, Siegel [3], p. 10, writes:

> Euler's penetrating proof continues to elicit admiration to this very day. What is not satisfying about Euler's proof is that one cannot tell a priori that the substitution (∗) will yield the required result. The addition theorems for the trigonometric functions become quite transparent when we bring in the connection with the exponential function, and it is reasonable to conjecture that the Euler addition theorem can also be proved more simply with the aid of function-theoretical arguments if we study thoroughly the relevant integrals as analytic functions in their full domains of definition. . . .

Legendre, Gauss, Eisenstein, Jacobi, and Weierstrass, who came after Euler, went very deeply into the analysis and arithmetic of elliptic integrals. Legendre spent a lifetime working out the theory of elliptic functions and wrote a monumental treatise on them, *Traité des fonctions elliptiques et des integrales eulériennes*. Gauss discovered the double periodicity of the function inverse to the lemniscatic integral

$$\int_0^z \frac{dx}{\sqrt{1-x^4}}$$

and developed a theory of what he called the lemniscatic sine and cosine functions. Jacobi, Eisenstein, and later Weierstrass developed a complete theory of elliptic functions and their applications to analysis and arithmetic.

The key step in the above derivation of the addition theorem is the fact that the transformation $u \longmapsto v$ leads to the differential equation

$$\frac{du}{\sqrt{1-u^4}} + \frac{dv}{\sqrt{1-v^4}} = 0.$$

It is clear from the very first paper that he wrote on this subject that Euler saw in the differential equations

$$m\frac{dx}{\sqrt{1-x^4}} = \pm n\frac{dy}{\sqrt{1-y^4}}$$

the veritable key to unlock the entire algebraic theory of elliptic integrals. He obtained what he called the *complete integral* or the *canonical equation*

$$c^2 x^2 y^2 + x^2 + y^2 = c^2 + 2xy\sqrt{1-c^4}$$

for the equation

$$\frac{dx}{\sqrt{1-x^4}} = \frac{dy}{\sqrt{1-y^4}},$$

c being an arbitrary constant. If we solve for y from this equation, we get

$$y = \frac{x\sqrt{1-c^4} \pm c\sqrt{1-x^4}}{1+c^2 x^2},$$

and this is nothing other than the transformation entering the statement of the addition theorem! If we now observe that $y = c$ when $x = 0$ and integrate the equation

$$\frac{dx}{\sqrt{1-x^4}} = \frac{dy}{\sqrt{1-y^4}}$$

from 0 to x, we get

$$\int_0^x \frac{dt}{\sqrt{1-t^4}} = \int_c^y \frac{dt}{\sqrt{1-t^4}};$$

adding

$$\int_0^c \frac{dt}{\sqrt{1-t^4}}$$

to both sides we get

$$\int_0^x \frac{dt}{\sqrt{1-t^4}} + \int_0^c \frac{dt}{\sqrt{1-t^4}} = \int_0^y \frac{dt}{\sqrt{1-t^4}}$$

where

$$y = \frac{x\sqrt{1-c^4} \pm c\sqrt{1-x^4}}{1+c^2 x^2}.$$

This discussion still falls short of explaining the mystery of how Euler saw that the transformations that preserve the differential form

$$\frac{dt}{\sqrt{1-t^4}}$$

would also lead to the addition theorem. We shall come to this point presently when we look at Euler's work through modern glasses, as suggested by Siegel.

Euler extended the determination of the canonical equation to the more general quartic $1 + mx^2 + nx^4$; the complete integral for

$$\frac{dx}{\sqrt{1+mx^2+nx^4}} = \frac{dy}{\sqrt{1+my^2+ny^4}}$$

was found by him to be

$$-nc^2x^2y^2 + x^2 + y^2 = c^2 + 2xy\sqrt{1 + mc^2 + nc^4}$$

where c is an arbitrary constant. Solving this for y we get

$$y = \frac{x\sqrt{P(c)} \pm c\sqrt{P(x)}}{1 - nc^2x^2}, \qquad P(t) = 1 + mt^2 + nt^4$$

from which one obtains the general addition theorem. Finally, Euler obtained for the general quartic

$$F(x) = A + 2Bx + Cx^2 + 2Dx^3 + Ex^4$$

the canonical equation in the *symmetric* form

$$\Phi(x, y) = 0$$

where

$$\Phi(x, y) = \alpha + 2\beta(x + y) + \gamma(x^2 + y^2) + 2\delta xy + 2\varepsilon xy(x + y) + \zeta x^2 y^2 = 0.$$

Given F, he determined the coefficients of Φ in terms of a single parameter, so that $\Phi = 0$ is the complete integral of

$$\frac{dx}{\sqrt{F(x)}} = \frac{dy}{\sqrt{F(y)}},$$

namely,

$$\frac{dx}{\sqrt{F(x)}} = \frac{dy}{\sqrt{F(y)}} \iff \Phi(x, y) = 0.$$

Once again, this would lead to the general addition theorem.

From the modern perspective it is best to view Euler's work in terms of the theory of smooth projective algebraic curves of genus 1 or, equivalently, the theory of compact Riemann surfaces of genus 1. A complete discussion would take us too far afield, so I shall just restrict myself to a few brief remarks; the interested reader should look up [3], Ch. 1, and [4], pp. 242-252, 296-307.

If F is a quartic polynomial with distinct roots, the affine curve

$$y^2 = F(x)$$

can be viewed as a complex manifold of dimension 1. By adding points at infinity (two of them in fact) in the classical manner prescribed by Weierstrass, we obtain the compact Riemann surface Γ associated to the algebraic function given by the equation $y^2 = F(x)$. x and y become meromorphic functions on Γ satisfying the relation $y^2 = F(x)$; the map

$$x : \Gamma \longrightarrow S = \mathbf{C} \cup \{\infty\}$$

exhibits Γ as a two-sheeted ramified cover of the Riemann sphere S; and the field of meromorphic functions on Γ is precisely the field of all rational functions of x and y. Since $y^2 = F(x)$, these are the functions of the form $R(x) + yS(x)$ where R, S are rational functions of x. We refer to x and y as the affine coordinates. Let $\Gamma^\times$ be the affine curve $y^2 = F(x)$ viewed inside Γ. The basic fact is that Γ has genus 1; i.e., it is an *elliptic curve*. The space of holomorphic 1-forms on Γ, i.e., differentials of the first kind, then has dimension 1. If we put

$$\omega = \frac{dx}{y},$$

then it can be verified that ω is actually holomorphic on Γ. One then begins by considering its integral

$$\int_A^M \omega = g(M)$$

where the integral is taken over some piecewise smooth path from A to M on Γ. Keeping A fixed for the moment, we may view g as a many-valued function of M since $g(M)$ depends on the path from A to M, but the difference between two evaluations of $g(M)$ is the integral of ω over a closed path, called a *period*. The periods form a *lattice* L in the complex plane $\mathbf{C}$, and it is the central fact of the theory that the single-valued map

$$g : \Gamma \longrightarrow \mathbf{C}/L$$

is an *analytic diffeomorphism*. One can thus invert it and carry over the additive group structure on $\mathbf{C}/L$ to Γ. Γ thus becomes a compact complex abelian Lie group, with A as its identity element. A has been arbitrary so far, but we shall choose it to be $(0, 1)$ later. Γ is an algebraic group; i.e., the group operations are rational in the coordinates x, y. As a complex Lie group of dimension 1, Γ will admit a holomorphic 1-form invariant under translations. This form has to be a multiple of ω as the space of holomorphic forms is of dimension 1. So ω itself is translation invariant.

Write the sum of two points $M, N \in \Gamma$ as

$$M \dotplus N.$$

Then, if the paths are appropriately chosen, we have

$$\int_A^M \omega + \int_A^N \omega = \int_A^{M \dotplus N} \omega.$$

This is the abstract addition theorem. The map

$$M, N \longmapsto M \dotplus N$$

is holomorphic from $\Gamma \times \Gamma$ to Γ, and so becomes rational in x, y. In other words, $M \dotplus N$ can be rationally expressed in terms of the affine coordinates of M and N. *Euler's addition theorem in the general case is thus an explicit description of the group structure on Γ, i.e., the explicit evaluation of the rational map*

$$M, N \longmapsto M \dotpm N$$

that gives the group structure on Γ; the multiplication formulae correspond to the maps

$$M \longmapsto M \dotplus M \dotplus M \dotplus \ldots \dotplus M \quad (n \text{ times}).$$

Once this framework is in place we can explain the role played by the complete integral for the differential equation

$$\omega(x, dx) = \omega(x', dx').$$

To this end we start by determining the maps $\Gamma \longrightarrow \Gamma$ that preserve ω. The situation is entirely group theoretic and is described by the following lemma.

LEMMA. *If G is a real or complex connected Lie group, then G is isomorphic to the group of analytic diffeomorphisms $G \longrightarrow G$ that preserve all left invariant 1-forms; i.e., every analytic diffeomorphism of G that preserves each of the left invariant 1-forms is itself a left translation by an element of G.*

PROOF. If $S(G \longrightarrow G)$ fixes all left invariant 1-forms, it fixes all left invariant vector fields. Since these vector fields integrate to *right translations* of G, we see that S commutes with all right translations. If r_x (resp. ℓ_x) is right (resp. left) translation by x and $S1 = a$, then

$$Sx = Sr_x 1 = r_x S1 = r_x a = ax = \ell_a x$$

so that $S = \ell_a$.

In the simple case in front of us we can prove this in an elementary manner. Recall that A is the identity element of Γ. If S takes A to B and T is the translation that takes B back to A, then TS preserves ω and fixes A. So we may assume that S itself fixes A and prove that it is the identity map. Now $d\omega = 0$. To see this, in some local coordinate z at A, we have

$$\omega = f(z)dz, \qquad f(A) \neq 0,$$

locally where f is holomorphic. We may then write

$$f = \frac{dg}{dz}$$

where g is holomorphic at 0, and g will be unique if we require that $g(0) = 0$. Then $\omega = dg$ locally. But S preserves ω and takes dg to $d(g \circ S)$, and $S(A) = A$. So,

$$dg = d(g \circ S), \quad g(A) = (g \circ S)(A) = 0,$$

from which, by the uniqueness of g, we get $g \circ S = g$. Now, $g(0) = 0$ and $(dg/dz)(0) \neq 0$, so that g is a local coordinate. Since it is fixed by S, it is immediate that S is the identity near A. Hence S is the identity transformation.

The fact that every transformation that preserves ω is nothing but the translation by some element of Γ, i.e., is of the form

$$M \longmapsto M' = M \dotplus B$$

for some $B \in \Gamma$, makes it very transparent why the complete integral of the differential equation

$$\omega(x, dx) = \omega(x', dx')$$

gives the transformations that lead to the addition theorem.

It remains to use this general theory to obtain Euler's formulae. This is done in full detail in [4], pp. 300-304, and I shall give just a sketch of the calculations. We take A to be the point $(0, 1)$ and fix it. Let $B \in \Gamma$ and let θ be the automorphism

$$M \longmapsto M \dotplus B, \qquad \theta(x, y) = (x', y')$$

of Γ. Then x' and y' are rational functions of x and y and so can be written in the form $R(x) + yS(x)$ where R and S are rational functions of x. We need a lemma.

THEOREM. *There exists an involutive automorphism of Γ that interchanges x and x'.*

PROOF. The map $\sigma((x, y) \longmapsto (x, -y))$ extends to an automorphism of Γ and changes ω to $-\omega$. Since $M \longmapsto \dot{-}M$ also has the same properties, it follows that for some $C_0 \in \Gamma$,

$$\sigma(M) = -M \dotplus C_0 \qquad (M \in \Gamma).$$

The automorphism $\sigma\theta$ is then involutive and is easily verified to exchange x and x'.

Write $x' = R(x) + yS(x)$ as above. Let $\Phi(x, x')$ be the irreducible polynomial such that $\Phi(x, x') = 0$ is the minimal equation of x' over the field of rational functions of x. Φ is unique up to a scalar multiple and can be written as

$$\Phi(x, x') = P_0(x)x'^2 + 2P_1(x)x' + P_2(x)$$

where the P_i are polynomials of degree ≤ 2. It follows from the above lemma that Φ is *symmetric* in x and x'. Thus we have an *a priori* justification of the fact that Euler always took his canonical equation as symmetric.

Let us now assume that

$$F(x) = 1 + mx^2 + nx^4.$$

Let τ be the involutive automorphism of Γ given by

$$\tau(x, y) = (-x, -y), \qquad \tau(M) = M^*.$$

Then τ preserves ω and so

$$\tau(M) = M \dot{+} B_0$$

for some $B_0 \in M$. Thus τ commutes with θ and so $\theta(M^*) = \theta(M)^*$. This implies that $\Phi(-x, -x') = 0$, and so, from the uniqueness of Φ we get

$$\Phi(x, x') = \alpha + \gamma(x^2 + x'^2) + 2\delta xx' + \zeta x^2 x'^2.$$

Thus

$$P_0(x) = \gamma + \zeta x^2, \quad P_1(x) = \delta x, \quad P_2(x) = \alpha + \gamma x^2.$$

We shall now determine the constants $\alpha, \gamma, \dots$ above. Let $B = (a, b)$. Then, as the x-coordinates of A and $\theta(A) = B$ are 0 and a respectively, we must have $\Phi(0, a) = 0$, giving $\alpha = -\gamma a^2$. Now since $x' = R + yS$ we must have $R \pm yS$ as the roots of $\Phi = 0$, so that

$$R = -\frac{P_1}{P_0}, \quad R^2 - FS^2 = \frac{P_2}{P_0}.$$

Hence

$$P_1^2 - P_0 P_2 = S^2 P_0^2 F.$$

Writing F as u/v where u, v are polynomials without a common factor, we have

$$v^2(P_1^2 - P_0 P_2) = u^2 F.$$

In particular u and v have the same degree. But it is clear from the above that v^2 must divide F, and so, as we are assuming that F has no multiple roots, v, hence u also, must be constant. So

$$P_1^2 - P_0 P_2 = \rho^2 F$$

where ρ is a nonzero constant (it may depend on B). This gives

$$\alpha\gamma = -\rho^2, \quad \zeta\gamma = -n\rho^2, \quad \delta^2 = \gamma^2 + \alpha\zeta + m\rho^2.$$

So γ cannot be 0 and so we may normalize it by taking it as -1. Thus

$$\alpha = a^2 = \rho^2, \quad \zeta = na^2, \delta^2 = 1 + ma^2 + na^4 = b^2.$$

This determines Φ:

$$\Phi(x, x') = a^2 - (x^2 + x'^2) + 2\delta xx' + na^2 x^2 x'^2.$$

Solving for x' from the equation $\Phi(x, x') = 0$ we get

$$x' = \frac{\pm\delta x \pm ay}{1 - na^2 x^2}.$$

If we now remember that (i) for $x = 0$ one has $x' = a$ and (ii) the expression for x' does not change when we interchange (x, y) with (a, b) $(M \dot{+} B = B \dot{+} M)$, we see that both ambiguous signs are $+$ and hence

$$x' = \frac{bx + ay}{1 - na^2 x^2} = \frac{x\sqrt{1 + ma^2 + na^4} + a\sqrt{1 + mx^2 + nx^4}}{1 - na^2 x^2},$$

which is Euler's formula. However Euler had nothing like the theory we are using. All he had when he started were the scattered calculations of Fagnano on the lemniscate, and so this achievement of his is absolutely remarkable, even by his own standards.

One final remark: one can construct the above theory of the elliptic curves and integrals more carefully, keeping track of the fields in which the coefficients defining the elliptic curve lie. Such a treatment will lead to a complete *arithmetico-geometric* approach to diophantine equations of genus 1. In his late years Euler treated many diophantine equations of genus 1, in particular obtaining rational solutions using the addition theorem applied to some given solutions. The reader should consult [4], pp. 252-256, for a modern view of this aspect of Euler's work.

In the modern era it is natural to look for the higher-dimensional versions of the above theory. This is a very rich part of mathematics that is still far from being exhausted. The higher-dimensional analogue of the elliptic curves are *abelian varieties*, which are certain special compact connected complex manifolds having a projective imbedding; the word special here has one of two equivalent meanings: either that they are tori, i.e., of the form $\mathbf{C}^g / L$ where L is a lattice (of rank $2g$ in $\mathbf{C}^g$), or they are compact connected complex *Lie groups*. A compact connected complex Lie group is necessarily abelian, and the exponential map is a *morphism* so that it is of the form $\mathbf{C}^g / L$ as defined above. By a theorem of Chow any projective imbedding of such a variety is algebraic, and the addition map $A, B \longmapsto A \dot{+} B$ is rational. The deeper study of rational maps that define an algebraic group was begun by Weil,[*] but there remains much to be done to bring the explicit aspects of the higher-dimensional theory to resemble what Euler did.

2.4. Calculus of variations

Problems involving the determination of curves with specified maximum or minimum properties go back to ancient times. However it is fair to say that the treatment of such problems with calculus or some similar technique began with Fermat. In his studies on the propagation of light in inhomogeneous media Fermat postulated around 1650 the principle that light always travels along paths that make the transit time the least. Fermat obtained the laws of reflection and refraction from this principle (see [6], Chs. 26 and 27, for a beautiful treatment of geometrical optics based on the Fermat principle).

A significant landmark in the history of such problems was the challenge of Johann Bernoulli in 1696 to mathematicians to solve what he called the *Brachystochrone* problem: given two points A, B in the vertical plane, to find the path that a point mass, moving under gravity, takes from A to reach B in the least possible time. This problem vastly extends the Fermat principle; indeed, whereas Fermat always considered only paths involving only finitely many distinct homogeneous

---

[*]A. Weil, *On algebraic groups of transformations*, Amer. Jour. of Math., 77 (1955), 355-381; *Collected Papers*, Vol. II, 197-233.

media in each of which the path was along straight lines (this assumption will reduce problems to ordinary calculus), Bernoulli's problem involved a continuously varying parameter, namely the height. Many people solved this problem, including Leibniz, Johann Bernoulli himself, his brother Jacob, and Newton (anonymously, and apparently in a few hours). The solutions appeared in the May 1697 issue of the *Acta Eruditorium*. Johann Bernoulli's solution used the device of dividing the vertical plane into horizontal slices in each of which he applied the Snell law of refraction. Historically, the brachystochrone problem with its many variations and their solutions played an important role in the origin and development of the subject of calculus of variations, especially at the hands of Euler.

It is almost certain that Euler's interest in such problems dates back to his days with Johann Bernoulli. However it was during his first St. Petersburg stay that he began to think seriously about such problems, and he started publishing his results around 1738. From the beginning he was interested in a systematic theory that would encompass whole classes of problems, thus going beyond the special questions that were the fashion in the earlier era. He was also, characteristically, not one to indulge in challenges and counterchallenges that were typical of the earlier era, but went about quietly developing the general theory. This he did in his magnificent book *Methodus inveniendi lineas curvas maximi minimive proprietate gaudentes sive solution problematis isopermietrici latissimo sensu accepti, Opera Omnia,* I-24, written in 1744. Constantin Carathéodory, who was the editor of this volume (published in 1952), called it one of the most beautiful books in mathematics. In this book Euler systematically described a large number of problems of various types and developed a general *geometric* approach to their solution.

The simplest of such problems is to find a curve $y(x)$ that makes the integral

$$\int_a^b F(x, y, y')$$

a maximum or a minimum, the curves being required to have the end points fixed, i.e.,

$$y(a) = A, \quad y(b) = B$$

being fixed. Euler derived the famous differential equation that the solution curve should satisfy, namely,

$$\frac{\partial F}{\partial y} = \frac{d}{dx} \frac{\partial F}{\partial y'}.$$

Notice that this is a second-order differential equation with *boundary conditions* that specify the solution curve at the end points a and b. He also treated the same problem with the additional constraint

$$\int_a^b G(x, y, y')dx = k$$

where k is a constant. These were called *isoperimetric* because of the analogy with the classical problem of maximizing the area under a curve when the perimeter is held constant. For these he found the principle that it is equivalent to the earlier problem but with F replaced by $F + \kappa G$ where κ is a suitable constant to be determined. He also generalized these problems to the situation when F and G depend on $x, y, y', y^{(2)}, \ldots, y^{(n)}$. In the simple case when there are no constraints except that the end points are fixed, he obtained the differential equation

$$\frac{\partial F}{\partial y} - \frac{d}{dx}\frac{\partial F}{\partial y'} + \frac{d^2}{dx^2}\frac{\partial F}{\partial y^{(2)}} - \cdots + (-1)^n \frac{d^n}{dx^n}\frac{\partial F}{\partial y^{(n)}} = 0$$

with the boundary conditions

$$y(a) = A_0, \ y'(A) = A_1, \ldots, \ y^{(n-1)} = A_{n-1},$$
$$y(b) = B_0, \ y'(b) = B_1, \ldots, \ y^{(n-1)}(b) = B_{n-1}.$$

For deriving these equations his technique was to start with a possible curve $y(x)$ and vary $y(t)$ at each point t and equate the variation to 0. This technique, which is very geometric, was cumbersome to work with when the constraints became complicated. To be sure, because of the novelty of the subject and the generality that Euler strove for, there were many gaps in his treatment. For instance, in applying this method to classical problems, the fact that the solution curve is a maximum or a minimum would have to be verified separately.

The *methodus inveniendi* contained other remarkable things: the invariance of the Euler equations under coordinate transformations and the treatment of the *principle of least action* for the dynamics of point masses. This latter is generally attributed to Maupertius, but there is no doubt about Euler's priority (see [9]). This paper of Euler is the first indication after Fermat's work that the equations governing the physical world are really variational equations. Nowadays this has been elevated to a central over-arching principle in all of physics.

The subject took a huge jump in 1755, when Lagrange, then a young mathematician from Turin, wrote a letter to Euler in which he outlined a new concept, that of *variational derivative*, that enabled him to treat in an algebraic manner all the problems that Euler studied in *Methodus inveniendi* with much greater ease and simplicity ([7], Vol. XIV). Euler immediately recognized the power and depth of Lagrange's ideas and dropped his own methods in favor of Lagrange's. He named the subject as the *calculus of variations*, a name that has been retained to this day. His papers written after 1755 are to be found in *Commentationes analyticae ad calculum variationum pertinentes, Opera Omnia*, I-25. Langrange's variational derivative was a new kind of derivative which he denoted by δ. It was new because it was truly a derivative in infinitely many variables. It is perhaps fair to say that the concept of variational derivative was the first true indication that the calculus of variations is really the use of differential calculus on *infinite dimensional* spaces like the space of paths to determine the extrema of functionals defined on them. Lagrange then wrote two fundamental papers based on his ideas, treating the isoperimetric problems more completely than Euler by the method of what is now called *Lagrange multipliers*. Although its genesis was already in *Methodus inveniendi*, it became a systematic tool only with Lagrange's work.

In the problem of extrema for

$$J[y] = \int_a^b F(x, y, y')dx,$$

the variational derivative δJ is to be viewed as the linear form

$$\delta J : h \longmapsto \delta J[h] = \int_a^b (F_y h + F_{y'} h')dx$$

where h varies over the linear space of functions h with $h(a) = h(b) = 0$. Indeed, if y_α is a curve in the space of paths, namely, a family of curves $y(x, \alpha)$ passing through A and B with $y(x, 0) = y(x)$, we define

$$h(x) = \left(\frac{d}{d\alpha}\right)_{\alpha=0} y(x, \alpha)$$

as the *tangent direction* for the family at the curve $y(x)$, and

$$\delta J[h] = \left(\frac{d}{d\alpha}\right)_{\alpha=0} \int_a^b F(x, y(x, \alpha), y'(x, \alpha)).$$

It is thus the analog in path space of the directional derivative in the calculus of functions of many variables. The Euler equations then simply express the fact that for an extremum the variational derivatives should be 0 in all tangent directions, i.e., $\delta J[h] = 0$ for all h as above. Thus they are precisely the analog in path space of the classical conditions for a critical point.

After Euler and Lagrange, the calculus of variations became a major discipline in mathematics and attracted the attention of many. Legendre introduced the fundamental idea of the *second variation* in an attempt to decide when the solution to the Euler-Lagrange equations is a minimum. In analogy with what happens in ordinary calculus, the second variation at a solution curve $y(x)$ is the *Hessian form* of the functional J; i.e., it is the quadratic functional

$$\delta^2 J[h] = \frac{1}{2} \int_a^b (F_{yy}h^2 + 2F_{yy'}hh' + F_{y'y'}h'^2)dx.$$

Integrating by parts and using $h(a) = h(b) = 0$ we get

$$\int_a^b 2F_{yy'}hh'dx = -\int_a^b \left(\frac{d}{dx}F_{yy'}\right)h^2 dx$$

from which it follows that

$$\delta^2 J[h] = \int_a^b (Ph'^2 + Qh^2)dx$$

where

$$P = \frac{1}{2}F_{y'y'}, \quad Q = \frac{1}{2}\left(F_{yy} - \frac{d}{dx}F_{yy'}\right).$$

The second variation is thus a *quadratic form*, and it is natural to expect that the condition for $y(x)$ to be a minimum is that this quadratic form should be *positive definite*. Legendre's work led to the necessary condition (*Legendre condition*) that $F_{y'y'} \geq 0$ at each point of the solution curve, but his attempts to prove the sufficiency of the *strengthened Legendre condition* that $F_{y'y'} > 0$ should be satisfied at all points of the solution curve $y(x)$ for $\delta^2 J$ to be positive definite did not succeed, although his work introduced many techniques that would be important in subsequent work. Actually, no *local condition* on the curve $y(x)$ can settle this question. It was left to Jacobi to find additional necessary conditions for this question with his concept of *conjugate points*. Jacobi's great paper of 1836 ([8], Vol. IV, pp. 39–55) did not contain any proofs, and his successors spent years working them out. Jacobi's work, which contained many new ideas, pointed very clearly to the global nature of the problems of the calculus of variations. The Jacobi necessary condition, namely that for $\delta^2 J$ to be positive definite there should be no point in $(a, b]$ conjugate to a, together with the strict positivity of $F_{y'y'}$, was eventually proved by Weierstrass to be sufficient for the extremal $y(x)$ to be a minimum. Of course, here as in the usual calculus, it is a question of *local* minimum (or maximum), i.e., minimum in a neighborhood of $y(x)$ in path space.

Already in the determination of the solution curve one should notice that the boundary conditions that enter are not of the usual initial value problem type where the conditions are imposed at only one point. For the problems of the calculus of variation the solution curve has to be globally determined by solving a boundary value problem, and the general existence theorems that are needed are more subtle than those for the initial value problems. The global nature of the problems got highlighted further with the discovery of conjugate points and their role. The subject was finally put on a firm and rigorous foundation when Weierstrass lectured on this subject in 1879 and added fundamental discoveries of his own. The lectures of Weierstrass were published as part of his *Collected Papers*. For a very nice and detailed historical discussion of the entire subject, the reader should look into [9]. For a discussion of some of the aspects of the initial work of Euler and Lagrange, see [10]. For a good introduction to the subject itself from the modern point of view, the reader should consult [11], especially Chapter 5, for a beautiful discussion of the question of when the second variation is positive definite. The second variation, being a quadratic form in a Hilbert space, has an *index*, namely the number of eigenvalues of the corresponding symmetric operator which are < 0, a fact emphasized by Hilbert. The index could well be positive and finite, thus providing the analog of *saddle points* in path space. Their importance was revealed only later with Morse's work [12].

The calculus of variations was the object of intense interest for mathematicians who came afterwards, such as Mayer, Hilbert, Weyl, and many others. Then Morse took it to another level, bringing in global topology as a fundamental ingredient, and created a new area called *Morse theory*. Morse theory, with ramifications in geometry and topology, has become a central part of mathematics and physics. For the work of Morse see [12] and [13], and in the latter, the introduction by Bott.

Just like his work on elliptic integrals, Euler's ideas in the calculus of variations marked the creation of a new subject which is still active and has, over the years, revealed connections with topology, geometry, and mathematical physics that are nowhere close to being exhausted. In this area also Euler was thus the great pioneer.

2.5. Number theory

Euler worked on number theory all his life, expressing repeatedly his delight in thinking about number theoretic questions. Although he was well aware of the indifference of his contemporaries to his arithmetical works, it remained a favorite subject for him throughout his life. Unlike in analysis and its applications where his work often represented a finished masterpiece, in number theory it was just a beginning. His profound insights unearthed the problems to be studied, and his tremendous powers of calculation and ability to see patterns led to conjectures which occupied mathematicians who came after him for a long time.

In the century before Euler, Fermat was almost the only mathematician who was deeply interested in the theory of numbers. Fermat of course had a lifelong interest in number theory and was the first mathematician to think systematically about questions in that area, but many of his great contemporaries like Huygens could hardly conceal their lack of interest in Fermat's arithmetic concerns. The situation did not change when Euler took up the torch that Fermat passed on. Even Euler's close friend Daniel Bernoulli had on occasion expressed the opinion that there were many other topics more worthy of Euler's genius.

Nevertheless Euler's work changed all of this. One can say that it was Euler's work that made number theory an important branch of mathematics. The diversity of the problems he studied, the techniques he created for their solutions, the tremendous range and depth of his numerical calculations, and the elegance of his results made number theory beautiful and attracted the greatest mathematicians who came after him: Lagrange, Legendre, and above all, Gauss. Gauss of course reached extraordinary heights, and after him there was no doubt that number theory had become a central area of mathematics with connections to geometry, analysis, and algebra. In modern times this has become even more pronounced, and the connections have extended even to physics. The beginning of this entire story goes back to Euler's work.

Problems involving integers had exerted a fascination for mathematicians from ancient times. The Babylonians had a deep understanding of Pythagorean triplets, namely, triplets of positive integers (x, y, z) with $z^2 = x^2 + y^2$. Diophantus was concerned with finding integral and rational solutions of equations with integer coefficients, and his famous book *Arithmetica* has survived from the ancient times. The Indians made a systematic and profound study of the diophantine equations of the first degree as well as those of the form $x^2 - Ny^2 = \pm 1$ where N is a positive integer. They knew that the set of solutions of the latter type had a multiplicative structure and knew also a method, the so-called *Cakravala* or *cyclic method*, which allowed them to calculate the basic solution from which all others could be obtained. Although they did not possess proofs for their discoveries in any modern sense of the word, they computed many difficult examples leading to really huge numbers, and their writings displayed complete confidence in the validity of their methods [4], [5].

The first mathematician who looked at number theoretic problems systematically and tried to obtain proofs in the modern sense of the word was Fermat. However he had great difficulties in writing his arguments down, and his results were to be found only in his correspondence with other mathematicians or savants and in his annotated copy of Bachet's edition of Diophantus's *Arithmetica*. Part of the reason for this was that there were no models for Fermat to follow. Fermat tried to interest some of his great contemporaries like Pascal and Huygens to collaborate with him in writing up his number theoretic investigations but did not succeed. He was thus almost the only person working on number theory in all of the seventeenth century. After Fermat's death there was great concern that his results would be forgotten, and so his son Samuel Fermat brought out in 1679 a publication of the annotated Bachet edition and some parts of the Fermat correspondence.

Very soon after he joined the St. Petersburg Academy, Euler became acquainted with Christian Goldbach. Their acquaintance soon grew into a close friendship. Goldbach remained a close friend and patron of Euler, and Euler had great regard and affection for him throughout his life. Goldbach was an amateur mathematician who had great interest in number theory. It was he who, in a letter to Euler in 1742, made the conjecture that *every even number $n > 4$ is the sum of 2 odd primes,* a conjecture that has not been settled to this day. He brought Euler's attention to results of Fermat that he had heard about and was the initial catalyst for Euler's lifelong interest in number theoretic questions. Euler communicated to him his results and conjectures on number theory. P.-H. Fuss's *Correspondance...* lists 177 letters between Euler and Goldbach from 1729 to 1764 [1].

In his very first letter to Euler in 1729, Goldbach asks Euler in a postscript, one that Weil calls "fateful"([4], p. 172), if he (Euler) knew of Fermat's assertion that all numbers of the form $2^{2^n} + 1$ are primes. Although he did not respond to Goldbach's suggestion at first, by 1730 Euler had become deeply interested in number theory and started studying Fermat's writings, at least those which were available to him. Eventually Euler was able to give proofs of almost all of Fermat's assertions, literally starting from scratch and developing the tools as he went along. He then went further, creating new results and introducing new themes that completely changed the landscape of number theory and paved the way for his great successors: Lagrange, Legendre, Gauss, Dirichlet, Jacobi, and others. Euler's number theoretic investigations occupy just 4 volumes (I-2 through I-5) out of the 70 plus volumes of *Opera Omnia*, but they are responsible for the emergence of number theory as a mature and independent mathematical discipline. His work was the essential stepping-stone for everyone who came after him.

The following partial list of some of the themes Euler worked on in number theory is already an indication of the tremendous range of Euler's arithmetical research. Even if he had done nothing else, his number theoretic work alone would have assured him of immortality. For more details see [4] and the article of A.O. Gel'fond [14] and other references that I shall cite later on while discussing individual topics.

Fermat primes

Fermat's little theorem
Sums of two squares

Representation of primes by binary quadratic forms

Sums of four squares

Diophantine equations

Fermat's last theorem

Quadratic reciprocity

Zeta and multizeta values

Euler products

Distribution of primes

Large primes and primality testing

Partition of integers

Continued fractions

Algebraic and transcendental numbers

The Fermat primes and the little Fermat theorem. Let us put

$$F_n = 2^{2^n} + 1.$$

The F_n are called *Fermat numbers*, and the first few are:

$$F_0 = 3$$
$$F_1 = 5$$
$$F_2 = 17$$
$$F_3 = 257$$
$$F_4 = 65537$$
$$F_5 = 4294967297$$
$$F_6 = 18446744073709551617$$

A *Fermat prime* is a number in the sequence (F_n) which is prime. Fermat had expressed in his letters his belief that all the F_n are primes. Although F_i $(0 \le i \le 4)$ are primes, Euler showed in 1732 in the very first paper he published in number theory that F_5 was not a prime and was in fact divisible by 641. For F_6 we have the prime factor 274177 as was shown by Landry in 1880. Once we guess that 641 is a potential divisor for $2^{32} + 1$, it is trivial to verify that

$$2^{32} + 1 = 4294967297 = 641 \times 6700417,$$

but the idea to look for 641 as a potential divisor was Euler's observation. It was based on Fermat's little theorem, which we shall talk about presently. Indeed, if p is an odd prime dividing $2^{32} + 1$, we have $2^{32} \equiv -1 \bmod p$ so that $2^{64} \equiv 1 \bmod p$. If x is the order of 2 mod p, then x must divide 64 but x cannot divide 32, and so $x = 64$. But then $p - 1$ is divisible by 64, a consequence of Fermat's little theorem. So $p = 64k + 1$, and the primes of this form are 193, 257, 577, 641, ... It is not difficult to show that the Fermat numbers are mutually prime. The fact that they increase very rapidly suggests that only a finite number of them are primes, i.e., that the number of the Fermat primes is finite. However this has not been settled yet.

Throughout his life Euler was interested in finding very large primes and seldom missed an opportunity to do so. I shall give examples of this in discussing his work on representing numbers as sums of two squares. By considering prime divisors of numbers of the form $a^n \pm b^n$ he discovered some very large (for his time) primes, such as $2^{31} - 1 = 2147483647$. With computer help, much larger primes of this form have been found these days.

Congruences and Fermat's little theorem. Fermat's little theorem asserts that if p is a prime and a is not divisible by p, then $a^{p-1} - 1$ is divisible by p. Euler gave a proof of this and generalized it to nonprime moduli. For any integer $N > 1$ he introduced the function $\varphi(N)$ which is the number of integers $< N$ and prime to N; for instance, $\varphi(p) = p - 1$ for a prime p. Then Euler's generalization is that if a is prime to N, then $a^{\varphi(N)} - 1$ is divisible by N. Euler established that $\varphi(N)$ is multiplicative, i.e., $\varphi(NN') = \varphi(N)\varphi(N')$ for mutually prime N, N', and also that $\varphi(p^\alpha) = p^{\alpha-1}(p-1)$. Hence φ is completely determined. Euler's prolonged concern with this question ultimately led to the modern view where we introduce the ring of residue classes mod N, namely the ring $\mathbf{Z}_N = \mathbf{Z}/N\mathbf{Z}$, and the *group* G_N of its *units*, namely, invertible elements. The elements of G_N are the residue classes of integers prime to N. Then the order of G_N is $\varphi(N)$, and Euler's result is that for any $a \in G_N$ we must have $a^{\varphi(N)} = 1$ in G_N, which is a special case of the result in elementary group theory that says that if G is a group of order g, then $a^g = 1$ for any element a of G. It took Euler some time to prove his results, and one can say that these results of Euler are the beginnings of the structure theory of finite abelian groups.

Although $a^{\varphi(N)} = 1$ in G_N, it could happen that $a^d = 1$ for some $d < \varphi(N)$. The smallest such d is the *order* of a and it divides $\varphi(N)$. a is a *primitive root* if its order is exactly $\varphi(N)$. To say that primitive roots mod N exist is to say that G_N is cyclic. Primitive roots exist if $N = p$ is a prime, but Euler could not prove this. This had to wait for Gauss. It can be shown that primitive roots mod N exist if and only if

$$N = 2^r (r = 1, 2), p^s (p \text{ a prime } > 2, s \geq 1), 2p^s (p \text{ a prime } > 2, s \geq 1).$$

Sums of squares, quadratic reciprocity, and more. Whatever was available of Fermat's work and correspondence indicated that he had made a profound study of the representations of numbers as sums of two squares, and Euler naturally gravitated to the task of understanding the theorems that Fermat had asserted (without proofs). It took many years before Euler was able to obtain the proofs of these statements of Fermat. Indeed, at one point in 1742, when he still did not have proofs for Fermat's results on sums of two squares, Euler even wrote to Clairaut in Paris asking that a search be made for the writings of Fermat. The search proved negative and so Euler had to work out the proofs himself!

Let us make some definitions. By a *representation* of n is meant a choice of integers u, v such that $n = u^2 + v^2$; the representation is *nontrivial* if $uv \neq 0$ and *proper* if it is nontrivial and u, v are mutually prime. We write $r_2(n)$ for the number of representations of n. In counting the number of representations we count as distinct all representations obtained from one by change of signs or order. Thus for $n = 585$ we have $585 = 3^2 + 26^2 = 12^2 + 21^2$, so that $r_2(585) = 16$. If we agree not to distinguish between representations that differ from one another by changes of sign and order, we obtain $r_{\text{ess}}(n)$, the number of essential representations. The basic result is the following.

THEOREM. *An integer $n > 0$ has a representation as a sum of two squares if and only if all its prime factors of the form $4m + 3$ have even exponents in the unique factorization of n. If $r_1, r_2, \ldots, r_k$ are the exponents of the primes of the form $4m + 1$ in the unique factorization of n, the number $r_{\text{ess}}(n)$ of essential representations of n is given by*

$$r_{\text{ess}}(n) = \begin{cases} \frac{1}{2}(1 + r_1)(1 + r_2) \ldots (1 + r_k) & \text{if some } r_j \text{ are odd} \\ \frac{1}{2}[(1 + r_1) \ldots (1 + r_k) + 1] & \text{if all } r_j \text{ are even.} \end{cases}$$

In particular, an odd prime is representable if and only if it is of the form $4m + 1$; and for an integer $n = 4m + 1(m > 0)$ to be a prime it is necessary and sufficient that it have a representation that is unique up to changes of sign and order and, moreover, that this representation be proper.

A key to the proofs of these results is the fact the numbers with representations are closed under multiplication, as is clear from the identity

$$(a^2 + b^2)(c^2 + d^2) = (ac \mp bd)^2 + (ad \pm bc)^2.$$

Once this is noted, everything comes down to knowing which primes have representations; the basic step is to show that for an odd prime p,

$$p|x^2 + y^2, (x, y) = 1 \Longleftrightarrow p \equiv 1 (\mathrm{mod}\ 4).$$

This fact and descent arguments form the core of the proofs of Euler and presumably of Fermat also. For a detailed discussion of the proofs of Euler, see [4]. Notice that the above is equivalent to saying that -1 is a quadratic residue mod p. In terms of the Legendre symbol this is the statement that

$$\left(\frac{-1}{p} \right) = 1,$$

and the statement that this happens if and only if $p \equiv 1$ mod 4 is the first supplement of the law of quadratic reciprocity:

$$\left(\frac{-1}{p} \right) = (-1)^{\frac{p-1}{2}}.$$

Euler used the last statement of the theorem as a device for discovering large primes. Indeed, if one has a table of squares and N is a large integer, one then has to check if $N - u^2 = v^2$ uniquely and then if u, v are mutually prime. It was thus he discovered that 82421, 100981, and 2626557 are primes, while 1000009, 233033, and 32129 are not. Here $32129 = 95^2 + 152^2$ is a unique representation, but it is not proper.

For many reasons, some of which are practical (for some people) like writing codes which are hard to break, the problem of testing whether a given number is prime has attracted a great deal of attention. The fundamental question is whether for testing a number N to be prime an algorithm can be developed which works in $O((\log N)^r)$ time. This has been done now, thanks to Agarwal, Kayal, and Saxena, who confounded all the experts by coming up with a sure-fire algorithm that succeeds in $O((\log N)^{7.5}(\log \log N)^r)$ time and that moreover is based on extremely simple ideas (see [16], [17]).

The view from the theory of quadratic fields. From our own perspective, the work of Fermat and Euler on sums of two squares can be viewed as the complete theory of divisibility and prime factorization in the Gaussian field $K = \mathbf{Q}(i)$ where $i = \sqrt{-1}$, and its ring of integers $R = \mathbf{Z}[i]$. It was Gauss who introduced this field and the idea of doing arithmetic in it and who knew that its Euclidean nature permitted unique factorization in it.

In particular R is a principal ring; i.e., all ideals of R are principal. The units of R are the elements $\pm 1, \pm i$, and for any ideal in R its generator is unique up to multiplication by a unit. The primes of R are obtained by factorizing the rational primes in R. For any nonzero integer $n = u + iv \in R$ it is always possible to multiply it by a unit so that $u > 0, v \geq 0$; this normalization is unique. In particular one can normalize the primes in R in this manner. Then any $0 \neq x \in R$ has a unique factorization

$$x = \varepsilon \pi_1^{e_1} \ldots \pi_k^{e_k}$$

where ε is a unit and the π_j are distinct normalized primes. We have $2 = -i(1+i)^2$ so that $1 + i$ is the prime above 2.

The key fact is that if $p \equiv 3$ mod 4, then p stays prime in R, but if $p \equiv 1$ mod 4, then $p = \pi\pi^*$ where π is a normalized prime and π^* is the complex conjugate of π. These follow from the quadratic character of -1 mod p and the fact that R

admits a unique factorization. Indeed, if $p = 4m+3$ and p splits, then $p = u^2 + v^2$, so that 3 is a sum of two squares mod p, which is impossible. If $p = 4m + 1$, then -1 is a quadratic residue mod p so that $p|x^2 + 1$, thus $p|(x+i)(x-i)$. If p stays prime, then unique factorization implies that p must divide one of $x \pm i$, which is a contradiction, since p cannot divide 1.

We write p for rational primes $4m+1$ and q for primes $4m+3$. If $c = a+ib \in R$ and N denotes the norm map from R to $\mathbf{Z}$, then $n = N(c) = cc^* = a^2 + b^2$. Thus the positive integers with a representation are those in $N(R)$. If

$$n = 2^a p_1^{r_1} \ldots p_k^{r_k} q_1^{c_1} \ldots q_r^{c_r} = uu^* \qquad (u \in R),$$

then, writing

$$u = \varepsilon (1+i)^\alpha \pi_1^{\beta_1}(\pi_1^*)^{\gamma_1} \ldots \pi_k^{\beta_k}(\pi_k^*)^{\gamma_k} q_1^{d_1} \ldots q_r^{d_r}$$

where ε is a unit, we find that

$$a = \alpha, \quad r_j = \beta_j + \gamma_j, \quad c_\ell = 2d_\ell.$$

Thus, for n to have a representation the c_ℓ should all be even, and then the number of u such that $n = uu^*$ is just

$$r_2(n) = 4\Pi_{j=1}^k(1 + r_j).$$

Moreover

$$\frac{1}{4}r_2(n) = \#\{ \text{ ideals in } R \text{ of norm } = n\} = \Pi_{j=1}^k(1 + r_j).$$

Thus, for the zeta function of K defined as

$$\zeta_K(s) = (1 - 2^{-s})^{-1}\Pi_p(1 - p^{-s})^{-2}\Pi_q(1 - q^{-2s})^{-1},$$

we have

$$\zeta_K(s) = \frac{1}{4}\sum \frac{r_2(n)}{n^s} = \zeta(s)L(s)$$

where $\zeta(s)$ is the usual Riemann zeta function and

$$L(s) = 1 - \frac{1}{3^s} + \frac{1}{5^s} - \cdots = \sum_{n \geq 1} \frac{\omega(n)}{n^s} = \Pi_{r'}(1 - \omega(r)r^{-s})^{-1}.$$

Here ω is the quadratic character mod 4 given by

$$\omega(n) = \begin{cases} 0 & \text{if } n \text{ is even} \\ (-1)^{\frac{n-1}{2}} & \text{if } n \text{ is odd.} \end{cases}$$

We have $\omega(mn) = \omega(m)\omega(n)$.

One can derive an alternative description of $r_2(n)$. Let $d_j(n)$ be the number of divisors of n of the form $4m + j (j = 1, 3)$. Let

$$\delta(n) = d_1(n) - d_3(n) = \sum_{d|n} \omega(d).$$

Then δ is *multiplicative*; i.e., $\delta(nn') = \delta(n)\delta(n')$ whenever n, n' are mutually prime. Indeed, if F is any multiplicative function on the positive integers, then

$$G(n) = \sum_{d|n} F(d)$$

is again multiplicative. Now $\delta(2^a) = \delta(q^{2s}) = 1$, while $\delta(p^r) = 1 + r$. From the unique factorization of n we conclude that

$$\delta(n) = \prod_{j=1}^{k}(1 + r_j)$$

so that

$$r_2(n) = 4\delta(n).$$

Finally,

$$r_{\text{ess}}(n) = \begin{cases} \frac{1}{2}\delta(n) & \text{if } n \neq 2u^2, \neq v^2 \\[2mm] \frac{1}{2}[\delta(n) + 1] & \text{if } n = 2u^2, v^2. \end{cases}$$

Fermat was also concerned with results for the representability of integers as $u^2 + Nv^2$ for $N = 2, 3$, and even $N = -2$. Of course in this last case the units, i.e., the solutions of $X^2 + NY^2 = \pm 1$, are infinitely many and the theory has a different flavor. We shall treat only the case when $N > 0$. Euler worked out the cases $N = 2, 3$, which Fermat considered, and went much further. Since

$$x^2 + Ny^2 = \text{Norm}(x + y\sqrt{-N})$$

where Norm is the norm from $\mathbf{Q}(\sqrt{-N})$ to $\mathbf{Q}$ and since Norm is multiplicative, the key issue comes down to representability of primes as $x^2 + Ny^2$. When $N = 2, 3$ the Euler-Fermat results for an odd prime p not dividing N are

$$p = x^2 + 2y^2 \Longleftrightarrow p \equiv 1, 3 \bmod 8$$
$$p = x^2 + 3y^2 \Longleftrightarrow p \equiv 1 \bmod 3.$$

Quadratic reciprocity. Euler did not stop with these special cases but had the goal of determining which odd primes p not dividing N can be represented as $x^2 + Ny^2$ where N can be *any* positive integer. He did not succeed in this, but his journey took him far beyond Fermat and led him to the discovery of quadratic reciprocity and even into conjectures that can be properly understood only with the help of class field theory. In his paper *Theoremata circa divisores numerorum in hac forma paa $\pm$ qbb contentorum* ([2], I-2, 194-222), Euler stated a number of remarkable Theorems and Annotations (which were really conjectures supported by experimental evidence), which revealed how far he had gone into unchartered waters. He started writing in 1748-1750 a treatise entitled *Tractatus de numerorum doctrina capita sedecim quae supersunt* ([2], I-5, 182-283), but left it unfinished because he could not settle many questions, and it was published only in 1849. In this book he formulated his conjectures on representability of primes as $x^2 + Ny^2$ for $N = 27, 64$ that brought in cubic and biquadratic residues mod p.

The quadratic reciprocity law, as formulated by Legendre and proved by Gauss, states that for two distinct odd primes p, q,

$$\left(\frac{q}{p}\right) = (-1)^{\frac{p-1}{2}\frac{q-1}{2}}\left(\frac{p}{q}\right),$$

with the two supplements

$$\left(\frac{-1}{p}\right) = (-1)^{\frac{p-1}{2}}, \qquad \left(\frac{2}{p}\right) = (-1)^{\frac{p^2-1}{8}}.$$

The main statement is equivalent to

$$\left(\frac{q}{p}\right) = \left(\frac{p^*}{q}\right), \qquad p^* = (-1)^{\frac{p-1}{2}} p.$$

In his work on primes dividing $x^2 + Ny^2$ for $N = 2, 3$, Euler had already worked out the supplements of the quadratic reciprocity law and had discovered that for distinct odd primes p, q, the key issue was to write the condition

$$\left(\frac{q}{p}\right) = 1$$

when p varies in terms of congruences with respect to the fixed modulus q. His numerical work and his genius for recognizing patterns led him to the conjecture that

$$\left(\frac{q}{p}\right) = 1 \iff p \equiv \pm\beta^2 \bmod 4q, \quad \beta \text{ odd} .$$

It is remarkable but not difficult to show that this is equivalent to the usual form in which quadratic reciprocity is stated. But Euler was after the case when q was replaced by an *arbitrary* positive integer. He knew that for p an odd prime not dividing N,

$$p \mid x^2 + Ny^2 \iff \left(\frac{-4N}{p}\right) = 1,$$

and his great goal was to rewrite this condition in terms of congruences for p mod $4N$. He discovered that the above can be transformed as the condition that p lies in a *subgroup* of the residue class group mod $4N$. Euler's conjectures singled out rather precisely a subgroup K_N of the group G_{4N} of residue classes mod $4N$ prime to $4N$ such that for an odd prime p not dividing N,

$$\left(\frac{-4N}{p}\right) = 1 \iff [p] \in K_N.$$

Nowadays the subgroup K_N is obtained as the kernel of the homomorphism of G_N into $\{\pm 1\}$ which is the *Jacobi symbol*. Perhaps he came to these discoveries rather late in his life when he had gone blind and had to rely on his assistants to do calculations for him. In any case he was unable to take the steps needed to prove these conjectures, and his work was finished by Gauss. For detailed discussions of the evolution of Euler's ideas on quadratic reciprocity there are many excellent accounts, especially those of [4], [18], and [19].

Representability of primes p in the form $p = x^2 + Ny^2$ for general N. There is a second line of thought in Euler's work, namely, that the full story of representability of primes as $x^2 + Ny^2$ for arbitrary $N > 0$ is not contained in quadratic reciprocity but goes very much beyond it. It is in fact tied up with the theory of abelian extensions of imaginary quadratic fields, thus leading one directly to modern number theory at its most beautiful. Let us now look somewhat more closely into this aspect of Euler's work.

In the paper *Theoremata* and his treatise *Tractatus* Euler stated the following conjectures:

$$p = x^2 + 5y^2 \Longleftrightarrow p \equiv 1, 9 \bmod 20$$

$$2p = x^2 + 5y^2 \Longleftrightarrow p \equiv 3, 7 \bmod 20.$$

$$p = \begin{cases} x^2 + 14y^2 \\ 2x^2 + 7y^2 \end{cases} \Longleftrightarrow p \equiv 1, 9, 15, 23, 25, 39 \bmod 56$$

$$3p = x^2 + 14y^2 \Longleftrightarrow p \equiv 3, 5, 13, 19, 27, 45 \bmod 56.$$

$$p = x^2 + 27y^2 \Longleftrightarrow p \equiv 1 \bmod 3 \text{ and } 2 \text{ is a cubic residue mod } p$$

$$p = x^2 + 64y^2 \Longleftrightarrow p \equiv 1 \bmod 4 \text{ and } 2 \text{ is a biquadratic residue mod } p.$$

We shall, as before, use the theory of primes in imaginary quadratic fields to elucidate the questions of representability of primes as $x^2 + Ny^2$. For an arbitrary $N > 0$, whether or not $\mathbf{Q}(\sqrt{-N})$ is Euclidean (as was the case when $N = 2, 3$), one has, for p an odd prime not dividing N,

$$p = x^2 + Ny^2 \Longrightarrow \left(\frac{-N}{p} \right) = 1 \Longrightarrow p \text{ splits in } \mathbf{Q}(\sqrt{-N}).$$

Here splitting means that

$$(p) = \mathfrak{p}\mathfrak{p}^*$$

where $\mathfrak{p}$ is a prime ideal in $\mathbf{Q}(\sqrt{-N})$ and $\mathfrak{p}^*$ is its complex conjugate. If $\mathbf{Q}(\sqrt{-N})$ is Euclidean or, more generally, if the class number of $\mathbf{Q}(\sqrt{-N})$ is 1, then $\mathfrak{p}$ above is a principal ideal and we can write $p = \mathrm{Norm}(u)$ for an integer u in $\mathbf{Q}(\sqrt{-N})$; we can take u to be a generator of $\mathfrak{p}$ for instance. If N is square free and $N \not\equiv 3 \bmod 4$, then $u \in \mathbf{Z}[\sqrt{-N}]$ and so we can complete the above chain of implications to get

$$p = x^2 + Ny^2 \Longleftrightarrow \left(\frac{-N}{p} \right) = 1.$$

This already takes care of the cases $N = 1, 2$, for then $\mathbf{Q}(\sqrt{-N})$ is Euclidean. But if $N \equiv 3 \bmod 4$, then we can say only that

$$p = \mathrm{Norm}(u), \qquad u \in \mathbf{Z}\left[\frac{1 + \sqrt{-N}}{2} \right]$$

so that

$$p = \left(a + \frac{b}{2} \right)^2 + N \left(\frac{b}{2} \right)^2.$$

For $N = 3$ this gives $p = a^2 + ab + b^2$, but a and b cannot both be even, and then $p = x^2 + 3y^2$; indeed,

$$b = 2\beta \Longrightarrow p = (a + \beta)^2 + 3\beta^2, a = 2\alpha \Longrightarrow p = (b + \alpha)^2 + 3\alpha^2$$

$$a = 2\alpha + 1, b = 2\beta + 1 \Longrightarrow p = (\alpha - \beta)^2 + 3(\alpha + \beta + 1)^2.$$

If $N = 7$, $p = a^2 + ab + 2b^2$; then $b = 2\beta$ must be even and $p = (a + \beta)^2 + 7\beta^2$. Thus the cases $N = 3, 7$ are also taken care of, with

$$p = x^2 + 7y^2 \Longleftrightarrow p \equiv 1, 2, 4 \bmod 7.$$

However there still remain the cases

$$N = 11, 19, 43, 67, 163$$

where the class number of $\mathbf{Q}(\sqrt{-N})$ is 1.

For arbitrary N, the criterion

$$\left(\frac{-N}{p}\right) = 1$$

is only necessary and not sufficient for the representability $p = x^2 + Ny^2$ because this condition is only equivalent to the splitting of p in $\mathbf{Q}(\sqrt{-N})$, whereas representability requires in addition that the factors of p be principal prime ideals with generators of the form $a + b\sqrt{-N}$. Now the ring of integer in $\mathbf{Q}(\sqrt{-N})$ need not be $\mathbf{Z}[\sqrt{-N}]$, and, even if it is, R can have prime ideals that are not principal; i.e., the class number of $\mathbf{Q}(\sqrt{-N})$ may be > 1. For the representability of primes as $x^2 + 5y^2$ the condition

$$\left(\frac{-5}{p}\right) = 1 \Longleftrightarrow (-1)^{\frac{p-1}{2}}\left(\frac{p}{5}\right) = 1$$

gives

$$p \equiv 1, 3, 7, 9 \bmod 20,$$

but Euler's conjectures split these classes into two subsets, with only $p \equiv 1, 9$ as being equivalent to $p = x^2 + 5y^2$, while for $p \equiv 3, 7$, he has $2p = x^2 + 5y^2$. Now $x^2 + 5y^2 \equiv x^2$ or $x^2 + 5 \bmod 20$ from which it follows that primes $p \equiv 3, 7 \bmod 20$ cannot be of the form $x^2 + 5y^2$. But the fact that for $p \equiv 1, 9$, p is representable as $x^2 + 5y^2$ lies much deeper. Let us introduce the *Hilbert class field* L over $\mathbf{Q}(\sqrt{-5})$. Since the class number of $\mathbf{Q}(\sqrt{-N})$ is 2, L is quadratic over $\mathbf{Q}(\sqrt{-N})$. All ideals of $\mathbf{Q}(\sqrt{-N})$ become principal in L. Write $Rp = \mathfrak{p}\mathfrak{p}^*$ where $\mathfrak{p}, \mathfrak{p}'$ are primes in $\mathbf{Q}(\sqrt{-N})$. If $\mathfrak{p}$ splits in L, it follows that p splits completely in L and the primes into which it factorizes in L are all principal. Thus p will be the norm of an integer of L, hence the norm of an integer in $\mathbf{Q}(\sqrt{-5})$.

It turns out that

$$L = \mathbf{Q}(\sqrt{-1}, \sqrt{-5}) \subset \mathbf{Q}(\zeta) = M$$

where

$$\zeta = e^{i\pi/10}$$

is the primitive 20th root of unity. Since M is abelian over $\mathbf{Q}$, it follows that L is also abelian over $\mathbf{Q}$, and hence the complete splitting of p in L can be decided by congruences mod 20. The Galois group of M over $\mathbf{Q}$ is the group G_{20} of residue classes mod 20 which are prime to 20, and L is the fixed field of [19]. For a prime p not dividing 20 to split completely in L is equivalent to saying that the subgroup of G_{20} fixing the ideal (p) in M is just $\{[1], [19]\}$. The known factorization laws of rational primes in M then imply that this is the case when $p \equiv 1, 9 \bmod 20$.

If N is square free and $\not\equiv 3 \bmod 4$, one can again introduce the Hilbert class field L of $\mathbf{Q}(\sqrt{-N})$ and argue that $p = x^2 + Ny^2$ if and only if p splits completely in L. But now, although L is Galois over $\mathbf{Q}$, it may not be abelian, and so the complete splitting of p in L can no longer be reduced to congruences mod $4N$. This is the reason why extra conditions are needed. In [18] Cox obtains the following criterion for representability $p = x^2 + Ny^2$ in this case: let $L = \mathbf{Q}(\sqrt{-N})(\alpha)$ where α is a real algebraic integer (such α exist), and let $f_N \in \mathbf{Z}[X]$ be its (monic) minimal polynomial; then

$$p = x^2 + Ny^2 \iff \begin{cases} \left(\frac{-N}{p}\right) = 1 \text{ and} \\\\ f_N(x) = 0 \text{ has a solution in integers mod } p. \end{cases}$$

For $N = 14$, a case considered by Euler, the class number of $\mathbf{Q}(\sqrt{-4})$ is 4. In this case it can be shown that $\alpha = \sqrt{2\sqrt{2} - 1}$ generates the Hilbert class field of $\mathbf{Q}(\sqrt{-14})$, and hence we have

$$p = x^2 + 14y^2 \iff \begin{cases} p \equiv 1, 9, 15, 23, 25, 39 \bmod 56 \text{ and} \\\\ (x^2 + 1)^2 = 8 \text{ has a solution in integers mod } p. \end{cases}$$

The cases $N = 27, 64$ studied by Euler—and even when N is square free, the cases $N \equiv 3 \bmod 4$—are not covered by the criterion described above. In these cases, $\mathcal{O}_N = \mathbf{Z}[\sqrt{-N}]$ is only an *order* in $\mathbf{Q}(\sqrt{-N})$, not its maximal order, which is the full ring of integers in the field. In these cases the Hilbert class field has to be replaced by the *ring class field* corresponding to the order $\mathcal{O}_N$. In [18] the reader can find the proof of the generalization of the above result, which consists of replacing the Hilbert class field by the ring class field. In particular, for $N = 27$ the ring class field is $\mathbf{Q}(\sqrt{-27})(\sqrt[3]{2})$, while for $N = 64$, the ring class field is $\mathbf{Q}(\sqrt{-27})(\sqrt[4]{2})$, leading to the conjectured results of Euler. Moreover it is also shown in [18] how these ring class fields are constructed using various devices including complex multiplication. These explicit constructions are part of the Kronecker *Jugendtraum*. More about such matters in Chapter 6.

Polygonal numbers and expression of numbers as sums of polygonal numbers. Not every number is a sum of 3 squares; for instance, 7 is not a sum of 3 squares. Fermat wrote in letters to Mersenne, Pascal, and others that *every number is a sum of 3 triangular numbers, 4 squares, 5 pentagonal numbers, 6 hexagonal numbers, and so on* (we count 0 as a polygonal number). The result for the triangular numbers, namely that every positive integer is the sum of at most 3 triangular numbers, was proved by Gauss. In his *Tagebuch*, which was a diary that Gauss kept in his early years to record his mathematical discoveries, there is the cryptic entry dated July 10, 1796 (Gauss was 19 then) (see [20]):

$$\text{EUREKA. num} = \Delta + \Delta + \Delta.$$

Cauchy proved the general result starting from 5 pentagonals in 1815. Euler took up the result involving 4 squares, namely that every positive integer is a sum of at most 4 squares (of integers). He finally was able to prove that every positive integer is a sum of at most 4 squares of *rational numbers* but could not remove the restriction to rationals. Then Lagrange took the decisive final step and proved the four-square theorem in complete generality. After seeing Lagrange's proof, Euler gave his own proof, which was a beautiful modification of his arguments for the two-square results of Fermat, as well as a great improvement of Lagrange's argument.

Analytic methods for number of representations. Concerning the four-square theorem it is interesting that Euler, with his great insight into power series and generating functions, suggested in a letter to Goldbach that the most natural way to prove it is to prove that all coefficients of

$$(1 + x + x^4 + x^9 + \ldots)^4$$

are positive. One can say that this remark marks the birth of the use of analytic methods in the problem of representing numbers as sums of squares and in fact in number theory as a whole. Jacobi indeed proved the four-square theorem by Euler's method, but working not with

$$1 + x^4 + x^9 + \ldots$$

but with his theta function

$$\vartheta(x) = 1 + 2x^4 + 2x^9 + \cdots = \sum_{n=-\infty}^{\infty} x^{n^2}.$$

He thus obtained an explicit expression for $r_4(n)$ and indeed for $r_s(n)(s = 2, 4, 6, 8)$.

The method of Jacobi already applies to the two-square problem and indeed to the problem of counting the number of representations of integers as sums of an arbitrary number of squares. Let

$$\vartheta(x) = 1 + 2x + 2x^4 + 2x^9 + \cdots = \sum_{n+-\infty}^{+\infty} x^{n^2}$$

be the Jacobi theta function as above. Then

$$\vartheta(x)^s = 1 + \sum_{n=1}^{\infty} r_s(n)x^n$$

where $r_s(n)$ is the number of representations of n as a sum of s integer squares, with the agreement that we count as different all representations obtained from one by changes of signs and permutations; this agrees with the definition of $r_2(n)$ given above. Jacobi obtained by this method formulae for $r_s(n)$ for $s = 2, 4, 6, 8$. For $s = 2$ Jacobi obtained the remarkable identity

$$(1 + 2x^2 + 2x^4 + 2x^9 + \ldots)^2 = 1 + 4\left(\frac{x}{1-x} - \frac{x^3}{1-x^3} + \frac{x^5}{1-x^5} + \cdots\right),$$

yielding

$$\sum_{n=1}^{\infty} r_2(n)x^n = 4\left(\frac{x}{1-x} - \frac{x^3}{1-x^3} + \frac{x^5}{1-x^5} + \cdots\right).$$

The formula

$$r_2(n) = 4\delta(n) = 4\sum_{d|n} w(d)$$

follows immediately.

For $s = 4$ Jacobi proved that

$$\left(1 + 4\left(\frac{x}{1-x} - \frac{x^3}{1-x^3} + \frac{x^5}{1-x^5} - \cdots\right)\right)^2 = 1 + 8\sum_{n\geq 1, n\neq 0 \bmod 4} n\frac{x^n}{1-x^n};$$

[8], Vol. I, [21] [22]. Hence

$$(1 + 2x + 2x^4 + 2x^9 + \ldots)^4 = 1 + 8\sum_{n\geq 1, n\neq 0 \bmod 4} n\frac{x^n}{1-x^n}.$$

This implies that

$$r_4(n) = 8 \sum_{d|n,\, d\not\equiv 0 \bmod 4} d.$$

It is easy to see that

$$r_4(n) = \begin{cases} 8\sum_{d|n} d & \text{if } n \text{ is odd} \\[2ex] 24\sum_{d|n,\, d \text{ odd}} d & \text{if } n \text{ is even.} \end{cases}$$

Clearly $r_4(n) > 0$ for all n. It must be noted that in calculating $r_4(n)$ directly one must add 0's to always have four squares. Thus, for $n = 10$, the sum of odd divisors of 10 is $1 + 5 = 6$, so that Jacobi's formula gives $r_4(10) = 144$. On the other hand, we have $10 = 2^2 + 2^2 + 1^2 + 1^2 = 3^2 + 1^2 + 0^2 + 0^2$. Allowing for changes of signs and order, the first gives rise to 96 and the second to 48, giving 144 altogether.

A galaxy of mathematicians continued these analytic methods, such as Hardy, Hardy-Littlewood, Mordell, Ramanujan, and many others. Starting from Jacobi the end result is the representation of $r_s(n)$ in terms of elementary arithmetical functions.

The theory of representations of numbers as sums of squares suggests the corresponding problem of representing a number as a sum of higher powers. In 1782, Waring stated in his *Meditationes Algebraicae* without proof that every number is a sum of 9 cubes, 19 fourth powers, and so on. Hilbert proved more than 100 years after Waring's assertion that for every k there is an integer $g(k)$ such that every number is a sum of at most $g(k)$ k^{th} powers. The analytic problem would then be the determination of the coefficients $r_{k,s}(n)$ defined by

$$1 + \sum_{n=1}^{\infty} r_{k,s}(n)x^n = \left(1 + 2\sum_{n=1}^{\infty} x^{n^k}\right)^s.$$

The asymptotic analysis of the $r_{k,s}(n)$ was developed by Hardy and Littlewood in a famous series of papers [23].

In the 1930's, Siegel generalized the problem of sums of squares in a far-reaching manner and developed the theory of representations of one quadratic form by another in arbitrary dimensions. Siegel's work and its subsequent reformulation and generalizations by Tamagawa and Weil represent a great climax of these themes.

As mentioned above, the result that any number is a sum of three triangular numbers was proved in [24]. The equation

$$n = \frac{x(x+1)}{2} + \frac{y(y+1)}{2} + \frac{z(z+1)}{2}$$

becomes

$$8n + 3 = (2x+1)^2 + (2y+1)^2 + (2z+1)^2$$

and so it becomes a question of proving that any number $\equiv 3 \bmod 8$ is a sum of three squares (which have to be necessarily odd), the form in which Gauss proved this result. Gauss's method was algebraic. For recent treatments using the analytical method, see Shimura [27]. Among other things Shimura obtains the number of representations of a number by k triangular and pentagonal numbers. Here by a pentagonal number we mean a number of the form

$$\frac{k(3k-1)}{2} \qquad (k = 0, \pm 1, \pm 2, \dots)$$

which is more inclusive than the classical definition which considers only $k > 0$.

Partitions of numbers. In how many ways can an integer N be written as a sum of m integers or m distinct integers? This question was posed to Euler by the Berlin mathematician Phillip Naude. This triggered some beautiful papers from Euler. Euler immediately introduced formal infinite products which were generating functions for the partition functions. Thus, let $p(n)$ be the number of partitions of n, i.e., the number of ways of writing n as a sum of positive integers, where we ignore the order of the summands; then

$$\sum_{n=1}^{\infty} p(n)x^n = \frac{1}{(1-x)(1-x^2)(1-x^3)\dots}.$$

Similarly

$$(1+x)(1+x^2)(1+x^3)\dots$$

is the generating function of the number of partitions of n into distinct parts, while

$$\frac{1}{(1-x)(1-x^2)(1-x^3)\dots(1-x^m)}$$

generates partitions of n into parts not exceeding m. Euler obtained many results about such partition functions in his paper *De partitione numerorum* ([2], I-2, 254-294). The whole theory is very pretty, and there are results which are not trivial to prove directly by combinatorial methods but are almost trivial if we use generating functions. For instance the obvious identity

$$(1+x)(1+x^2)(1+x^3)\dots = \frac{1-x^2}{1-x}\frac{1-x^4}{1-x^2}\frac{1-x^6}{1-x^3}\dots$$
$$= \frac{1}{(1-x)(1-x^3)(1-x^5)\dots}$$

shows that the *number of partitions of n into distinct parts is equal to the number of partitions of n into odd parts.*

Euler's pentagonal number theorem. In the course of this work Euler came across the infinite product

$$(1-x)(1-x^2)(1-x^3)\dots.$$

At first he contented himself with just an evaluation of a large number of coefficients which led him to a conjecture of the expansion involving the pentagonal numbers. Eventually he succeeded in proving the remarkable formula ([2], I-3, 472-479)

$$\prod_{k=1}^{\infty}(1-x^k) = 1 + \sum_{k=1}^{\infty}(-1)^k \left(x^{\frac{k(3k-1)}{2}} + x^{\frac{k(3k+1)}{2}} \right) = \sum_{-\infty}^{\infty}(-1)^k x^{\frac{k(3k-1)}{2}}.$$

Since the $\frac{k(3k-1)}{2}$ are the *pentagonal numbers* this theorem is called the *pentagonal number theorem* of Euler. It is certainly one of the most beautiful formulae in the theory of elliptic modular functions, and Jacobi obtained it in the course of his investigations in the theory of elliptic functions. For combinatorial proofs, see [21] [25]. The proof in [21] is Franklin's proof; it is by clever counting arguments and establishes the equivalent formula

$$p_e(n) - p_o(n) = \begin{cases} (-1)^k & \text{if } n = \frac{k(3k\pm 1)}{2} \\ \\ 0 & \text{if otherwise,} \end{cases}$$

where $p_e(n)$ (resp. $p_o(n)$) is the number of partitions of n into an even (resp. odd) number of distinct parts. The proof of Andrews [25] is a modern exposition and generalization of Euler's original proof. Andrews also gives a proof of Fine's result, namely,

$$\pi_e(n) - \pi_o(n) = \begin{cases} 1 & \text{if } n = \frac{k(3k+1)}{2} \\ -1 & \text{if } n = \frac{k(3k-1)}{2} \\ 0 & \text{if otherwise,} \end{cases}$$

where $\pi_e(n)$ (resp. $\pi_o(n)$) is the number of partitions of n into distinct parts, the largest of which is even (resp. odd).

I shall sketch Andrews' argument which follows and generalizes Euler's. The starting point is the identity of Euler:

$$\prod_{n=1}^{\infty}(1 + a_n) = 1 + \sum_{n=1}^{\infty} a_n(1 + a_1)(1 + a_2)\ldots(1 + a_{n-1}).$$

Thus

$$\prod_{n=1}^{\infty}(1 - q^n) = 1 - \sum_{n=1}^{\infty}(1 - q)(1 - q^2)\ldots(1 - q^{n-1})q^n.$$

Andrews now introduces

$$f(x, q) = 1 - \sum_{n=1}^{\infty}(1 - xq)(1 - xq^2)\ldots(1 - xq^{n-1})x^{n+1}q^n$$

so that

$$f(1, q) = \prod_{n=1}^{\infty}(1 - q^n).$$

By direct manipulation which takes after Euler's calculation one obtains the functional equation

$$f(x, q) = 1 - x^2 q - x^3 q^2 f(xq, q).$$

Calculation of a few terms shows that f has the following form:

$$f(x, q) = 1 + \sum_{n=1}^{\infty}(-1)^n \left(x^{3n-1}q^{a_n} + x^{3n}q^{b_n}\right)$$

where $(a_n), (b_n)$ are two sequences of integers. Substituting in the functional equation gives

$$a_n - a_{n-1} = 3n - 2(a_1 = 1) \qquad b_n - b_{n-1} = 3n - 1(b_1 = 2)$$

from which we get that the a_n and b_n are the pentagonal numbers

$$a_n = \frac{n(3n-1)}{2}, \qquad b_n = \frac{n(3n+1)}{2}.$$

Thus

$$f(x, q) = 1 + \sum_{n=1}^{\infty}(-1)^n \left(x^{3n-1}q^{\frac{n(3n-1)}{2}} + x^{3n}q^{\frac{n(3n+1)}{2}}\right).$$

For $x = 1$ we get Euler's theorem. By taking $x = -1$ Andrews gets the result of Fine quoted above.

The deepest work on partitions after Euler came from Ramanujan. Hardy [22] gives a masterful account of this. Ramanujan obtained the remarkable congruences

$$p(5m + 4) \equiv 0 \bmod 5$$
$$p(7m + 5) \equiv 0 \bmod 7$$
$$p(11m + 6) \equiv 0 \bmod 11.$$

In fact he discovered some beautiful identities which implied these congruences:

$$\sum_{m=0}^{\infty} p(5m + 4)x^m = 5\frac{\left(\prod_{k \geq 1}(1 - x^{5k})\right)^5}{\left(\prod_{k \geq 1}(1 - x^k)\right)^6}$$

and

$$\sum_{m=0}^{\infty} p(7m + 5)x^m = 7\frac{\left(\prod_{k \geq 1}(1 - x^{7k})\right)^3}{\left(\prod_{k \geq 1}(1 - x^k)\right)^4}$$
$$+ 49x\frac{\left(\prod_{k \geq 1}(1 - x^{7k})\right)^7}{\left(\prod_{k \geq 1}(1 - x^k)\right)^8}.$$

An entirely new theme in the theory of partitions was introduced by Hardy and Ramanujan when they asked how large $p(n)$ can be for large n. The method that they developed led them to a remarkable asymptotic formula for $p(n)$, which was then refined by Rademacher into an even more remarkable *exact* formula for $p(n)$. It is a chapter in the theory of modular forms. The formula of Rademacher is

$$p(n) = \frac{1}{\pi\sqrt{2}} \sum_{k=1}^{\infty} A_k(n)k^{1/2}\psi_q(n)$$

where

$$\psi_k(n) = \frac{d}{dn}\left(\frac{\sinh C\lambda_n/k}{\lambda_n}\right)$$

with

$$C = \pi\sqrt{\frac{2}{3}}, \qquad \lambda_n = n - \frac{1}{24},$$

and $A_k(n)$ is a certain trigonometric sum given in Selberg's form by

$$A_k(n) = \sqrt{(k/3)} \sum_{(3j^2+j)/2 \equiv -n \bmod k} (-1)^j \cos\frac{6j + 1}{6k}\pi.$$

The original form of $A_k(n)$ of Ramanujan and Rademacher was much more complicated and involved certain 24th roots of unity [22], [26], [23]. Subsequently, Hardy, in collaboration with Littlewood, used what is nowadays known as the Hardy-Ramanujan-Littlewood circle method, born out of the Hardy-Ramanujan paper on $p(n)$, to obtain the solutions to some of the most difficult asymptotic problems in additive number theory [23].

Euler products. We shall discuss this aspect of Euler's work in greater detail when we take up his contributions to infinite series and products. Here we just mention his product expansion for the zeta function

$$\zeta(s) = 1 + \frac{1}{2^2} + \frac{1}{3^s} + \cdots$$

given by

$$\zeta(s) = \prod_p (1 - p^{-s})^{-1}$$

where the product is over all the primes or in his notation

$$\zeta(s) = \frac{2^s.3^s.5^s.\cdots}{(2^s - 1).(3^s - 1).(5^s - 1)\cdots}.$$

At $s = 1$ the product becomes infinite, and Euler used this to give an analytic demonstration of the fact that the set of primes is infinite. Actually Euler proved the more general result

$$\sum_{p \le x} \frac{1}{p} \simeq \log \sum_{n \le x} \frac{1}{n}$$

which is one of the earliest results on the distribution of primes. Euler's idea of getting asymptotic statements about distribution of primes from the product formula for the zeta function and its behavior as $s \to 1 + 0$ was the starting point for Dirichlet's epoch-making work in 1837 on primes in arithmetic progressions. For any multiplicative function χ on the integers with absolute values 1 Dirichlet introduced

$$L(s : \chi) = \prod_p \frac{1}{1 - \chi(p)p^{-s}} = \sum_{n \ge 1} \frac{\chi(n)}{n^s}$$

where the product is taken over all the primes. Already Euler had considered these for some real functions χ which are periodic for some small periods. Analyzing the behavior of the functions $L(s : \chi)$ as $s \to 1 + 0$, Dirichlet was able to prove the existence of infinitely many primes in arithmetic progressions. The series that Dirichlet introduced are now called *Dirichlet series*, and the products such as the ones above are known as *Euler products*. They have occupied a central position in number theory ever since. We shall discuss them in a little more detailed manner in Chapter 6.

The distribution of primes. Let

$$\pi(x) := \#\{p \mid p \text{ a prime}, \ p \le x\}.$$

The determination of $\pi(x)$ at least asymptotically for large x was an important goal for number theorists after Gauss had conjectured that

$$\pi(x) \simeq \frac{x}{\log x} \qquad (x \to \infty),$$

and it was proved towards the end of the 19th century by Hadamard and de la Vallée Poussin independently; it is known as the *prime number theorem*. Before this Riemann had already understood that the key to the study of the distribution of primes is the behavior of the zeta function in the complex plane; indeed, it must have been for such a study that Riemann proved the analytic continuation of the zeta function. This was clearly a significant step forward, because for Euler and certainly for Dirichlet it was necessary to treat the zeta function only as a function of the real variable s. The proofs of Hadamard and de la Vallée Poussin used deep properties of entire functions, and ultimately their proofs depended on the fact that the zeta function does not vanish on the line $\Re(s) = 1$. It was only in 1948 that *elementary* proofs (in the sense that complex function theory is not used) for the prime number theorem were discovered by Erdös and Selberg independently. A more transparent proof from the analytical point of view was given by Wiener when he viewed it as a Tauberian theorem. I shall discuss it in Chapter 5.

The functional equation of the zeta function is

$$\zeta(1-s) = 2^{1-s}\pi^{-s}\cos\left(\frac{s\pi}{2}\right)\Gamma(s)\zeta(s)$$

as was proved by Riemann. But a hundred years before Riemann, Euler had verified this at all integer values of s and even for many fractional values, and there was no doubt in his mind of its validity for all s. However before Euler could do this, he had to satisfy himself that the series expressing $\zeta(1-m)$ for $m = 2, 3, \ldots$, which are divergent, could be summed by some procedure. He used his favorite method, one that has since come to be known as *Abel summation*, namely,

$$\zeta(s) = \lim_{x \to 1-0} \sum_{n \geq 1} \frac{x^n}{n^s},$$

to compute $\zeta(s)$ for $s < 0$. The function $\Gamma(s)$ is the Gamma function, which for integer values $m + 1(m \geq 0)$ coincides with $m!$. It was also discovered by Euler. One must admire the sureness of his judgment and intuition in this matter.

Diophantine equations. Equations for which one asks for solutions in integers or rationals are called *Diophantine* after Diophantus, who was an ancient Greek mathematician and was the first to formulate such problems. Fermat was interested his whole life about the impossibility of the solutions in integers of equations like

$$x^3 + y^3 = z^3$$
$$x^4 + y^4 = z^2$$
$$x^n + y^n = z^n$$

the last of which, the so-called *Fermat's Last Theorem* (FLT), is the one about which he made his celebrated remark that he has a truly wonderful proof that it has no solutions in integers when $n > 2$ but the margin of the book he was reading was insufficient for giving it. He did give a proof for the case $n = 4$; in fact, the second of the above equations is more general and Fermat gave a proof in that case. It was Euler who proved FLT for the case $n = 3$. Although Euler's proof was incomplete at one point, it could be fixed by using results that Euler himself had proved at that time. The significance of Euler's proof was that it made use of number theory in complex numbers, specifically by introducing $\mathbf{Q}[\sqrt{-3}]$, thus beginning the era in which algebraic number fields became an important player in arithmetic. The trail blazed by Euler in this problem was subsequently followed by Kummer when he created modern algebraic number theory and applied it to FLT successfully in many cases. I shall discuss Kummer's discoveries in Chapter 6. As we all know, Wiles solved FLT in 1995.

In addition to the Fermat equation Euler made many contributions to the solution of the so-called Pells's equation,

$$x^2 - Ny^2 = 1.$$

Here the basic problem is to find the fundamental solution. He knew the relationship between the fundamental solution to the continued fraction associated to $\sqrt{N}$ as

well as the periodicity of the continued fraction for quadratic irrationalities. He
computed the fundamental solutions for a whole range of numerical values of N.
Eventually it was left to Lagrange to prove the fundamental results of the theory.

Euler also treated the more general equations of the form

$$y^2 = ax^2 + bx + c$$

where a, b, c are integers with $a > 0$ and not a square and, more generally, equations
of the form

$$Ax^2 + 2Bxy + By^2 + Dx + Ey + F = 0$$

where the discriminant $\Delta := B^2 - AC > 0$ and not a square. Finally, we have
mentioned during our discussion of Euler's work on elliptic integrals that late in his
life Euler worked out many diophantine equations of genus 1.

Zeta and multizeta values. We shall take a more detailed look at this theme
in the next chapter and so we shall be very brief at this time. His first stay at St.
Petersburg was notable for his determination of the value of

$$\zeta(2) = 1 + \frac{1}{2^2} + \frac{1}{3^2} + \cdots.$$

His method was an audacious, almost reckless, use of Newton's theorem on the
symmetric functions of the roots of polynomials in the context of infinite power
series; he succeeded in determining the exact value of $\zeta(2)$. In a letter to Daniel
Bernoulli he communicated his formulae

$$\zeta(2) = \frac{\pi^2}{6} \qquad \zeta(4) = \frac{\pi^4}{90}.$$

Although there were objections (from Bernoulli and others) to Euler's method in
deriving these results, the numerical calculations essentially confirmed his results,
and Euler's reputation as a mathematician of the first rank was established. Euler
was aware that some of the objections to his derivation were legitimate, and he
continued to work on making his arguments more solid, at least by the standards
of his era (actually, as we shall see later, even by modern standards, once we have
the notion of uniform convergence). He succeeded in getting such a solid proof in
1742 when he proved the infinite product expansion

$$\frac{\sin x}{x} = \prod_{n=1}^{\infty} \left(1 - \frac{x^2}{n^2\pi^2}\right).$$

Once he had this, the use of Newton's theorem became legitimate. He then went
on to prove the remarkable formula

$$\zeta(2k) = 1 + \frac{1}{2^{2k}} + \frac{1}{3^{2k}} + \cdots = \frac{(-1)^{k-1}B_{2k}2^{2k}}{2(2k)!}\pi^{2k}$$

where the B_m are the *Bernoulli numbers* given as the coefficients of the expansion

$$\frac{z}{e^z - 1} = 1 - \frac{z}{2} + \sum_{k=1}^{\infty} B_{2k}\frac{z^{2k}}{(2k)!}.$$

From the infinite product expansion for $\sin x$, which incidentally appears already in
his 1734/35 paper on the value of $\zeta(2)$, he obtained by logarithmic differentiation

the remarkable partial fraction expansion

$$\pi \cot \pi x = \frac{1}{x} + \sum_{n=1}^{\infty} \left(\frac{1}{x+n} + \frac{1}{x-n} \right).$$

However, in spite of all his efforts, Euler could not make any progress in the problem of finding the sum

$$\zeta(3) = 1 + \frac{1}{2^3} + \frac{1}{3^3} + \cdots$$

nor the more general sums

$$\zeta(2k+1) = 1 + \frac{1}{2^{2k+1}} + \frac{1}{3^{3k+1}} + \cdots.$$

These remain unknown to this day. The only result is that of Roger Apéry [28], who succeeded in proving in 1978 that $\zeta(3)$ is an irrational number. Recently Rivoal [29] has proved that an infinite number of the $\zeta(2k+1)$ are irrational. Notice however how weak these results are when compared to Euler's explicit evaluation of $\zeta(2k)$.

More than 30 years after these results Euler returned to them and introduced the *multizeta values*. In a remarkable paper written in 1775 Euler introduced the double zeta series

$$\zeta(p,q) = \sum_{p>q>0} \frac{1}{n^p m^q} \qquad (p \geq 2, q \geq 1)$$

and obtained a number of beautiful identities involving them, such as

$$\zeta(2,1) = \zeta(3).$$

The double and the more general multizeta values have generated intense interest and have been shown to be related to some very significant objects in topology and geometry. There is a whole set of conjectures regarding them which, if proved, will significantly advance our knowledge of the zeta values at the odd integers.

Algebraic and transcendental numbers. Throughout his life Euler was interested in the arithmetic nature of special numbers like e; $\log 2$; Euler's constant, which he denoted by C, although nowadays we write γ for it; and so on. His continued fraction for e allowed Lambert to prove in 1758 that e is irrational. He prepared extensive tables for natural logarithms and observed that the natural logarithm of a rational number to a rational base must be transcendental. Thus if a and b are rational, then $a^x = b$ can be true only for transcendental x. The problem of the arithmetic nature of a^b when a, b are both algebraic was one of the famous problems of Hilbert and was solved by Gel'fond and Schneider [15]. It is remarkable that Euler had thought about such questions long before Hilbert and even long before the work of Dedekind and Cantor that clarified the nature of the real number system.

Notes and references

[1] P. -H. Fuss, *Correspondance Mathématique et Physique de quelques célèbres géomètres du XVIIIeme siècle*, Vols I, II, Johnson Reprint Corporation, New York, 1968.

[2] L. Euler, *Opera Omnia*, Series I, Birkhäuser, Boston and Basel.

[3] C. L. Siegel, *Topics in Complex Function Theory*, Vol. I, Wiley-Interscience, 1969.

[4] A. Weil, *Number Theory: An Approach through History from Hammurapi to Legendre*, Birkhäuser, 1984.

[5] V. S. Varadarajan, *Algebra in Ancient and Modern Times*, AMS and Hindustan Book Agency, 1998.

[6] R. P. Feynman, *Lectures in Physics*, Vol. I, Addison-Wesley, 1963.

[7] J. L. Lagrange, *Oeuvres*, Gauthier-Villars, 1867-1892.

[8] C. G. J. Jacobi, *Werke*, Chelsea, 1969.

[9] H. H. Goldstine, *A History of the Calculus of Variations from the 17th through the 19th century*, Springer-Verlag, 1980.

[10] C. G. Fraser, *Isoperimetric problems in the variational calculus of Euler and Lagrange*, Historia Mathematica, 19 (1992), 4-23.

[11] I. M. Gel'fand, and S. V. Fomin, *Calculus of Variations*, Translated by R. A. Silverman, Prentice Hall, 1963.

[12] M. Morse, *Calculus of Variations in the Large*, Amer. Math. Soc., 1934.

[13] M. Morse, *Marston Morse Selected Papers*, Springer-Verlag, 1981.

[14] A. O. Gel'fond, *The role of Euler's work in the development of number theory*, pp. 80-95, in *Leonhard Euler* (in Russian), dedicated to Euler's 250th anniversary, Akademia Nauk SSSR, 1958.

[15] *Mathematical developments arising from Hilbert's problems*, Proc. Symposia in Pure Math., Vol. XXVIII, AMS, 1976.

[16] M. Agarwal, N. Kayal, and N. Saxena, *PRIMES is in P*, Preprint.

[17] A. Granville, *It is easy to determine whether a given integer is a prime*, Bull. Amer. Math. Soc., 42 (2005), 3-38.

[18] D. A. Cox, *Primes of the form $x^2 + ny^2$*, Wiley, 1989.

[19] H. M. Edwards, *Euler and quadratic reciprocity*, Math. Magazine, 56 (November 1983), 285-291.

[20] J. J. Gray, *A commentary on Gauss's mathematical diary, 1796-1814, with an English translation*, Expositiones Mathematicae, 2 (1984), 97-130. (For the original, see Gauss, *Werke*, Vol. X, 1, 1917, 485-574.)

[21] G. H. Hardy and E. M. Wright, *An Introduction to the Theory of Numbers*, Oxford, 1968.

[22] G. H. Hardy, *Ramanujan*, Chelsea, 1940.

[23] G. H. Hardy, *Collected Papers of G. H. Hardy*, Vol. I, Oxford, 1966.

[24] C. F. Gauss, *Disquisitiones Arithmeticae*, Chelsea, 1965.

[25] G. E. Andrews, *Euler's pentagonal number theorem*, Math. Magazine, 56 (November 1983), 279-284.

[26] H. Rademacher, *Collected Papers of Hans Rademacher*, Vol. II, MIT Press, 1974.

[27] G. Shimura, *Inhomogeneous quadratic forms and triangular numbers*, Amer. Jour. Math., 126 (2004), 191-214.

[28] R. Apery, *Irrationalité de $\zeta(2)$ et $\zeta(3)$*, Astérisque, 61 (1979), 11-13.

[29] T. Rivoal, *La fonction zeta de Riemann prend une infinité de valeurs irrationelles aux entiers impairs*, arXiv: math. NT/0008051 v1 2000.

CHAPTER 3

Zeta Values

3.1. Summary

Before Euler's work, infinite series and products had occurred only sporadically in mathematics and nearly always in an isolated manner. With Euler this situation changed dramatically. He was the first and greatest master of infinite series and products and created the first general theory dealing with them. Certainly the geometric series

$$\sum_{n=1}^{\infty} \frac{1}{R^n}$$

for special values such as $R = 2, 4$ goes back to the Greeks. For example, the quadrature of the parabola due to Archimedes depends on the formula

$$\frac{4}{3} = 1 + \frac{1}{4} + \frac{1}{4^2} + \dots.$$

The series

$$\sum_{n=1}^{\infty} \frac{1}{n} = 1 + \frac{1}{2} + \frac{1}{3} + \dots$$

appears in the work of Nicholas of Orseme (1323-1382), and Pietro Mengoli (1625-1686) seems to have posed [1] the problem of finding the sum of the series

$$\sum_{n=1}^{\infty} \frac{1}{n^2} = 1 + \frac{1}{4} + \frac{1}{9} + \dots.$$

The series

$$\frac{\pi}{4} = 1 - \frac{1}{3} + \frac{1}{5} - \dots$$

was discovered by Leibniz, although it appears to have been known to Gregory. We shall follow current universal usage and write

$$\zeta(s) = \sum_{n=1}^{\infty} \frac{1}{n^s} = 1 + \frac{1}{2^s} + \frac{1}{3^s} + \dots$$

following Riemann's notation in his famous paper on the distribution of primes. Many famous mathematicians of the 17th and 18th centuries, including the Bernoulli brothers Jacob and Johann, worked on the problem of determining $\zeta(2)$, which came to be known as the Basel problem. However its solution eluded all these mathematicians. Even the approximate computation of its sum, accurate to several decimal places, proved troublesome because of the slow convergence of the series.

Enter Euler. First in 1731 he discovered a brilliant transformation of the series for $\zeta(2)$. He proved that [2]

$$\zeta(2) = (\log 2)^2 + 2 \sum_{n=1}^{\infty} \frac{1}{n^2 . 2^n}$$

which is rapidly convergent because of the factors 2^n in the denominator. Now, from the Taylor series for $\log(1 - x)$ we have

$$-\log(1 - x) = x + \frac{x^2}{2} + \frac{x^3}{3} + \cdots = \sum_{n=1}^{\infty} \frac{x^n}{n}$$

and so, taking $x = 1/2$,

$$\log 2 = \frac{1}{2} + \frac{1}{8} + \frac{1}{24} + \cdots = \sum_{n=1}^{\infty} \frac{1}{n . 2^n}.$$

The geometric nature of the terms again allows an accurate computation of the series from a small number of terms. Euler did this and found that

$$\zeta(2) = 1.644934\ldots.$$

The problem of numerical evaluation of $\zeta(n)$ for higher values of n probably led Euler to his discovery of his famous *Euler-Maclaurin summation formula*. He announced it in 1732 [3a] and discussed it more elaborately in 1736 in [3b] and in his book [3c] on differential calculus. From then on this was his preferred tool for very accurate numerical summation of series. It was by using it that he was able to show that [3b], [3d]

$$\zeta(2) = 1.64493406684822643647\ldots$$
$$\zeta(3) = 1.202056903159594\ldots$$
$$\zeta(4) = 1.08232323371113819\ldots.$$

So matters stood till suddenly and unexpectedly, around 1735, Euler had a stroke of inspiration that led him to the exact value of $\zeta(2), \zeta(4)$, etc. [4]. Here is the translation by A. Weil of the opening lines of Euler's paper ([5], p. 261): *So much work has been done on the series $\zeta(n)$ that it seems hardly likely that anything new about them may still turn up. ... I, too, in spite of repeated efforts, could achieve nothing more than approximate values for their sums. ... Now, however, quite unexpectedly, I have found an elegant formula for $\zeta(2)$, depending on the quadrature of a circle [i.e., upon π].* Across a gulf of centuries, this passage, and indeed the whole paper, still conveys the excitement Euler must have felt on his discovery. Euler's method was an audacious, one might say, even reckless, generalization of Newton's theorem on the symmetric functions of the roots of a polynomial to a power series. By using it Euler succeeded in determining the exact value of $\zeta(2)$

and $\zeta(4)$. In a letter to Daniel Bernoulli he communicated his formulae

$$1 + \frac{1}{2^2} + \frac{1}{3^2} + \cdots = \frac{\pi^2}{6} \qquad 1 + \frac{1}{2^4} + \frac{1}{3^4} + \cdots = \frac{\pi^4}{90}.$$

Although questions were raised by the Bernoulli brothers and others regarding the validity of Euler's method in deriving these results, the numerical calculations essentially confirmed his results, and Euler's reputation as a mathematician of the first rank was established. Euler was aware that the objections to his derivation were grave and legitimate and so he continued to work on meeting these objections and making his arguments more solid, at least by the standards of his era (actually, as we shall see later, even by modern standards once we have the notion of uniform convergence). He succeeded in doing this around 1742 [6] when he proved the infinite product expansion

$$\frac{\sin x}{x} = \prod_{n=1}^{\infty} \left(1 - \frac{x^2}{n^2 \pi^2}\right)$$

from which, by a legitimate application of Newton's theorem generalized to the context of power series, he could justify completely the values calculated earlier in [4]:

$$\zeta(2) = \frac{\pi^2}{6}$$

$$\zeta(4) = \frac{\pi^4}{90}$$

$$\zeta(6) = \frac{\pi^6}{945}$$

$$\zeta(8) = \frac{\pi^8}{9450}$$

$$\zeta(10) = \frac{\pi^{10}}{93555}$$

$$\zeta(12) = \frac{691\pi^{12}}{6825 \times 93555}$$

It is here that he remarks that the product expansions show that the equations of infinite degree that he had used earlier had no roots other than the obvious real ones (see p. 146 in [6] and Weil [5], p. 271). The appearance of 691 might perhaps have suggested to him that there was a connection with the *Bernoulli numbers*, which he had already encountered in his work on the summation formula. He then went on to prove the beautiful formula [7]

$$\zeta(2k) = 1 + \frac{1}{2^{2k}} + \frac{1}{3^{2k}} + \cdots = \frac{(-1)^{k-1}B_{2k}2^{2k}}{2(2k)!}\pi^{2k}$$

where the B_m are the *Bernoulli numbers* given as the coefficients of the expansion

$$\frac{z}{e^z - 1} = 1 - \frac{z}{2} + \sum_{k=1}^{\infty} B_{2k}\frac{z^{2k}}{(2k)!}.$$

These would follow from the striking partial fraction expansion

$$\pi \cot \pi x = \frac{1}{x} + \sum_{n=1}^{\infty} \left(\frac{1}{x+n} + \frac{1}{x-n}\right)$$

which itself is a consequence of his infinite product for $\sin x$, as was pointed out to him by Nicolaus Bernoulli.[*] Euler knew this and had in fact two different proofs of this, and in any case it already appeared implicitly in his paper.

In addition to the zeta series Euler also considered the L-series

$$L(2k+1) = 1 - \frac{1}{3^{2k+1}} + \frac{1}{5^{2k+1}} - \cdots.$$

For these he obtained the formulae

$$L(2k+1) = \frac{1}{2^{2k+2}(2k)!} E_{2k} \pi^{2k+1}$$

where the E_{2k} are the *Euler numbers* which are defined by

$$\sec z = 1 + \sum_{k=1}^{\infty} \frac{E_{2k} z^{2k}}{(2k)!}.$$

These would follow from the equally striking partial fraction expansion

$$\frac{\pi}{\sin \pi x} = \frac{1}{x} + \sum_{n=1}^{\infty} (-1)^n \left(\frac{1}{n+x} - \frac{1}{n-x} \right).$$

All these results were given a majestic exposition in his great book [3d].

These infinite product and partial fraction expansions were the beginning of a glorious era in the global theory of complex functions. On the one hand, in the hands of Weierstrass, Mittag–Leffler, Hadamard and others it would lead to the golden age of entire function theory. Hadamard used his work to get a proof of the prime number theorem (at the same time as de la Vallée Poussin). On the other hand, at the hands of Riemann, whose work was preceded by that of Eisenstein and Gauss, it would lead to function theory on compact complex manifolds and dominate much of twentieth century geometry. In a third direction it would lead, at the hands of Dirichlet, Kronecker, and their modern successors (not to mention Eisenstein and Gauss again), to the arithmetic theory of abelian extensions of algebraic number fields. It is thus extraordinary that Euler's work could be thought of as the springboard for his successors. In our own time the themes initiated by Riemann have perhaps obscured the possibility that the theory of entire functions has still many deep secrets waiting to be discovered.

Let us return to the zeta values. Euler gave another proof, perfectly unobjectionable by the standards of his (or our) day and completely elementary, for his formula $\zeta(2) = \pi^2/6$. He started from the formula

$$\frac{1}{2}(\arcsin x)^2 = \int_0^x \frac{\arcsin t}{\sqrt{1-t^2}} dt.$$

Taking $x = 1$ gives $\pi^2/8$ for the left-hand side. On the other hand, the integral on the right can be evaluated by expanding $\arcsin t$ as a power series and integrating term by term. The result is

$$\frac{\pi^2}{8} = 1 + \frac{1}{3^2} + \frac{1}{5^2} + \cdots = \sum_{n=1}^{\infty} \frac{1}{(2n+1)^2}$$

which leads to

$$\zeta(2) = \frac{\pi^2}{6}$$

---

[*]See [1] of Ch. 2, Notes and References.

since
$$1 + \frac{1}{3^2} + \frac{1}{5^2} + \cdots = \zeta(2) - \frac{1}{2^2}\zeta(2) = \frac{3}{4}\zeta(2).$$

However, in spite of all his efforts, Euler could not make any progress in the problem of finding the sum
$$1 + \frac{1}{2^3} + \frac{1}{3^3} + \cdots$$

nor the more general sums
$$1 + \frac{1}{2^{2k+1}} + \frac{1}{3^{3k+1}} + \cdots.$$

He gave some wonderful formulae for $\zeta(3)$ such as
$$1 + \frac{1}{3^3} + \frac{1}{5^3} + \cdots = \frac{\pi^2}{\log 2} + 2 \int_0^{\pi/2} x \log \sin x \, dx$$

(see [1] for a detailed discussion). Amazingly, the problem of evaluating the $\zeta(2k+1)$ has remained intractable to this day.

It is remarkable, but fitting, that the next step in the story of zeta values was taken by Euler himself. In a beautiful paper [9] written in 1775 more than 30 years after his great discoveries on zeta values, he introduced the analogues of the zeta series in two variables, more precisely the series
$$\zeta(a, b) = \sum_{n > m > 0} \frac{1}{n^a m^b}.$$

Here a and b are positive integers and $a \geq 2$ so that the series is convergent. He obtained several striking formulae for these double zeta values among which
$$\zeta(2, 1) = \zeta(3)$$

stands out. It must be noted that the above definition of $\zeta(a, b)$ differs from Euler's, who summed over $n \geq m$; thus
$$\zeta_{\text{Euler}}(a, b) = \zeta(a, b) + \zeta(a + b).$$

The previous formula then appears in [8] as
$$\zeta_{\text{Euler}}(2, 1) = 2\zeta(3).$$

It is very clear from his paper that his goal was to express, for each given positive integer $r \geq 3$, the $\zeta(a, b)$ with $a + b = r$ as a rational linear combination of $\zeta(r)$ and the $\zeta(p)\zeta(r - p)$ $(2 \leq p \leq r - 2)$. He succeeded in many cases but not all.

It is obvious that Euler's idea can be generalized to higher dimensions. The *multizeta values* (MZV) are defined by
$$\zeta(a_1, \ldots, a_r) = \sum_{n_1 > \cdots > n_r > 0} \frac{1}{n_1^{a_1} \ldots n_r^{a_r}}$$

where the a_i are positive integers and $a_1 \geq 2$ to ensure convergence. Actually one can use a summation process to define the MZV even when there is divergence.

The first new result after Euler was Roger Apéry's stunning proof in 1978 that $\zeta(3)$ is irrational ([28] of Ch. 2.). Going beyond Apéry, T. Rivoal has now proved that an infinite number of the $\zeta(2k + 1)$ must be irrational, although no specific $\zeta(2k + 1)$ is known to be irrational other than $\zeta(3)$ ([29] of Ch. 2).

In recent years the MZV have generated tremendous interest, and people have shown them to be connected to some very deep aspects of algebraic geometry and arithmetic. Some remarkable conjectures have been proposed, but the evaluations of the MZV appear out of reach today. If the conjectures are true, they would imply that

$$\pi, \zeta(3), \zeta(5), \zeta(7), \ldots,$$

are algebraically independent over $\mathbf{Q}$; even this is a far cry from any exact evaluation of the $\zeta(2k+1)$.

3.2. Some remarks on infinite series and products and their values

In Euler's time there was no habit of precise definitions and this led to enormous and intense discussions regarding infinite series and their sums in which nonmathematicians, occasionally even clergymen, participated. For instance such a great mathematician as Leibniz could not decide whether the series

$$1 - 1 + 1 - 1 + 1 - 1 + \ldots$$

had a sum. He thought that as the partial sum s_n is 1 for odd n and 0 for even n, the sum should be $1/2$. But since there was no precise definition of what a sum ought to be, there was confusion and disagreement. Euler had a much clearer idea than his contemporaries on what one means when one says that such and such a series has the sum s. In fact precisely because his ideas were close to ideas in our own time he was able to make a substantial theory of *divergent series* and make serious use of them. His 1760 paper on divergent series signals the birth of the modern theory of divergent series (see Chapter 5).

The problems of dealing with infinite series fall into several categories.

(1) To decide if the series is convergent.
(2) If convergent, to determine the exact value of the sum.
(3) In case evaluation of the sum proves impossible, to try to evaluate it to a high degree of accuracy by summing up to a certain limit; here if the terms a_n decrease to 0 slowly, as is often the case, transformations are needed before approximate calculation can be attempted. Euler discovered many of these transformations.
(4) To try to associate a "sum" to a divergent series. Euler was the first mathematician to do this systematically.

In 1674 Leibniz had calculated that

$$\frac{\pi}{4} = 1 - \frac{1}{3} + \frac{1}{5} - \frac{1}{7} + \frac{1}{9} - \ldots$$

where the terms are the reciprocals of odd numbers with signs that alternate. There were other series of this type, such as

$$\frac{\pi}{2\sqrt{2}} = 1 + \frac{1}{3} - \frac{1}{5} - \frac{1}{7} + \frac{1}{9} + \frac{1}{11} - \ldots$$

which was attributed to Newton by Euler, and

$$\log 2 = 1 - \frac{1}{2} + \frac{1}{3} - \frac{1}{4} + \ldots$$

which follows from the Taylor series for $\log(1+x)$, namely

$$\log(1+x) = x - \frac{x^2}{2} + \frac{x^3}{3} - \ldots$$

by taking $x = 1$.

For a series of *positive* terms there are only two possibilities: either the series is convergent or divergent to ∞. If

$$0 \leq a_n \leq b_n,$$

then the series $\sum_n a_n$ converges if the series $\sum_n b_n$ converges, while the series $\sum b_n$ diverges if the series $\sum a_n$ diverges. This principle leads to the various tests of convergence that were in vogue in the 19th century. The choice of $b_n = r^n$ leads to the ratio and root tests which are not conclusive; if we choose $b_n = n^{-s}$, it leads to a conclusive test at least for series where $a_n = f(n)$ for a rational function f. This is the Gauss test. All this became clear only in the 19th century. For a while people believed that one could write down a *universal test* of convergence, but the work of *Abel, Dini, Pringsheim,* and others showed that this is impossible. For instance, given any series $\sum a_n$ of positive terms which is convergent, there is another series $\sum b_n$ which is convergent, but such that

$$\lim_{n \to \infty} \frac{a_n}{b_n} = 0,$$

or, as one would say, *the series $\sum_n b_n$ converges more slowly than the series $\sum_n a_n$.* Similarly, given any divergent series of positive terms one can find a divergent series which diverged more slowly. For a beautiful account of the classical work on series, see Knopp's famous book [10].

Let us look at the series

$$1 + \frac{1}{2^s} + \frac{1}{3^s} + \cdots.$$

Since x^{-s} is decreasing for $x \geq 1$ we get

$$\frac{1}{(n+1)^s} < \int_n^{n+1} \frac{1}{x^s} dx < \frac{1}{n^s}.$$

For $s > 1$ we have convergence since

$$1 + \frac{1}{2^s} + \cdots + \frac{1}{n^s} < 1 + \int_1^\infty \frac{dx}{x^s} = 1 + \frac{s}{s-1},$$

while for $s = 1$ we have logarithmic divergence because

$$\frac{1}{2} + \cdots + \frac{1}{n} < \int_1^n \frac{dx}{x} = \log n < 1 + \frac{1}{2} + \cdots + \frac{1}{n},$$

hence divergence for $s \leq 1$. In the convergent case if

$$R_N = \sum_{n=N+1}^\infty \frac{1}{n^s}.$$

Then

$$\int_{N+1}^\infty \frac{1}{x^s} dx < R_N < \int_N^\infty \frac{1}{x^s} dx$$

which gives the error estimate

$$\frac{1}{s-1} \frac{1}{(N+1)^{s-1}} < R_N < \frac{1}{s-1} \frac{1}{N^{s-1}}.$$

In other words, *the error R_N committed in stopping after N terms lies between*

$$\frac{1}{s-1} \frac{1}{(N+1)^{s-1}} \quad and \quad \frac{1}{s-1} \frac{1}{N^{s-1}}.$$

Thus, when $s = 2$, the error R_N lies between $\frac{1}{N+1}$ and $\frac{1}{N}$. To get accuracy up to 6 decimals we thus have to take a million terms. This should give an idea of the difficulty of numerical computation of these series, difficulties that Euler often overcame with his famous summation formula.

Examples of series whose sums could be evaluated were already known when Euler started working on these things, and he added many more to this list. We mention a few of these:

$$\frac{\pi}{4} = 1 - \frac{1}{3} + \frac{1}{5} - \frac{1}{7} \cdots$$

$$\log 2 = 1 - \frac{1}{2} + \frac{1}{3} - \frac{1}{4} \cdots$$

$$\frac{\pi}{2\sqrt{2}} = 1 + \frac{1}{3} - \frac{1}{5} - \frac{1}{7} + \frac{1}{9} + \frac{1}{11} - \cdots$$

$$\frac{1}{3}\left(\log 2 + \frac{\pi}{\sqrt{3}}\right) = 1 - \frac{1}{4} + \frac{1}{7} - \cdots.$$

For the first two we integrate $(1 + x^2)^{-1}$ and $(1 + x)^{-1}$ from 0 to 1. The last two summations, obtained by integrating $(1 + x^2)/(1 + x^4)$ and $1/(1 + x^3)$ respectively from 0 to 1, are less elementary and follow from Euler's general expression for such integrals in terms of circular functions (cf. infra). The series for $\log 2$ is not good for numerical evaluation. If we take $x = -1/2$ in the series for $\log x$ we get the series for $\log 2$ which Euler used for his very accurate computation of $\log 2$ correct to several decimal places:

$$\log 2 = \sum_{n=1}^{\infty} \frac{1}{n.2^n}.$$

The geometric nature of the series ensures its rapid convergence.

Unlike series of positive terms the series of terms of arbitrary signs do not behave well under various transformations. It was Riemann who really clarified the true nature of series of terms with arbitrary signs. For the series

$$a_1 + a_2 + \ldots$$

let us define

$$a_n^+ = \frac{|a_n| + a_n}{2}, \qquad a_n^- = \frac{|a_n| - a_n}{2}.$$

Let us consider the partial sums of the three series

$$\sum_n a_n, \qquad \sum_n a_n^{\pm}.$$

Suppose that the series $\sum_n a_n$ is convergent to the sum s. Then for absolute convergence it is necessary and sufficient that $s_n^{\pm}$ have finite limits $s^{\pm}$ and $s = s^+ - s^-$. For nonabsolute or conditional convergence it is necessary and sufficient that $s_n^{\pm} \to \infty$ but $s_n^+ - s_n^- \to s$. This shows that conditional convergence is very delicate and that it depends on the difference of two sequences going to ∞ in such a way that their difference goes to a finite limit. Riemann used this to show that an absolutely convergent series can be *rearranged* in any manner without affecting its convergence or its sum, but a conditionally convergent series can be rearranged to have any real number as its sum or to diverge to $\pm\infty$ or to oscillate with prescribed upper and lower limits. This is the famous *Riemann rearrangement theorem*. If the a_n are complex or more generally vectors in a finite dimensional vector space, the set of sums obtained by rearrangement is an affine subspace; in the complex case it is either a point or a line or the whole complex plane.

The evaluations above would not be completely understood till Abel and Weierstrass laid the foundations of the theory of convergent series of functions, especially power series. For instance, the series for $(1 + x^2)^{-1}$ converges only in $(0, 1)$, and so to get the Leibniz result one has to proceed in two stages. In the first stage we integrate from 0 to y and get

$$\arctan y = y - \frac{y^3}{3} + \frac{y^5}{5} - \dots \qquad (0 < y < 1).$$

This already requires the use of the uniform convergence of the power series. Then we let y approach 1. The left side goes to $\pi/4$, but in the right side we have to pass to the limit termwise, and this time it is more delicate as there is no uniform convergence to fall back on. This was done first by Abel. In any case these things were beyond what was understood and available in the 18th century.

For completeness let us sketch the proofs of both steps mentioned above. Consider a power series

$$f(x) = a_0 + a_1 x + a_2 x^2 + \dots + a_n x^n + \dots \qquad (0 < x < 1)$$

convergent in $(0, 1)$. If $0 < x \leq r < r' < 1$, we have $|a_n r'^n| \leq C$ for a constant C and hence

$$|a_n x^n| \leq C \left(\frac{r}{r'}\right)^n$$

showing uniform convergence in $[0, r]$. Term by term integration and differentiation are thus permitted in any interval $[0, y]$ with $y < 1$. Abel's result is more delicate and depends on *Abel's partial summation*, the discrete version of integration by parts, which was probably developed by Abel to deal with series of terms whose signs are not fixed. It asserts that if the series $\sum_n a_n$ is convergent, then

$$\lim_{x \to 1-} f(x) = \sum_n a_n.$$

Here the existence of the limit on the left side is part of the assertion. The partial summation formula is

$$\sum_{0 \leq r \leq n} a_r b_r = \sum_{0 \leq r \leq n-1} s_r (b_r - b_{r+1}) + s_n b_n, \qquad s_r = \sum_{0 \leq j \leq r} a_j.$$

This leads to

$$\sum_{0 \leq r \leq n} a_r x^r = (1 - x) \sum_{0 \leq r \leq n-1} s_r x^r + s_n x^n.$$

Since s_n is bounded we may let $n \to \infty$ to get, for $0 < x < 1$,

$$f(x) = (1 - x)g(x), \qquad g(x) = \sum_{r \geq 0} s_r x^r = \frac{s}{1 - x} + \sum_{r \geq 0} (s_r - s)x^r.$$

Hence, writing $c_r = s_r - s$ we get

$$f(x) = s + (1 - x)h(x), \qquad h(x) = \sum_{r \geq 0} c_r x^r.$$

Now $c_r \to 0$ and so, for any $\delta > 0$ we can find $k > 1$ so that $|c_r| \leq \delta$ for $r > k$. Then

$$|(1 - x)h(x)| \leq (1 - x) \sum_{1 \leq r \leq k} |c_r| x^k + \delta$$

so that

$$\limsup_{x \to 1-} |(1-x)h(x)| \le \delta.$$

Letting $\delta \to 0$ we see that $(1-x)h(x) \to 0$ as $x \to 1-$. This is Abel's theorem.

When the series $\sum_n a_n$ does not converge but $\lim_{x \to 1-} f(x)$ exists and has the value s, we may *define* the sum of the series $\sum_n a_n$ as s. This is the definition of *Abel summability*, but it goes back to Euler, for whom it was a favorite method of summation. He used it extensively.

Infinite products. We have already seen that infinite product expansions of the circular functions form the center piece of Euler's discoveries on the zeta values. With this in mind let us review a few elementary facts about infinite products. We shall consider only infinite products of the form

$$\prod_{n=1}^{\infty}(1+a_n), \qquad (a_n \in \mathbf{C},\ a_n \to 0 \text{ as } n \to \infty).$$

Then $|a_n| \le 1/2$ for all n sufficiently large so that we may assume that $|a_n| \le 1/2$ for all n. The product is said to be *convergent* with the value p if

$$p_n = \prod_{k=1}^{n}(1+a_k) \to p \ne 0.$$

We write

$$\prod_{n=1}^{\infty}(1+a_n) = p.$$

It is easy to see that

$$\prod_{n=1}^{\infty}(1+a_n) \quad \text{and} \quad \sum_{n=1}^{\infty}\log(1+a_n)$$

have the same behavior, log being the principal branch at 1, and the sum s of the series is related to p by $p = e^s$. Moreover,

$$\sum_{n=1}^{\infty}|a_n| < \infty \iff \sum_{n=1}^{\infty}|\log(1+a_n)| < \infty.$$

To see this, note that for $|x| \le 1/2$,

$$|\log(1+x)-x| \le \frac{|x|^2}{2}(1+|x|+|x|^2+\dots) \le \frac{|x|^2}{1-|x|} \le \frac{1}{2}|x|$$

so that

$$\frac{1}{2}|x| \le |\log(1+x)| \le \frac{3}{2}|x|.$$

The assertion is then clear.

3.3. Evaluation of $\zeta(2)$ and $\zeta(4)$

We have seen that

$$\frac{1}{n+1} < R_n = \frac{1}{(n+1)^2} + \frac{1}{(n+1)^2} + \dots < \frac{1}{n}.$$

So if one wants to calculate $\zeta(2)$ accurately up to 6 decimals we must take $n > 10^6$; i.e., we must take a million terms. This must have been daunting even to Euler, who was an indefatigable calculator. In his very first contribution to the subject in

1731, entitled *De Summatione Innumerabilium Progressionum*, I-14, 25-41, Euler discovered a transformation of this series that allowed him to calculate $\zeta(2)$ to 6 decimal places, and even more accurately had he wanted to do so. Euler proved that

$$\zeta(2) = \log u \log(1 - u) + \sum_{n=1}^{\infty} \frac{u^n}{n^2} + \sum_{n=1}^{\infty} \frac{(1 - u)^n}{n^2} \qquad (0 < u < 1).$$

In particular, taking $u = 1/2$, we get

$$\zeta(2) = (\log 2)^2 + \sum_{n=1}^{\infty} \frac{1}{n^2 . 2^{n-1}}.$$

Euler's argument is the following. Start with

$$-\frac{\log(1 - x)}{x} = 1 + \frac{x}{2} + \frac{x^2}{3} + \cdots.$$

Integrating from 0 to 1 we have,

$$\zeta(2) = \int_0^1 \frac{-\log(1 - x)}{x} dx.$$

We split the integration from 0 to u and u to 1. So for any u with $0 < u < 1$,

$$\zeta(2) = \int_0^u \frac{-\log(1 - x)}{x} dx + \int_u^1 \frac{-\log(1 - x)}{x} dx.$$

In the second integral we change x to $1 - x$ to get

$$\zeta(2) = \int_0^u \frac{-\log(1 - x)}{x} dx + \int_0^{1-u} \frac{-\log x}{1 - x} dx = I_1 + I_2.$$

For I_1 we expand $\log(1 - x)$ as a series and integrate term by term from 0 to u. So

$$I_1 = \sum_{n=1}^{\infty} \frac{u^n}{n^2}.$$

To evaluate I_2 we integrate by parts first to get

$$I_2 = \log x \log(1 - x)\Big|_0^{1-u} + \int_0^{1-u} \frac{-\log(1 - x)}{x} dx.$$

The second term is like I_1. For the first term note that

$$\log(1 - x) = -x - \frac{x^2}{2} - \cdots \sim -x \qquad (x \to 0)$$

so that

$$\log x \log(1 - x) \sim -x \log x \to 0 \qquad (x \to 0).$$

Hence we obtain

$$\zeta(2) = \log u \log(1 - u) + \sum_{n=1}^{\infty} \frac{u^n}{n^2} + \sum_{n=1}^{\infty} \frac{(1 - u)^n}{n^2},$$

which is Euler's result. For numerical evaluation Euler used the formula for $\log 2$ obtained earlier,

$$\log 2 = \sum_{n=1}^{\infty} \frac{1}{n . 2^n}.$$

Using this Euler computes

$$(\log 2)^2 = 0.480453\ldots, \qquad \sum_{n=1}^{\infty} \frac{1}{n^2 . 2^{n-1}} = 1.164481\ldots$$

to obtain

$$\zeta(2) = 1.644934\ldots.$$

The above transformation used by Euler is the first instance of the appearance of the *dilogarithm*. Let

$$\mathrm{Li}_2(x) := \int_0^x \frac{-\log(1-t)}{t}\,dt = \int_{x > t_1 > t_2 > 0} \frac{dt_1\,dt_2}{t_1(1-t_2)}.$$

Then

$$\mathrm{Li}_2(x) = \sum_n \frac{x^n}{n^2}, \qquad \mathrm{Li}_2(1) = \zeta(2).$$

The discussion above is essentially the proof of the functional equation

$$\mathrm{Li}_2(x) + \mathrm{Li}_2(1-x) = -\log x \log(1-x) + \mathrm{Li}_2(1) \qquad (0 < x < 1)$$

which also is in Euler (loc. cit.). The dilogarithm has obvious generalizations to several variables, and these play an important role in the theory of MZV.

In addition to $\zeta(s)$ Euler worked with several other functions N, M, L given by

$$N(s) = 1 + \frac{1}{3^s} + \frac{1}{5^s} + \cdots = \sum_{n=1}^{\infty} \frac{1}{(2n-1)^s}$$

$$M(s) = 1 - \frac{1}{2^s} + \frac{1}{3^s} - \frac{1}{4^s} + \frac{1}{5^s} - \cdots = \sum_{n=1}^{\infty} (-1)^{n-1} \frac{1}{n^s} \qquad (s > 0)$$

$$L(s) = 1 - \frac{1}{3^s} + \frac{1}{5^s} - \frac{1}{7^s} + \cdots = \sum_{n=1}^{\infty} (-1)^{n-1} \frac{1}{(2n+1)^s} \qquad (s > 0).$$

Unlike $\zeta(s)$ and $N(s)$ which are defined only for $s > 1$, the functions M and L are defined for $s > 0$. Moreover

$$N(s) = (1 - 2^{-s})\zeta(s)$$
$$M(s) = (1 - 2^{1-s})\zeta(s).$$

The other function $L(s)$ is the first instance of what Dirichlet would later introduce as an "L-function associated with a character". In this case the character is essentially the function ω defined on the odd integers by

$$\omega(n) = (-1)^{(n-1)/2}.$$

It is an easy verification that ω is *multiplicative*, i.e.,

$$\omega(mn) = \omega(m)\omega(n) \qquad (m, n \text{ odd numbers}).$$

One can then write

$$L(s) = \sum_{m=1}^{\infty} \frac{\omega(2m+1)}{(2m+1)^s} \qquad (s > 0).$$

Notice that ω has period 4, using which we can extend it to all integers by requiring that

$$\omega(2m) = 0, \qquad \omega(n) = \omega(n+4).$$

Then
$$L(s) = \sum_{n=1}^{\infty} \frac{\omega(n)}{n^s}.$$

Euler preferred to work with $L(s)$ and $M(s)$ because of their convergence for $s > 0$, while $N(s)$ and $\zeta(s)$ converged only for $s > 1$.

The accurate evaluation of $\zeta(2)$ must have motivated Euler to look for a general method of summing series. He succeeded in this attempt when he discovered what is now known as the *Euler Maclaurin summation formula*. This was around 1732, and Maclaurin discovered it in 1742. While Euler's priority was unquestioned, he made no attempt to deny Maclaurin his share of the discovery, which was quite typical of him. With this formula he computed the values $\zeta(k)$ for even k up to 26. These evaluations would serve him well when objections were raised about his first proof of the exact even zeta values.

Euler's first proof that $\zeta(2) = \pi^2/6$ **and** $\zeta(4) = \pi^4/90.$ Around 1735 Euler suddenly found a method to evaluate $\zeta(2)$ exactly. He found that
$$\zeta(2) = \frac{\pi^2}{6}, \qquad \zeta(4) = \frac{\pi^4}{90}.$$

He was quite excited about his discovery and communicated it to his friend Daniel Bernoulli and to other mathematicians. We shall now take a look at his method which was given in the paper entitled *De Summis Serierum Reciprocarum*, I-14, 73-86.

Euler's method, as we mentioned in §3.1, was based on a generalization of Newton's theorem on symmetric functions of the roots of a polynomial, to the case when the polynomial is replaced by a power series (he called this an *infinite equation*). Let us first recall Newton's theorem. Let $f(s)$ be a polynomial of degree k in the variable s and let $a, b, \ldots, q, r$ be its roots. Then
$$f(s) = L(s - a)(s - b)\ldots(s - q)(s - r)$$

where L is the coefficient of s^k in $f(s)$. Newton's theorem asserts that any symmetric polynomial of the roots can be expressed as a polynomial of the coefficients of f. To prepare us for Euler's generalization of this to polynomials of infinite degree, let us rewrite f *starting from the constant term* (normalized to be 1) as
$$1 - \alpha s + \beta s^2 - \cdots \pm \rho s^k = \left(1 - \frac{s}{a}\right)\left(1 - \frac{s}{b}\right)\ldots\left(1 - \frac{x}{q}\right)\left(1 - \frac{s}{r}\right).$$

From this, expanding the right side and equating coefficients of the powers of x on both sides, we get
$$\alpha = a + b + \ldots r, \quad \beta = ab + ac + \cdots + qr, \quad \gamma = abc + \cdots + pqr$$

and so on. Here $\beta, \gamma, \ldots$ are the sums of products of the roots taken $2, 3, \ldots$ at a time. If
$$S_k = a^k + b^k + \cdots + r^k,$$

then Newton's formulae give
$$S_2 = \alpha^2 - 2\beta, \quad S_3 = \alpha^3 - 3\alpha\beta + 3\gamma, \quad S_4 = \alpha^4 - 4\alpha^2\beta + 4\alpha\gamma + 2\beta^2 - 4\delta$$

and so on. When the polynomial is replaced by a power series
$$f(s) = 1 - \alpha s + \beta s^2 - \gamma s^3 + \delta s^4 - \varepsilon s^5 + \ldots$$

one has to now allow for the fact that the roots will be infinite in number. Let us write them as $A, B, C, D, \ldots$. Then, imitating the case of the polynomials, Euler wrote

$$1 - \alpha s + \beta s^2 - \gamma s^3 + \cdots = \left(1 - \frac{s}{A}\right)\left(1 - \frac{s}{B}\right)\left(1 - \frac{s}{C}\right)\left(1 - \frac{s}{D}\right)\cdots.$$

If now one takes over (as Euler did) the Newton formulae for sums of powers of the roots, one obtains

$$\alpha = \frac{1}{A} + \frac{1}{B} + \frac{1}{C} + \frac{1}{D} + \ldots, \quad \beta = \frac{1}{AB} + \frac{1}{AC} + \frac{1}{BC} + \cdots$$

and so on. The formulae for $S_1, S_2, S_3, \ldots$ remain unchanged. Thus

$$S_2 = \frac{1}{A^2} + \frac{1}{B^2} + \cdots = \alpha^2 - 2\beta.$$

There is no obstacle to writing down expressions for the sums of higher powers of the roots. Following Euler let us write

$$P = \frac{1}{A} + \frac{1}{B} + \ldots, \quad Q = \frac{1}{A^2} + \frac{1}{B^2} + \ldots, \quad R = \frac{1}{A^3} + \frac{1}{B^3} + \ldots$$
$$S = \frac{1}{A^4} + \frac{1}{B^4} + \ldots, \quad T = \frac{1}{A^5} + \frac{1}{B^5} + \ldots, \quad V = \frac{1}{A^6} + \frac{1}{B^6} + \ldots$$

and so on. To make the calculations easier Euler wrote down expressions for these that are *recursive*. Thus

$$P = \alpha, \ Q = P\alpha - 2\beta, \ R = Q\alpha - P\beta + 3\gamma$$
$$S = R\alpha - Q\beta + P\gamma - 4\delta, \ T = S\alpha - R\beta + Q\gamma - P\delta + 5\varepsilon$$

and so on.

Euler applied this theory to the case

$$f(s) = 1 - \frac{\sin s}{y}$$

where y is a parameter in the sense that various values can be given to it later. Since

$$\sin s = s - \frac{s^3}{3!} + \frac{s^5}{5!} - \frac{s^7}{7!} + \ldots$$

we have

$$1 - \frac{s}{y} + \frac{s^3}{3!\,y} - \frac{s^5}{5!\,y} + \cdots = \left(1 - \frac{s}{A}\right)\left(1 - \frac{s}{B}\right)\left(1 - \frac{s}{C}\right)\left(1 - \frac{s}{D}\right)\cdots.$$

In this case the earlier formulae reduce to

$$\alpha = \frac{1}{y}, \ \beta = 0, \ \gamma = -\frac{1}{6y}, \ \delta = 0$$

and so on. Thus

$$P = \frac{1}{y}, \ Q = \frac{1}{y^2}, \ R = \frac{1}{y^3} - \frac{1}{2y}, \ S = \frac{1}{y^4} - \frac{2}{3y^2}$$

and so on.

$y = 1$: The function $1 - \sin s$ has the roots

$$\frac{\pi}{2}, \frac{\pi}{2}, -\frac{3\pi}{2}, -\frac{3\pi}{2}, \frac{5\pi}{2}, \frac{5\pi}{2}, -\frac{7\pi}{2}, -\frac{7\pi}{2}, \ldots$$

where the repetitions express the fact that these are all *double roots*. Writing $q = \pi/2$, the reciprocals of the roots are

$$\frac{1}{q}, \frac{1}{q}, -\frac{1}{3q}, -\frac{1}{3q}, \frac{1}{5q}, \frac{1}{5q}, -\frac{1}{7q}, -\frac{1}{7q}, \frac{1}{9q}, \frac{1}{9q}, \ldots$$

The formula for the sum of roots gives

$$1 = 2 \times \left(\frac{1}{q} - \frac{1}{3q} + \frac{1}{5q} - \frac{1}{7q} + \cdots \right)$$

which leads to Leibniz's result

$$\frac{\pi}{4} = 1 - \frac{1}{3} + \frac{1}{5} - \frac{1}{7} + \cdots.$$

The formula for the sums of squares of roots gives

$$1 = 2 \times \left(\frac{1}{q^2} + \frac{1}{9q^2} + \frac{1}{25q^2} + \frac{1}{49q^2} + \cdots \right)$$

leading to

$$\frac{\pi^2}{8} = 1 + \frac{1}{3^2} + \frac{1}{5^2} + \frac{1}{7^2} + \cdots.$$

If we now observe that the right side is

$$N(2) = (1 - 2^{-2})\zeta(2)$$

we get

$$\zeta(2) = 1 + \frac{1}{2^2} + \frac{1}{3^2} + \cdots = \frac{\pi^2}{6}.$$

This was Euler's proof.

One need not stop here of course. Going to the third and fourth powers we get

$$\frac{2}{q^3} \times \left(1 - \frac{1}{3^3} + \frac{1}{5^3} - \cdots \right) = \frac{1}{2}$$

$$\frac{2}{q^4} \times \left(1 + \frac{1}{3^4} + \frac{1}{5^4} + \cdots \right) = \frac{1}{3}.$$

These lead to

$$L(3) = \left(1 - \frac{1}{3^3} + \frac{1}{5^3} - \cdots \right) = \frac{\pi^3}{32}$$

$$N(4) = \left(1 + \frac{1}{3^4} + \frac{1}{5^4} + \cdots \right) = \frac{\pi^4}{96}.$$

Using the relation

$$N(4) = (15/16)\zeta(4)$$

we get

$$\zeta(4) = \frac{\pi^4}{90}.$$

Euler actually did not stop with these in his 1734/35 paper but went on to calculate

$$L(5) = \frac{5\pi^5}{1536}, \quad N(6) = \frac{\pi^6}{960}, \quad L(7) = \frac{61\pi^7}{184320}, \quad N(8) = \frac{17\pi^8}{161280}$$

and hence also

$$\zeta(6) = \frac{\pi^6}{945}, \quad \zeta(8) = \frac{\pi^8}{9450}.$$

In every case he found that the value is a rational multiple of an appropriate power of π. We shall return to this point later.

Further calculations. It was natural for Euler to ask what additional formulae could be obtained by this method if one takes other values for y. Euler did this and found that although nothing new was obtained for sums of higher powers, the formulae for the sum of the roots led to new results.

$y = 1/\sqrt{2}$: The roots of the equation $\sin s = 1/\sqrt{2}$ are $2k\pi + \pi/4$, i.e.,

$$\frac{\pi}{4}, \frac{3\pi}{4}, -\frac{5\pi}{4}, -\frac{7\pi}{4}, +\frac{9\pi}{4}, \frac{11\pi}{4}, \ldots$$

giving the formula

$$\frac{\pi}{2\sqrt{2}} = 1 + \frac{1}{3} - \frac{1}{5} - \frac{1}{7} + \frac{1}{9} + \frac{1}{11} - \cdots$$

which Euler attributed to Newton.

$y = \sqrt{3}/2$: The roots of the equation $\sin s = \sqrt{3}/2$ are $2k\pi + \pi/3$, i.e.,

$$\frac{\pi}{3}, \frac{2\pi}{3}, -\frac{4\pi}{3}, -\frac{5\pi}{3}, +\frac{7\pi}{3}, \frac{8\pi}{4}, -\frac{10\pi}{3}, -\frac{11\pi}{3}, \ldots$$

giving the formula

$$\frac{2\pi}{3\sqrt{3}} = 1 + \frac{1}{2} - \frac{1}{4} - \frac{1}{5} + \frac{1}{7} + \frac{1}{8} - \cdots.$$

The calculation of the sums of higher powers of the reciprocal integers by taking these various values of y does not yield additional formulae but provides confirmation of the method and the results obtained before.

This method works for all y. Write $y = \sin \sigma$. The roots of

$$1 - \frac{\sin s}{\sin \sigma} = 0$$

are

$$\sigma, \sigma \pm 2\pi, \sigma \pm 4\pi, \ldots$$

and

$$\pi - \sigma, \pi - \sigma \pm 2\pi, \pi - \sigma \pm 4\pi, \ldots.$$

Euler's argument then gives

$$1 - \frac{\sin s}{\sin \sigma} = \prod_{n=-\infty}^{\infty} \left(1 - \frac{s}{2n\pi + \sigma}\right)\left(1 - \frac{s}{2n\pi + \pi - \sigma}\right).$$

For convergence purposes this should be rewritten as

$$1 - \frac{\sin s}{\sin \sigma} = \left(1 - \frac{s}{\sigma}\right) \prod_{n=1}^{\infty} \left(1 - \frac{s}{2n\pi + \sigma}\right)\left(1 + \frac{s}{2n\pi - \sigma}\right)$$

$$\times \prod_{n=1}^{\infty} \left(1 - \frac{s}{(2n-1)\pi - \sigma}\right)\left(1 + \frac{s}{(2n-1)\pi + \sigma}\right).$$

We shall give a rigorous demonstration later, due to Euler himself. The formula for the sum of roots then gives

$$\frac{1}{\sin \sigma} = \frac{1}{\sigma} + \sum_{n=1}^{\infty} \left(\frac{1}{2n\pi + \sigma} - \frac{1}{2n\pi - \sigma} \right)$$
$$+ \sum_{n=1}^{\infty} \left(\frac{1}{(2n-1)\pi - \sigma} - \frac{1}{(2n-1)\pi + \sigma} \right)$$

which can be written as

$$\frac{1}{\sin \sigma} = \frac{1}{\sigma} + \sum_{n=1}^{\infty} (-1)^n \left(\frac{1}{n\pi + \sigma} - \frac{1}{n\pi - \sigma} \right).$$

Writing πs for σ leads to

$$\frac{\pi}{\sin \pi s} = \frac{1}{s} + \sum_{n=1}^{\infty} (-1)^n \left(\frac{1}{n + s} - \frac{1}{n - s} \right).$$

Euler writes this as

$$\frac{\pi}{\sin \pi s} = \frac{1}{s} - \frac{1}{1 + s} + \frac{1}{1 - s} + \frac{1}{2 + s} - \frac{1}{2 - s} - \frac{1}{3 + s} + \frac{1}{3 - s} + \text{etc.}$$

Euler would return to this infinite partial fraction later and prove it by another method, based on integral calculus and using what Legendre would later call the Eulerian integrals. The sums evaluated earlier correspond to the cases $s = 1/2, 1/4, 1/6$. Going back to the infinite product and writing the expression for the sum of squares and cubes of the roots, we get

$$\frac{1}{s^2} + \sum_{n=1}^{\infty} \left(\frac{1}{(n + s)^2} + \frac{1}{(n - s)^2} \right) = \frac{\pi^2}{\sin^2 \pi s}$$

$$\frac{1}{s^3} + \sum_{n=1}^{\infty} (-1)^n \left(\frac{1}{(n + s)^3} - \frac{1}{(n - s)^3} \right) = \frac{\pi^3}{\sin^3 \pi s} - \frac{\pi^3}{2 \sin \pi s}.$$

The last case Euler considered was the case $y = 0$. The equation $1 - \sin s/y = 0$ will now be written as $\sin s = 0$ and on removing the trivial root $s = 0$ becomes

$$\frac{\sin s}{s} = 0.$$

We get the equation

$$\frac{\sin s}{s} = 1 - \frac{s^2}{3!} + \frac{s^4}{5!} - \frac{s^6}{7!} + \cdots = \left(1 - \frac{s}{A} \right) \left(1 - \frac{s}{B} \right) \left(1 - \frac{s}{C} \right) \left(1 - \frac{s}{D} \right) \cdots.$$

The roots are now

$$\pm \pi, \pm 2\pi, \pm 3\pi, \ldots$$

and so the product formula becomes

$$\frac{\sin s}{s} = \left(1 - \frac{s^2}{\pi^2} \right) \left(1 - \frac{s^2}{4\pi^2} \right) \left(1 - \frac{s^2}{9\pi^2} \right) \cdots = \prod_{n=1}^{\infty} \left(1 - \frac{s^2}{n^2 \pi^2} \right).$$

The method used above then gives the sums of *even* powers of the reciprocal integers, and Euler calculated them up to $\zeta(12)$. Thus he obtained

$$\zeta(2) = \frac{\pi^2}{6}$$

$$\zeta(4) = \frac{\pi^4}{90}$$

$$\zeta(6) = \frac{\pi^6}{945}$$

$$\zeta(8) = \frac{\pi^8}{9450}$$

$$\zeta(10) = \frac{\pi^{10}}{93555}$$

$$\zeta(12) = \frac{691\pi^{12}}{6825 \times 93555}.$$

He noticed that the values of $L(2k+1)$ and $N(2k)$ differ by a rational multiple of π, giving him the value of π as the ratio of two infinite series in many different ways, for example

$$\pi = \frac{25}{8} \frac{1 + \frac{1}{3^6} + \frac{1}{5^6} + \cdots}{1 - \frac{1}{3^5} + \frac{1}{5^5} - \cdots}$$

$$\pi = \frac{192}{61} \frac{1 - \frac{1}{3^7} + \frac{1}{5^7} - \cdots}{1 + \frac{1}{3^6} + \frac{1}{5^6} + \cdots}.$$

Objections to Euler's method. Euler lost no time in communicating his discoveries to his friends and other mathematicians. Some of them, Daniel and Nicholas Bernoulli for instance, raised objections to his method. The application of Newton's theorem to a context far more general than the original one was a troublesome point. A more fundamental objection was that the function $\sin s$ could have complex roots, apart from the obvious ones that Euler wrote down. Moreover, while a polynomial can be easily seen to be factorizable as a product of linear factors corresponding to its roots, the extension of such factorizations to transcendental functions such as $\sin s$ seemed too big a step and demanded justification. For instance, $\sin s/s$ and $e^s \sin s/s$ have the same real roots, but clearly both cannot have the same product formula. It was pointed out to him that if one uses an ellipse instead of a circle to define the function $\sin s$, the zeros would be the same but the function would be different and the same series would appear to have a different value, now depending on the lengths of the axes of the ellipse, which is absurd.[*] This objection could really be answered only after Weierstrass's work on functions with product representations and specified zeros. To these one can also add a more modern objection, one that would become clear after Riemann constructed his theory of rearrangements of a series that is not absolutely convergent, namely that in the case of such a series, the sum of the infinite series depends on the order of summation and so is really not a symmetric function of the roots. Thus the order in which the roots are taken, as well as the terms of the infinite products that enter Euler's proofs, becomes critical in a way that is not clear in his treatment. The

---

[*]See [1], Ch. 2., Vol. II, 433-435 (Daniel Bernoulli to Euler), and 681-689 (Nicolaus Bernoulli to Euler).

same objection could be raised against the product formula for $1 - \frac{\sin s}{\sin \sigma}$ and hence also the partial fraction for $\frac{\pi}{\sin \pi s}$.

Euler himself was aware of these shortcomings of his proof but was encouraged by two facts. First, in the argument starting with $y = 1$ the first step gave Leibniz's series for $\pi/4$, and the one with $y = 1/2\sqrt{2}$ gave Newton's formula, both of which had been proved by other methods. So even if $\sin s$ had other roots, the sum of their reciprocals must be 0. Second, the value $\pi^2/6$ for $\zeta(2)$ was clearly confirmed by numerical calculation so that the other roots, if any, do not contribute to the calculations. These were the reasons that persuaded him to publish and publicize his method. Nevertheless he recognized the validity of the objections and devoted the next few years to making the arguments as solid as possible.

The objection that the sum of the roots depends on the order in which the roots are enumerated was of course not one his contemporaries could have made. Nevertheless it disappears in Euler's proof based on the infinite product expansion

$$\frac{\sin s}{s} = \prod_{n=1}^{\infty} \left(1 - \frac{s^2}{n^2 \pi^2} \right)$$

that he wrote down in [6]. The absolute convergence of the product makes the justification of the application of Newton's theorem more routine.

The calculations that he did for the values of $\zeta(2k)$ and $L(2k+1)$ for the first few values of k must have convinced Euler that he should try to determine these values for all k. This he would do later.

3.4. Infinite products for circular and hyperbolic functions

After his first evaluation of $\zeta(2)$ and $\zeta(4)$ and $\zeta(2k)$ for some small values of k, Euler began his long campaign to consolidate his proofs. This took him the better part of the next few years. Finally, around 1742, he obtained the infinite product expansion for $\frac{\sin x}{x}$ that conclusively demonstrated that his results were correct. In this section we shall discuss Euler's proof of the infinite product expansions of the circular functions (I-14, 138-155). Although Euler worked only with real x, he considered both the hyperbolic and the circular functions so that in essence he was working over the complex plane. Eventually he gave a beautiful exposition of these results in his book I-8.

The starting point for Euler was the limit formula

$$e^x = \lim_{n \to \infty} \left(1 + \frac{x}{n} \right)^n.$$

He would write this as

$$e^x = \left(1 + \frac{x}{n} \right)^n$$

where n is an infinite magnitude, sometimes denoted by ι. He used the limit formula also for imaginary values of the exponent expressed in current notation as

$$e^{ix} = \lim_{n \to \infty} \left(1 + \frac{ix}{n} \right)^n.$$

From this one has

$$\sin x = \lim_{n \to \infty} \frac{\left(1 + \frac{ix}{n}\right)^n - \left(1 - \frac{ix}{n}\right)^n}{2i}$$

$$\cos x = \lim_{n \to \infty} \frac{\left(1 + \frac{ix}{n}\right)^n + \left(1 - \frac{ix}{n}\right)^n}{2}$$

as well as the companion formulae

$$\frac{e^x - e^{-x}}{2} = \sinh x = \lim_{n \to \infty} \frac{\left(1 + \frac{x}{n}\right)^n - \left(1 - \frac{x}{n}\right)^n}{2}$$

$$\frac{e^x + e^{-x}}{2} = \cosh x = \lim_{n \to \infty} \frac{\left(1 + \frac{x}{n}\right)^n + \left(1 - \frac{x}{n}\right)^n}{2}.$$

Euler's brilliant idea was to factorize the polynomials

$$\frac{\left(1 + \frac{ix}{n}\right)^n - \left(1 - \frac{ix}{n}\right)^n}{2i} \quad \text{and} \quad \frac{\left(1 + \frac{x}{n}\right)^n - \left(1 - \frac{x}{n}\right)^n}{2}$$

into quadratic factors and pass to the limit as $n \to \infty$; the infinite product expansions for $\sin x$ and $\sinh x$ follow immediately.

Factorization using roots of unity. The roots of the equation

$$T^n - 1 = 0$$

are

$$e^{\frac{2r\pi i}{n}} \qquad (r = 0, 1, 2, \ldots, n-1).$$

However this equation has real coefficients and so the roots can be arranged in pairs where each pair consists of conjugate quantities. Thus, if $n = 2p+1$, then the roots can be written as

$$e^{\frac{2r\pi i}{n}} \qquad (r = 0, \pm 1, \pm 2, \ldots, \pm p).$$

On the other hand

$$(T - e^{\frac{2r\pi i}{n}})(T - e^{\frac{-2r\pi i}{n}}) = T^2 - 2T \cos \frac{2r\pi}{n} + 1$$

from which we obtain the factorization

$$T^n - 1 = (T - 1) \prod_{k=1}^{p} \left(T^2 - 2T \cos \frac{2k\pi}{n} + 1 \right) \qquad (n = 2p+1).$$

Writing $T = X/Y$ and clearing of fractions we get

$$X^n - Y^n = (X - Y) \prod_{k=1}^{p} \left(X^2 - 2XY \cos \frac{2k\pi}{n} + Y^2 \right).$$

Let us now take

$$X = 1 + \frac{ix}{n}, \qquad Y = 1 - \frac{ix}{n}$$

and write

$$q_n(x) = \frac{\left(1 + \frac{ix}{n}\right)^n - \left(1 - \frac{ix}{n}\right)^n}{2ix}.$$

Then we get, since $X - Y = \frac{2ix}{n}$, after a little calculation,

$$q_n(x) = C_n \prod_{k=1}^{p} \left(1 - \frac{x^2}{n^2} \frac{1 + \cos \frac{2k\pi}{n}}{1 - \cos \frac{2k\pi}{n}} \right)$$

where C_n is a numerical constant. Since

$$q_n(x) = 1 + \ldots$$

while the product on the right side of the equation above takes the value C_n at $x = 0$, we must have $C_n = 1$. It is an elementary exercise to verify this directly also. Thus we finally obtain

$$q_n(x) = \prod_{k=1}^{p} \left(1 - \frac{x^2}{n^2} \frac{1 + \cos \frac{2k\pi}{n}}{1 - \cos \frac{2k\pi}{n}} \right).$$

It remains only to let n go to ∞. We know that

$$\lim_{n \to \infty} q_n(x) = \frac{e^{ix} - e^{-ix}}{2ix} = \frac{\sin x}{x}.$$

On the other hand let us examine the kth term of the product on the right side, *for a fixed k*. From the power series for $\cos u$ it follows that

$$\lim_{u \to 0} \cos u = 1, \qquad \lim_{u \to 0} \frac{1 - \cos u}{u^2} = \frac{1}{2}.$$

Hence

$$\lim_{n \to \infty} 1 + \cos \frac{2k\pi}{n} = 2, \qquad \lim_{n \to \infty} n^2 \left(1 - \cos \frac{2k\pi}{n} \right) = \frac{2k^2}{\pi^2}.$$

Thus

$$\lim_{n \to \infty} 1 - \frac{x^2}{n^2} \frac{1 + \cos \frac{2k\pi}{n}}{1 - \cos \frac{2k\pi}{n}} = 1 - \frac{x^2}{k^2 \pi^2}.$$

From this Euler deduced finally his famous infinite product expression:

$$\frac{\sin x}{x} = \prod_{k=1}^{\infty} \left(1 - \frac{x^2}{k^2 \pi^2} \right).$$

Clearly the same method will work when x is replaced by ix so that one obtains with virtually no change in the argument the formula

$$\frac{\sinh x}{x} = \frac{e^x - e^{-x}}{2x} = \prod_{k=1}^{\infty} \left(1 + \frac{x^2}{k^2 \pi^2} \right).$$

The product formulae for $\cos x$ and $\cosh x$ are derived in almost identical fashion. For $\cos x$ we can also proceed by first noting that

$$\sin 2x = 2 \sin x \cos x$$

so that

$$\cos x = \frac{\sin 2x / 2x}{\sin x / x}$$

from which we get at once the infinite product for $\cos x$:

$$\cos x = \prod_{k=1}^{\infty} \left(1 - \frac{4x^2}{(2k-1)^2 \pi^2} \right).$$

Similarly we get

$$\cosh x = \frac{e^x + e^{-x}}{2} = \prod_{k=1}^{\infty} \left(1 + \frac{4x^2}{(2k-1)^2 \pi^2} \right).$$

The direct derivation is equally straightforward. We factorize first $T^n + 1$. The roots of the equation

$$T^n + 1 = 0 \qquad (n = 2p)$$

are

$$e^{\frac{\pi i}{2p}}, e^{\frac{3\pi i}{2p}}, e^{\frac{5\pi i}{2p}}, \ldots, e^{\frac{(4p-1)\pi i}{2p}}$$

which can be arranged in p mutually conjugate pairs

$$e^{\pm\frac{\pi i}{2p}}, e^{\pm\frac{3\pi i}{2p}}, \ldots, e^{\pm\frac{(2p-1)\pi i}{2p}}.$$

We thus obtain

$$T^n + 1 = \prod_{k=1}^{p} \left(T^2 - 2T \cos \frac{(2k-1)\pi}{2p} + 1 \right).$$

As before we deduce from this the factorization

$$X^n + Y^n = \prod_{k=1}^{p} \left(X^2 - 2XY \cos \frac{(2k-1)\pi}{2p} + Y^2 \right).$$

Take now

$$X = 1 + \frac{ix}{n}, \qquad Y = 1 - \frac{ix}{n}$$

and proceed exactly as we do for the case of the sine to get for

$$p_n(x) = \frac{\left(1 + \frac{ix}{n}\right)^n + \left(1 - \frac{ix}{n}\right)^n}{2}$$

the factorization

$$p_n(x) = \prod_{k=1}^{p} \left(1 - \frac{x^2}{n^2} \frac{1 + \cos \frac{(2k-1)\pi}{n}}{1 - \cos \frac{(2k-1)\pi}{n}} \right) \qquad (n = 2p).$$

Letting $n \to \infty$, we get

$$\cos x = \prod_{k=1}^{\infty} \left(1 - \frac{4x^2}{(2k-1)^2 \pi^2} \right).$$

The factorization of $\cosh x$ is obtained in the same way with x replacing ix throughout.

The product formula for the function

$$1 - \frac{\sin s}{\sin \sigma}$$

can be obtained immediately from the above. Indeed, we write

$$1 - \frac{\sin s}{\sin \sigma} = \frac{2 \cos \frac{\sigma+s}{2} \sin \frac{\sigma-s}{2}}{\sin \sigma}$$

and use the products for sin and cos to get for it the expression

$$\left(1 - \frac{s}{\sigma} \right) \frac{\prod_{n=1}^{\infty} \left(1 - \frac{(\sigma+s)^2}{((2n-1)\pi)^2} \right) \prod_{n=1}^{\infty} \left(1 - \frac{(\sigma-s)^2}{(2n\pi)^2} \right)}{\prod_{n=1}^{\infty} \left(1 - \frac{\sigma^2}{((2n-1)\pi)^2} \right) \prod_{n=1}^{\infty} \left(1 - \frac{\sigma^2}{(2n\pi)^2} \right)}.$$

Factorizing and rearranging a little we get

$$1 - \frac{\sin s}{\sin \sigma} = \left(1 - \frac{s}{\sigma}\right) \prod_{n=1}^{\infty} \left(1 - \frac{s}{2n\pi + \sigma}\right)\left(1 + \frac{s}{2n\pi - \sigma}\right)$$

$$\times \prod_{n=1}^{\infty} \left(1 - \frac{s}{(2n-1)\pi - \sigma}\right)\left(1 + \frac{s}{(2n-1)\pi + \sigma}\right).$$

Notice that the individual products separately converge because the terms are $1 + O(n^{-2})$; this justifies the rearrangement that was done.

A similar method can be used to obtain infinite product formulae for

$$\frac{\cos s \pm \cos \sigma}{1 \pm \cos \sigma}, \qquad \frac{\sin s \pm \sin \sigma}{\sin \sigma}.$$

Thus, for example,

$$\frac{\cos s \pm \cos \sigma}{1 \pm \cos \sigma} = \prod_{n=1}^{\infty} \left(1 + \frac{s}{(2n-1)\pi - \sigma}\right)\left(1 - \frac{s}{(2n-1)\pi + \sigma}\right)$$

$$\times \prod_{n=1}^{\infty} \left(1 + \frac{s}{(2n-1)\pi + \sigma}\right)\left(1 - \frac{s}{(2n-1)\pi - \sigma}\right).$$

See pp. 132-134 in [3d] (where Euler writes v, g for our s, σ).

In this manner, all the results of Euler's initial paper [4] were put on a firm basis. The infinite product for $\frac{\sin x}{x}$, from which all other products and partial fractions can be deduced, is thus the cornerstone of Euler's entire theory.

Justification of Euler's proof from a later point of view. Actually Euler's derivation is almost unexceptionable. The only point that we would cavil at is the actual passage to the limit. For instance, in the case of $\sin x$, Euler has the product expression

$$q_n(x) = \prod_{k=1}^{n} (1 - a_k(n)x^2) \qquad (n = 2p + 1)$$

where

$$a_k(n) = \frac{1 + \cos \frac{2k\pi}{n}}{1 - \cos \frac{2k\pi}{n}}$$

and simply passes to the limit *term by term*. Since however the number of factors is also going to infinity, it is clear from our vantage point that an additional argument is needed. We shall see that this is taken care of by the principle of *normal convergence*, which did not really become clear till the work of Weierstrass in the 19th century on uniform convergence. Certainly in Euler's time this proof was accepted without question; in any case it is fundamentally sound except for the technical point of uniform convergence.

This phenomenon actually occurs already in Euler's proof of the limit formula

$$e^{ix} = \lim_{n \to \infty} \left(1 + \frac{ix}{n}\right)^n.$$

Indeed, expanding the right side by the binomial theorem we get

$$1 + ix + (1 - 1/n)\frac{(ix)^2}{2!} + \ldots (1 - 1/n)(1 - 2/n) \ldots (1 - (k-1)/n)\frac{(ix)^k}{k!} + \ldots$$

where the sum extends to n terms. Euler simply passed to the limit term by term.

We thus have the following situation. We have a sequence

$$a_1(n), a_2(n), \ldots, a_{k_n}(n) \qquad (k_n \to \infty)$$

and the corresponding sum and product

$$s_n = \sum_{k=1}^{k_n} a_k(n) \qquad p_n = \prod_{k=1}^{k_n} (1 + a_k(n)),$$

and we know that for *each fixed* k,

$$\lim_{n \to \infty} a_k(n) = a_k$$

exists. Under what circumstances are we then allowed to conclude that

$$\lim_{n \to \infty} s_n = \sum_{k=1}^{\infty} a_k, \qquad \lim_{n \to \infty} p_n = \prod_{k=1}^{\infty} (1 + a_k)?$$

We can actually simplify a little by defining $a_k(n)$ for all k by saying that it is 0 if $k > k_n$. Then the problem is to find out when we can interchange summation and passing to the limit, i.e., when we can conclude that

$$\lim_{n \to \infty} \sum_{k=1}^{\infty} a_k(n) = \sum_{k=1}^{\infty} \lim_{n \to \infty} a_k(n)$$

$$\lim_{n \to \infty} \prod_{k=1}^{\infty} (1 + a_k(n)) = \prod_{k=1}^{\infty} (1 + \lim_{n \to \infty} a_k(n)).$$

The notion of *normal convergence*, which was introduced and exploited by Weierstrass, allows us to take care of this point. We shall say that the series

$$\sum_{k=1}^{\infty} a_k(n)$$

of complex terms is *normally convergent* if there is a sequence (M_k) such that

    (i) $|a_k(n)| \leq M_k$ for all k, n;
    (ii) $\sum_{k=1}^{\infty} M_k$ is convergent.
   Note that

$$|a_k| \leq M_k$$

for all k. Actually it is enough if we have (i) above for all $k \geq k_0$ where k_0 is *independent* of n. In the case of the series, we have, for any N,

$$\left| \sum_{k} a_k(n) - \sum_{k} a_k \right| \leq \sum_{1 \leq k \leq N} |a_k(n) - a_k| + 2 \sum_{k > N} M_k$$

so that

$$\limsup_{n \to \infty} \left| \sum_{k} a_k(n) - \sum_{k} a_k \right| \leq 2 \sum_{k > N} M_k.$$

Letting $N \to \infty$ we get the result

$$\lim_{n \to \infty} \sum_{k} a_k(n) = \sum_{k} a_k.$$

For the product we work with the series

$$\sum_{k} \log(1 + a_k(n))$$

and use the estimate

$$|\log(1+x)| \le 2|x| \qquad (|x| \le 1/2).$$

Then

$$|\log(1+a_k(n))| \le 2M_k$$

and so we have normal convergence.

As a first application let us consider the limit formula

$$\lim_{n\to\infty} \left(1+\frac{z_n}{n}\right)^n = e^z \qquad (z_n \in \mathbf{C}, z_n \to z).$$

Then

$$\left(1+\frac{z_n}{n}\right)^n = 1 + z_n + \frac{n(n-1)}{2!}\frac{z_n^2}{n^2} + \frac{n(n-1)(n-2)}{3!}\frac{z_n^3}{n^3} + \cdots$$

which can be written as

$$\sum_{k=1}^{\infty} c_k(n)\frac{z_n^k}{k!}$$

where

$$c_k(n) = \begin{cases} \left(1-\frac{1}{n}\right)\left(1-\frac{2}{n}\right)\cdots\left(1-\frac{k}{n}\right) & \text{if } k \le n \\ 0 & \text{if } k > n \end{cases}.$$

If $|z_n| \le A$ for all n and we take $a_k(n) = c_k(n)z_n^k/k!$, the obvious estimate

$$|a_k(n)| \le \frac{A^k}{k!}$$

gives normal convergence and hence the limit formula. This argument is valid even when z_n is a complex *matrix* and is the basic fact in the theory of Lie groups.

We are now ready to justify Euler's passage to the limit when $n \to \infty$ term by term in the infinite products. We have a product of the form

$$\prod_{k=1}^{(n-1)/2} \left(1 \pm \frac{x^2}{n^2}\frac{1+\cos\frac{2k\pi}{n}}{1-\cos\frac{2k\pi}{n}}\right)$$

so that

$$a_k(n) = \pm\frac{x^2}{n^2}\frac{1+\cos\frac{2k\pi}{n}}{1-\cos\frac{2k\pi}{n}} \qquad (2k < n).$$

We now observe that for some constant $\alpha > 0$ we have

$$1 - \cos u \ge \alpha u^2 \qquad (0 < u \le \pi).$$

Indeed, we note that $(1-\cos u)/u^2 \to 1/2$ as $u \to 0$ and is never zero in $(0, \pi]$. Hence it must have a positive minimum α in $[0, \pi]$. Returning to $a_k(n)$ we then have, using the above lemma and remembering that $2k\pi/n \le \pi$,

$$\left|\frac{1+\cos\frac{2k\pi}{n}}{1-\cos\frac{2k\pi}{n}}\right| \le \frac{n^2}{\alpha}\frac{1}{2k^2\pi^2}$$

and hence

$$|a_k(n)| \le \frac{x^2}{2\alpha\pi^2}\frac{1}{k^2}$$

giving normal convergence. The other cases may be treated in an entirely similar fashion.

There is no difficulty in carrying this entire discussion over to complex values of the argument. Nowadays these results are indeed established for complex values of the argument, using periodicity arguments and Liouville's theorem that a bounded entire function is a constant. While this is admirable in itself, it is regrettable from the historical and pedagogical points of view. Euler's proof is beautiful, it displays the problem and its solution in the proper historical context, and moreover, the justification will take the student through the fine points of real analysis and show that the theorems that have been learned do not exist in a vacuum but serve a crucial purpose. There is no better demonstration of Euler's power and originality than this proof.

Symmetric functions of infinitely many variables. Once the product formulae have been established, it remains only to examine the Euler method of generalizing Newton's theorem. I follow an exposition of Cartier [11]. Write

$$P(t) = \prod_n (1 - t t_n) = 1 - \lambda_1 t + \lambda_2 t^2 - \cdots + (-1)^r \lambda_r t^r + \ldots$$

where the t_n are complex quantities and t is also a complex variable. Formally we have

$$\lambda_r = \sum_{n_1 > n_2 > \cdots > n_r > 0} t_{n_1} t_{n_2} \ldots t_{n_r}$$

which are reminiscent of the multiple zeta series. Let

$$S_r = \sum_{n>0} t_n^r.$$

Then logarithmic differentiation gives the relation

$$P'(t) + P(t) \sum_n \frac{t_n}{1 - t t_n} = 0.$$

Expanding both sides as power series in t we get the Newton formulae in the present context:

$$(*) \qquad S_r = \lambda_1 S_{r-1} - \lambda_2 S_{r-2} + \cdots + (-1)^{r-2} \lambda_{r-1} S_1 + (-1)^{r-1} r \lambda_r.$$

Changing t to $-t$ we have

$$\sum_r \lambda_r t^r = \exp\left(\sum_n \log(1 + t t_n)\right)$$

which leads to

$$\sum_r \lambda_r t^r = \exp\left(\sum_r (-1)^r \frac{S_r}{r} t^r\right).$$

This is an alternative way to summarize the relations $(*)$.

We now take

$$t_n = \frac{1}{n^2 \pi^2}, \qquad P(t) = \frac{\sin t}{t}.$$

Then

$$\pi^{2r} \lambda_r = \zeta(2, 2, \ldots, 2)(r \text{ arguments}), \qquad \pi^{2r} S_r = \zeta(2r).$$

Since

$$\frac{\sin t}{t} = 1 - \frac{t^2}{3!} + \cdots + (-1)^r \frac{t^{2r}}{(2r+1)!} + \ldots$$

we have

$$\zeta(2, 2, \ldots, 2) = \frac{\pi^{2r}}{(2r+1)!}, \qquad \lambda_r = \frac{1}{(r+1)!}.$$

and so the $\zeta(2r)$ can be calculated recursively. Indeed, if

$$\zeta(2r) = f_r \pi^{2r},$$

then

$$f_r = \sum_{p=1}^{r-1} (-1)^{p-1} \frac{f_{r-p}}{(2p+1)!} + (-1)^{r-1} \frac{r}{(2r+1)!}$$

from which it is immediate that the f_r are rational numbers.

The justification of these remarks is quite simple if we assume that

$$\sum_n |t_n| < \infty.$$

It is then a question of working with the product of N terms and letting $N \to \infty$. Let

$$P_N(t) = \prod_{n \le N} (1 - t t_n) = \sum_{n \ge 0} (-1)^r \lambda_r(N) t^r$$

where

$$\lambda_r(N) = \sum_{n_1 > n_2 > \cdots > n_r > 0} t_{n_1}(N) \ldots t_{n_r}(N), \qquad t_n(N) = \begin{cases} t_n & \text{if } n \le N \\ 0 & \text{if } n > N. \end{cases}$$

If

$$S_r(N) = \sum_{n>0} t_n(N)^r$$

we have, by the classical theory,

$$S_r(N) = \lambda_1(N) S_{r-1}(N) + \cdots + (-1)^r \lambda_{r-1}(N) S_1(N) + (-1)^{r+1} r \lambda_r(N).$$

To obtain $(*)$ it is enough to show that as $N \to \infty$,

$$\lambda_r(N) \to \lambda_r, \qquad S_r(N) \to S_r.$$

This is immediate using the criterion of normal convergence since

$$|t_{n_1}(N) \ldots t_{n_r}(N)| \le |t_{n_1}| \ldots |t_{n_r}|, \qquad |t_n(N)^r| \le |t_n|^r$$

and

$$\sum_{n_1 > \cdots > n_r > 0} |t_{n_1}| \ldots |t_{n_r}| < \infty, \qquad \sum_{n>0} |t_n|^r < \infty.$$

On the other hand, it is clear that $P(t)$ exists for all $t \in \mathbf{C}$ and $P_N(t) \to P(t)$ for all t. Moreover

$$|\lambda_r(N)| \le |\lambda_r| \le \frac{1}{r!} \sum_{n_1, n_2, \ldots, n_r > 0} |t_{n_1}| \ldots |t_{n_r}| = \frac{\left(\sum_n |t_n|\right)^r}{r!}.$$

Hence

$$\sum_r |\lambda_r| |t|^r < \infty, \qquad \sum_r (-1)^r \lambda_r(N) t^r \to \sum_r (-1)^r \lambda_r t^r$$

for all t, completing the proof that

$$P(t) = \sum_r (-1)^r \lambda_r t^r.$$

Other proofs. Euler himself had another very simple proof of the formula $\zeta(2) = \pi^2/6$. In a memoir which was forgotten but resurrected by Paul Stäckel in 1907

[12], Euler had given a direct proof of this. Euler's starting point is the power series for $\arcsin x$ which is obtained from the relation

$$\arcsin x = \int_0^x \frac{dt}{\sqrt{1-t^2}}$$

by expanding in powers of t and integrating term by term. The result is

$$\arcsin x = x + \sum_{k \geq 1} \frac{1.3.\ldots 2k-1}{2.4.\ldots 2k} x^{2k+1}.$$

An integration by parts now leads to the relation

$$\int_0^x \frac{\arcsin t}{\sqrt{1-t^2}} dt = (\arcsin x)^2 - \int_0^x \frac{\arcsin t}{\sqrt{1-t^2}} dt$$

so that

$$\frac{1}{2}(\arcsin x)^2 = \int_0^x \frac{\arcsin t}{\sqrt{1-t^2}} dt.$$

We now replace $\arcsin x$ in the integrand by its power series expansion and integrate term by term from 0 to 1. An integration by parts shows that if

$$I_n = \int_0^1 \frac{t^n}{\sqrt{1-t^2}} dt,$$

then

$$I_{n+2} = \frac{n+1}{n+2} I_n$$

from which the I_n can be calculated for all n. The result is

$$\frac{\pi^2}{8} = 1 + \frac{1}{3^2} + \cdots + \frac{1}{(2k+1)^2} + \cdots$$

which leads to Euler's formula. Euler tried to push this method for $\zeta(2k)$ for $k \geq 2$ but did not succeed. He even determined the power series for $(\arcsin x)^2$ in this attempt. Indeed, he wrote $y = (1/2)(\arcsin x)^2$ and from the integral representation obtained the differential equation

$$(1 - x^2)y'' - xy' = 1$$

from which he determined the coefficients of the expansion of y recursively and obtained

$$\frac{1}{2}(\arcsin x)^2 = \sum_{r \geq 1} \frac{(2.4.\ldots 2r)^2}{(2r)!} x^{2r}.$$

After mentioning his failure to push this method for the higher zeta values, he lists a few of these zeta values for the benefit of those who would try to work on this question.

Over the years there have been many proofs of Euler's formulae, but none that capture us with the imagination and originality that inform his proofs.

3.5. The infinite partial fractions for $(\sin x)^{-1}$ and $\cot x$. Evaluation of $\zeta(2k)$ and $L(2k+1)$

Already in 1734 when he determined the values of $\zeta(2k)$ for some small values of k, it must have become clear to Euler that

$$\zeta(2k) = 1 + \frac{1}{2^{2k}} + \frac{1}{3^{2k}} + \cdots$$

should be a rational multiple of π^{2k} for all k and that similarly the series

$$L(2k+1) = 1 - \frac{1}{3^{2k+1}} + \frac{1}{5^{2k+1}} - \cdots$$

(which he had summed for small values of k) should have sums which are rational multiples of π^{2k+1}. In his papers [6], [7] Euler succeeded in evaluating all these series in closed form in terms of what would later be called *Bernoulli numbers* and *Euler numbers*. He did this starting from infinite partial fraction expansions for $\frac{1}{\sin x}$ and $\cot x$.

The expansions in question are

$$\frac{\pi}{\sin s\pi} = \frac{1}{s} + \sum_{n=1}^{\infty} (-1)^n \left(\frac{1}{n+s} - \frac{1}{n-s} \right)$$

$$\pi \cot s\pi = \frac{1}{s} + \sum_{n=1}^{\infty} \left(\frac{1}{n+s} - \frac{1}{n-s} \right)$$

where the brackets are introduced to ensure absolute convergence of the series involved. The expressions are valid for $0 < s < 1$ and indeed whenever s is not an integer. The absolute convergence is easy to check since

$$\frac{1}{n-s} - \frac{1}{n+s} = \frac{2s}{n^2 - s^2} = c_n(s) \frac{1}{n^2}$$

where $c_n = 2s/(1 - s^2/n^2)$ is bounded as $n \to \infty$; the convergence is even normal as long as s varies in a closed interval which does not contain any integer. Euler would write these as

$$\frac{\pi}{\sin s\pi} = \frac{1}{s} + \frac{1}{1-s} - \frac{1}{1+s} - \frac{1}{2-s} + \frac{1}{2+s} + \frac{1}{3-s} - \frac{1}{3+s} - \cdots$$

$$\pi \cot s\pi = \frac{1}{s} + \frac{1}{1-s} - \frac{1}{1+s} + \frac{1}{2-s} - \frac{1}{2+s} + \frac{1}{3-s} - \frac{1}{3+s} - \cdots$$

not caring about the fine point of absolute convergence.

To obtain these the simplest proof is to use logarithmic differentiation of the infinite product for $\frac{\sin s}{s}$, as was noticed by Nicolaus Bernoulli* in a letter to Euler. From

$$\frac{\sin s}{s} = \prod_{n=1}^{\infty} \left(1 - \frac{s^2}{n^2 \pi^2} \right)$$

we have, taking logarithms,

$$\log \sin s - \log s = \sum_{n=1}^{\infty} \log(1 - s/n\pi) + \log(1 + s/n\pi) \qquad (0 < s < \pi)$$

---

*See [1], Ch. 2, Vol. II, 681-689.

from which differentiation gives

$$\cot s = \frac{1}{s} + \sum_{n=1}^{\infty} \left(\frac{1}{n\pi + s} - \frac{1}{n\pi - s} \right) \qquad (0 < s < \pi).$$

Replacing s by πs we get

$$\pi \cot s\pi = \frac{1}{s} + \sum_{n=1}^{\infty} \left(\frac{1}{n + s} - \frac{1}{n - s} \right) \qquad (0 < s < 1).$$

The justification of this process requires normal convergence. For the partial fraction for $\frac{1}{\sin s}$ we follow the same procedure and logarithmically differentiate the infinite product for $1 - \sin s / \sin \sigma$ obtained earlier, with respect to s at $s = 0$. Once again we use normal convergence to justify this process. We get

$$\frac{1}{\sin \sigma} = \frac{1}{\sigma} + \sum_{n=1}^{\infty} \left(\frac{1}{2n\pi + \sigma} - \frac{1}{2n\pi - \sigma} \right)$$

$$+ \sum_{n=1}^{\infty} \left(\frac{1}{(2n-1)\pi - \sigma} - \frac{1}{(2n-1)\pi + \sigma} \right)$$

which, on setting $\sigma = s\pi$, leads to

$$\frac{\pi}{\sin s\pi} = \frac{1}{s} + \sum (-1)^n \left(\frac{1}{n + s} - \frac{1}{n - s} \right).$$

The periodicity properties are more transparent if we write these as sums over the full set of integers. Thus

$$\pi \cot \pi s = PV \sum_{n\in\mathbf{Z}} \frac{1}{n + s}, \qquad \frac{\pi}{\sin \pi s} = PV \sum_{n\in\mathbf{Z}} (-1)^n \frac{1}{n + s},$$

where the notation PV indicates that the series in question are understood as

$$\lim_{N\to\infty} \sum_{n=-N}^{n=+N}.$$

Of course, as soon as we differentiate, the PV can be dropped as the series converge absolutely. The series for $\frac{\pi}{\sin \pi s}$ can also be obtained in another way. We logarithmically differentiate the product for $\cos \pi s$ or change s to $\frac{1}{2} - s$ in $\pi \cot \pi s$ to get the series

$$\pi \tan \pi s = PV \sum_{n\in\mathbf{Z}} \frac{1}{n + \frac{1}{2} - s}.$$

Adding the two series for $\pi \cot \pi s$ and $\pi \tan \pi s$, we get, after changing s to $\frac{1}{2}s$, the formula

$$\frac{\pi}{\sin \pi s} = PV \sum_{n\in\mathbf{Z}} (-1)^n \frac{1}{n + s}.$$

There are at least two other proofs for these formulae. One of these was given by Euler himself in [13] based on integrating certain rational functions that generalize the ones occurring in the Leibniz series. We shall give this in the next section. The other depends on the theory of what are now call the gamma and beta functions, or as Legendre called them, *Eulerian integrals of the first and second kind*. These integrals were first studied by Euler, and he knew their principal properties. We shall give this proof also in the next section. It uses nothing that Euler did not know.

Bernoulli numbers and evaluation of $\zeta(2k)$. Let

$$Q(s) = \pi \cot \pi s.$$

Then

$$Q(s) - \frac{1}{s} = \sum_{n=1}^{\infty} \left(\frac{1}{n+s} - \frac{1}{n-s} \right).$$

Now $Q(s) - s^{-1}$ is smooth at $s = 0$ and

$$\frac{1}{(r-1)!} \frac{d^{r-1}}{ds^{r-1}} \left(Q(s) - \frac{1}{s} \right) = \sum_{n=1}^{\infty} \left((-1)^{r-1} \frac{1}{(n+s)^r} - (r-1)! \frac{1}{(n-s)^r} \right).$$

Now if $r = 2k$ is even, we get

$$\frac{1}{(2k-1)!} \frac{d^{2k-1}}{ds^{2k-1}} \left(Q(s) - \frac{1}{s} \right) \Big|_{s=0} = -2\zeta(2k)$$

which was Euler's explicit formula. If we do the same with $r = 2k + 1$, we get 0.

Let us now take a closer look at the function Q. We have

$$Q(s) - \frac{1}{s} = \pi \cot \pi s - \frac{1}{s} = \frac{2i\pi}{e^{2i\pi s} - 1} + \pi i - \frac{1}{s}.$$

We now introduce the function

$$B(z) = \frac{z}{e^z - 1} - 1 + \frac{1}{2} z.$$

It is clear that $B(z)$ is analytic at $z = 0$ and its power series begins with the z^2 term. Moreover it is easy to check that

$$B(-z) = B(z)$$

so that only even powers of z occur in the expansion of $B(z)$. We may therefore write

$$B(z) = \sum_{k=1}^{\infty} \frac{B_{2k}}{(2k)!} z^{2k}.$$

The coefficients B_{2k} are the so-called *Bernoulli numbers* because they appeared first in the work of Jacob Bernoulli in his book *Ars Conjectandi*. There are many different conventions about writing them, and we have chosen one of them. They have extraordinary properties and occur in the most diverse contexts. They may be calculated in succession using the recursion formulae obtained from the identity

$$\left(1 - \frac{z}{2} + \frac{B_2 z^2}{2!} + \cdots + \frac{B_{2k} z^{2k}}{(2k)!} + \ldots \right) \left(1 + \frac{z}{2} + \ldots \frac{z^k}{(k+1)!} + \ldots \right) = 1.$$

The recursion formulae are

$$(2k+1) B_{2k} = - \sum_{r=1}^{r=k-1} \binom{2k+1}{2r} B_{2r} + k - \frac{1}{2}.$$

It follows easily from this by induction on k that the B_{2k} are all rational numbers. The first few of them are

$$B_2 = \frac{1}{6}, \quad B_4 = -\frac{1}{30}, \quad B_6 = \frac{1}{42}, \quad B_8 = -\frac{1}{30}, \quad B_{10} = \frac{5}{66}, \quad B_{12} = -\frac{691}{2730}.$$

Notice the presence of 691 in the expression for B_{12} which occurred in Euler's first evaluation of $\zeta(12)$.

Returning to the evaluation of $\zeta(2k)$ we find

$$Q(s) - \frac{1}{s} = \frac{2\pi i}{2\pi i s} B(2\pi i s) = 2\pi i \sum_{k=1}^{\infty} B_{2k} \frac{(2\pi i s)^{2k-1}}{(2k)!}.$$

This formula shows that $Q(s) - 1/s$ is smooth and its derivatives at 0 are just the coefficients of the power series that appears on the right side of this formula. Hence

$$-2\zeta(2k) = \frac{1}{(2k-1)!} \frac{d^{2k-1}}{ds^{2k-1}} \left(Q(s) - \frac{1}{s} \right) \Big|_{s=0} = \frac{(2\pi i)^{2k} B_{2k}}{(2k)!}$$

leading to

$$\zeta(2k) = \frac{(-1)^{k-1} 2^{2k-1} B_{2k}}{(2k)!} \pi^{2k}$$

which is Euler's celebrated formula.

One can also calculate the derivatives at $s = \pi/2$ instead of at $s = 0$. Changing s to $s + 1/2$ we have

$$R(s) = -\pi \tan \pi s = \frac{1}{s + (1/2)} + \sum_{n=1}^{\infty} \left(\frac{1}{n + (1/2) + s} - \frac{1}{n - (1/2) - s} \right).$$

As before we can write

$$R(s) = \pi i \left(\frac{e^{2\pi i s} - 1}{e^{2\pi i s} + 1} \right) = \pi i \left(1 - \frac{2}{e^{2\pi i s} + 1} \right).$$

Now

$$\frac{1}{e^z + 1} = \frac{1}{e^z - 1} - \frac{2}{e^{2z} - 1} = \frac{1}{2} + \sum_{k=1}^{\infty} \frac{(1 - 2^{2k}) B_{2k}}{(2k)!} z^{2k-1}$$

from which it follows easily that

$$R(s) = -2\pi i \sum_{k=1}^{\infty} \frac{(1 - 2^{2k}) B_{2k}}{(2k)!} (2\pi i s)^{2k-1}.$$

So

$$\frac{1}{(2k-1)!} \frac{d^{2k-1} R(s)}{ds^{2k-1}} \Big|_{s=0} = (-1)^{k-1} (1 - 2^{2k}) \frac{2^{2k} B_{2k}}{(2k)!} \pi^{2k}.$$

On the other hand we also have

$$\frac{1}{(2k-1)!} \frac{d^{2k-1} R(s)}{ds^{2k-1}} \Big|_{s=0} = -2^{2k} + \sum_{n=1}^{\infty} \frac{-2^{2k}}{(2n+1)^{2k}} - \sum_{n=1}^{\infty} \frac{2^{2k}}{(2n-1)^{2k}}$$

$$= -2^{2k+1} M(2k).$$

Hence

$$M(2k) = (1 - 2^{-2k})(-1)^{k-1} \frac{2^{2k-1} B_{2k}}{(2k)!} \pi^{2k}$$

which agrees with the formula

$$M(2k) = (1 - 2^{-2k}) \zeta(2k).$$

Euler numbers and the evaluation of $L(2k+1)$. For the $L(2k+1)$ we start with the partial fraction expansion

$$\frac{\pi}{\sin \pi s} = \frac{1}{s} + \sum_{n=1}^{\infty} (-1)^n \left(\frac{1}{n+s} - \frac{1}{n-s} \right).$$

Let

$$P(s) = \frac{\pi}{\sin \pi s}.$$

Then

$$\frac{1}{r!} \frac{d^r P(s)}{ds^r} = \frac{(-1)^r}{s^{r+1}} + (-1)^r \sum_{n=1}^{\infty} (-1)^n \left(\frac{1}{(n+s)^{r+1}} - (-1)^r \frac{1}{(n-s)^{r+1}} \right).$$

We now evaluate the derivatives at $s = \frac{1}{2}$. Now $P(s + \frac{1}{2}) = \pi \sec \pi s$. So

$$\frac{1}{2^{2k+1}(2k)!} \frac{d^{2k} \pi \sec \pi s}{ds^{2k}} \bigg|_{s=0}$$

is equal to

$$1 + \sum_{n=1}^{\infty} (-1)^n \left(\frac{1}{(2n+1)^{2k+1}} - \frac{1}{(2n-1)^{2k+1}} \right).$$

Let us write

$$\sec z = 1 + \sum_{k=1}^{\infty} \frac{E_{2k} z^{2k}}{(2k)!}.$$

The E_{2k} are the so-called *Euler numbers*. Then taking $z = \pi s$ in this expansion we get

$$\frac{1}{2^{2k+1}(2k)!} E_{2k} \pi^{2k+1} = 2L(2k+1).$$

Thus

$$L(2k+1) = \frac{1}{2^{2k+2}(2k)!} E_{2k} \pi^{2k+1}.$$

Let $E_0 = 1$. The Euler numbers satisfy the recursion formulae

$$E_{2k} = \binom{2k}{2} E_{2k-2} - \binom{2k}{4} E_{2k-4} + \cdots + (-1)^{k-2} \binom{2k}{2k-2} E_2 + (-1)^{k-1} E_0$$

and hence are all integers. Therefore $L(2k+1)$ is a rational multiple of π^{2k+1} for all k. The first few Euler numbers are

$$E_2 = 1, \ E_4 = 5, \ E_6 = 61, \ E_8 = 1385$$

which lead to the formulae*

$$L(3) = \frac{\pi^3}{32}, \ L(5) = \frac{5\pi^5}{1536}, \ L(7) = \frac{61\pi^7}{184320}, \ L(9) = \frac{1385\pi^9}{1024 \times 40320}.$$

Sums of series and cyclotomic extensions. Let P, Q be as before:

$$P(s) = \frac{\pi}{\sin \pi s}, \qquad Q(s) = \pi \cot \pi s,$$

and let

$$x = \sin \pi s, \qquad y = \cos \pi s.$$

---

*See [1], Ch. 2, Vol. I, 131-132, 200-205 (Euler to Goldbach), in which Euler discusses the partial fractions and writes down the approximate values (up to a huge number of decimal places) of $\zeta(n)$ for $2 \le n \le 16$.

Then

$$\frac{dx}{ds} = \pi y, \qquad \frac{dy}{ds} = -\pi x$$

while

$$P = \frac{\pi}{x}, \qquad Q = \frac{\pi}{x} y.$$

Then we can obtain the derivatives of P and Q successively and get the following:

$$-\frac{dP}{ds} = \frac{\pi^2}{x^2} y, \quad \frac{d^2 P}{ds^2} = \frac{\pi^3}{x^3}(y^2 + 1) \quad -\frac{d^3 P}{ds^3} = \frac{\pi^4}{x^4}(y^3 + 5y) \text{ etc.}$$

and

$$-\frac{dQ}{ds} = \frac{\pi^2}{x^2}, \quad \frac{d^2 Q}{ds^2} = \frac{\pi^3}{x^3} 2y \quad -\frac{d^3 Q}{ds^3} = \frac{\pi^4}{x^4}(4y^2 + 2) \text{ etc.}$$

Taking for s a rational value

$$s = \frac{p}{q}$$

we then obtain

$$\frac{(-1)^k}{k!q^{k+1}} P^{(k)}\left(\frac{p}{q}\right) = \sum_{n\in\mathbf{Z}} \frac{(-1)^n}{(qn+p)^{k+1}},$$

$$\frac{(-1)^k}{k!q^{k+1}} Q^{(k)}\left(\frac{p}{q}\right) = \sum_{n\in\mathbf{Z}} \frac{1}{(qn+p)^{k+1}}.$$

In particular

$$\frac{\pi^2 \cos\frac{p}{q}\pi}{q^2 \sin^2\frac{p}{q}\pi} = \frac{1}{p^2} - \frac{1}{(q-p)^2} - \frac{1}{(q+p)^2} + \frac{1}{(2q-p)^2} + \frac{1}{(2q+p)^2} - \cdots$$

$$\frac{\pi^2}{q^2 \sin^2\frac{p}{q}\pi} = \frac{1}{p^2} + \frac{1}{(q-p)^2} + \frac{1}{(q+p)^2} + \frac{1}{(2q-p)^2} + \frac{1}{(2q+p)^2} + \cdots.$$

Also

$$\frac{\pi}{q \sin\frac{p}{q}\pi} = \frac{1}{p} + \left(\frac{1}{q-p} - \frac{1}{q+p}\right) - \left(\frac{1}{2q-p} - \frac{1}{2q+p}\right) + \cdots$$

$$\frac{\pi \cos\frac{p}{q}\pi}{q \sin\frac{p}{q}\pi} = \frac{1}{p} - \left(\frac{1}{q-p} - \frac{1}{q+p}\right) - \left(\frac{1}{2q-p} - \frac{1}{2q+p}\right) - \cdots.$$

In [6] Euler takes various special values for p and q and evaluates the series explicitly for $k = 0, 1$. The series he obtains are actually Dirichlet series corresponding to various characters mod q and their variants. Thus, for $q = 3, p = 1, k = 0$ he gets

$$\frac{2\pi}{3\sqrt{3}} = \sum_{n=1}^{\infty} \frac{(-1)^{n-1}\chi(n)}{n}, \qquad \frac{\pi}{3\sqrt{3}} = \sum_{n=1}^{\infty} \frac{\chi(n)}{n}$$

where

$$\chi(n) = \begin{cases} +1 & \text{if } n \equiv 1 \bmod 3, \\ -1 & \text{if } n \equiv -1 \bmod 3, \\ 0 & \text{if otherwise.} \end{cases}$$

For $q = 3, p = 1, k = 1$ we get

$$\frac{2\pi^2}{27} = \sum_{n=1}^{\infty} \frac{(-1)^{n-1}\chi(n)}{n^2}, \qquad \frac{4\pi^2}{27} = \sum_{n=1}^{\infty} \frac{\chi_0(n)}{n^2}$$

where

$$\chi_0(n) = \begin{cases} +1 & \text{if } n \equiv \pm 1 \bmod n, \\ 0 & \text{if otherwise.} \end{cases}$$

For $q = 4, p = 1$ we have

$$\frac{\pi^2}{8\sqrt{2}} = \sum_{n=1}^{\infty} \frac{\chi_8(n)}{n^2}$$

where

$$\chi_8(n) = \begin{cases} +1 & \text{if } n \equiv \pm 1 \bmod 8 \\ -1 & \text{if } n \equiv \pm 3 \bmod 8 \\ 0 & \text{if otherwise.} \end{cases}$$

For $q = 6, p = 1$ we get

$$\frac{\pi^2}{6\sqrt{3}} = \sum_{n=1}^{\infty} \frac{\chi_{12}(n)}{n^2}$$

where

$$\chi_{12}(n) = \begin{cases} +1 & \text{if } n \equiv \pm 1 \bmod 12 \\ -1 & \text{if } n \equiv \pm 5 \bmod 12 \\ 0 & \text{if otherwise.} \end{cases}$$

These cases can be continued indefinitely. It is easy to take this discussion to the point where we conclude that

$$\sum_{n \in \mathbf{Z}} \frac{1}{(qn+p)^r} = f_r \pi^r, \qquad \sum_{n \in \mathbf{Z}} \frac{(-1)^n}{(qn+p)^r} = g_r \pi^r$$

where f_r and g_r are elements in the field R_q generated by all the qth roots of unity; they are thus certainly *cyclotomic numbers*. If we now take h to be a periodic function on the integers of period q, or, for odd q, $(-1)^n$ times such a periodic function, whose values are cyclotomic numbers, then

$$\sum_{n \in \mathbf{Z}} \frac{h(n)}{n^r} = g\pi^r$$

where g is a cyclotomic number. It would be almost a hundred years before Dirichlet would systematically introduce such series in his path-breaking memoirs on primes in residue classes and reveal the arithmetical meaning of some of these sums (class numbers), and even later than that when Kronecker would discover the true arithmetical significance of the cyclotomic extensions R_q (see Chapter 6).

The work of Jacobi and Weierstrass on elliptic functions. The trigonometric functions are essentially the only periodic functions of a real variable. Indeed, any period can be changed to period 1 by a change of scale, and any polynomial or rational expression in $\cos \pi s$ and $\sin \pi s$ is essentially the only periodic function with period 1. The partial fraction formula can be written as

$$(*) \qquad \pi \cot \pi s = PV \sum_{-\infty}^{\infty} \frac{1}{n+s}.$$

This way $(*)$ of writing the expansion as a principal value makes it almost obvious that the infinite series represents a function of period 1.

In the 19th century, when Jacobi and later Weierstrass began studying functions of a *complex* variable which are *doubly periodic,* for instance with periods 1 and τ

$(\tau = i, \tau = 1 + i\sqrt{2}$ are classical examples, the first of which goes back to Gauss), it was natural to think of partial fractions and infinite product expansions for them. These series and products would now be taken over the lattice of all periods in the complex plane. The Weierstrass $\wp$–function and the infinite product expansions of Weierstrass and Jacobi may thus be viewed as natural descendants of Euler's work.

3.6. Partial fraction expansions as integrals

In view of the pivotal role of the partial fractions for $\frac{\pi}{\sin \pi s}$ and $\pi \cot \pi s$ in summing the zeta and the L-series, it was natural for Euler to try to derive these expansions by other more direct methods. His method had two stages. In the first, already given in [6], one obtains the partial fraction expansions as integrals from 0 to 1 of differential forms that are generalizations of the form $(1 + x^2)^{-1}dx$ which was used in evaluating the Leibniz series. The second stage is then dedicated to the evaluation of the integrals.

In generalization of the Leibniz series for $\arctan x$ Euler starts with the series obtained by evaluating the integrals

$$\int_0^1 \frac{x^{p-1} + x^{q-p-1}}{1 + x^q}dx, \qquad \int_0^1 \frac{x^{p-1} - x^{q-p-1}}{1 - x^q}.$$

Here p, q are integers and we assume that

$$q > p > 0.$$

Note that the integrand in the first integral is continuous in the range $[0, 1]$. On the other hand, both the numerator and denominator of the integrand in the second integral are polynomials divisible by $1 - x$, and on removing this factor we obtain an integrand which is a ratio of two polynomials, the denominator being

$$(1 + x + x^2 + \cdots + x^{q-1})$$

which is ≥ 1 on $[0, 1]$ and so the integrand is again continuous on $[0, 1]$.

For the first integral we start from the series

$$\frac{1}{1 + x^q} = 1 - x^q + x^{2q} - x^{3q} + x^{4q} + \cdots$$

and multiply by x^{p-1} and x^{q-p-1} and integrate from 0 to 1 to get

$$\int_0^1 \frac{x^{p-1}}{1 + x^q}dx = \frac{1}{p} - \frac{1}{q+p} + \frac{1}{2q+p} - \frac{1}{3q+p} + \cdots$$

$$\int_0^1 \frac{x^{q-p-1}}{1 + x^q}dx = \frac{1}{q-p} - \frac{1}{2q-p} + \frac{1}{3q-p} - \cdots.$$

The integrals obviously exist separately since p, q are integers with $q > p > 0$ and the two series are convergent. We add the two series but group them in the following manner to get

$$\frac{1}{p} + \left(\frac{1}{q-p} - \frac{1}{q+p}\right) - \left(\frac{1}{2q-p} - \frac{1}{2q+p}\right) - \cdots.$$

This gives the formula

$$\int_0^1 \frac{x^{p-1} + x^{q-p-1}}{1 + x^q}dx = \frac{1}{p} + \sum_{n=1}^{\infty} (-1)^{n-1}\left(\frac{1}{nq-p} - \frac{1}{nq+p}\right).$$

We shall discuss later the proof that

$$\int_0^1 \frac{x^{p-1} + x^{q-p-1}}{1 + x^q}\,dx = \frac{\pi}{q\sin(p/q)\pi}.$$

Assuming this we obtain

$$\frac{\pi}{q\sin(p/q)\pi} = \frac{1}{p} + \sum_{n=1}^{\infty}(-1)^{n-1}\left(\frac{1}{nq - p} - \frac{1}{nq + p}\right).$$

Multiplying by q and replacing p/q by s we get

$$\frac{\pi}{\sin s\pi} = \frac{1}{s} + \sum_{n=1}^{\infty}(-1)^{n-1}\left(\frac{1}{n - s} - \frac{1}{n + s}\right) \qquad (0 < s < 1).$$

To be sure, s is a rational number, but both sides are continuous functions and so the above formula is valid for *all* s in the range $0 < s < 1$. The continuity of the infinite series is a consequence of normal convergence when s varies in any interval of the form $[a, 1 - a]$ where $0 < a < 1$. To verify normal convergence observe that $1 - s^2/n^2 = (1 - s/n)(1 + s/n) \geq a$ and hence

$$\left|\frac{1}{n - s} - \frac{1}{n + s}\right| = \frac{2s}{(1 - s^2/n^2)}\,\frac{1}{n^2} \leq \frac{2}{a}\,\frac{1}{n^2}.$$

For the second integral the calculations are the same, but the integral cannot be split into two terms since the individual integrals do not converge. So we first write the series for $(x^{p-1} - x^{q-p-1})/1 - x^q$ in the form

$$\frac{x^{p-1} - x^{q-p-1}}{1 - x^q} = x^{p-1} - \sum_{n=1}^{\infty}x^{nq-p-1} + \sum_{n=1}^{\infty}x^{nq+p-1}.$$

Hence integrating from 0 to y where $0 < y < 1$ we have

$$\int_0^y \frac{x^{p-1} - x^{q-p-1}}{1 - x^q}\,dx = \frac{y^p}{p} - \sum_{n=1}^{\infty}\left(\frac{y^{nq-p}}{nq - p} - \frac{y^{nq+p}}{nq + p}\right).$$

It is now a question of taking the limit as $y \to 1$ and passing to the limit term by the term on the right side. For this we must verify normal convergence of the series on the right when y varies in $[0, 1]$. Write

$$a_n(y) = \frac{y^{nq-p}}{nq - p} - \frac{y^{nq+p}}{nq + p}.$$

Then, for $0 \leq y \leq 1$,

$$\frac{da_n}{dy} = y^{nq-p} - y^{nq+p} = y^{nq-p}(1 - y^{2p}) \geq 0$$

showing that a_n is increasing in y and hence reaches its maximum at $y = 1$ and minimum at $y = 0$. So

$$0 \leq a_n(y) \leq a_n(1) = \frac{1}{nq - p} - \frac{1}{nq + p} \leq \frac{2p}{q^2 - p^2}\,\frac{1}{n^2}.$$

So we have proved normal convergence. Thus

$$\int_0^1 \frac{x^{p-1} - x^{q-p-1}}{1 - x^q}\,dx = \frac{1}{p} - \sum_{n=1}^{\infty}\left(\frac{1}{nq - p} - \frac{1}{nq + p}\right).$$

We shall later give a direct argument that

$$\int_0^1 \frac{x^{p-1} - x^{q-p-1}}{1 - x^q} dx = \frac{\pi \cot(p/q)\pi}{q}.$$

Hence we obtain, writing $s = p/q$,

$$\pi \cot s\pi = \frac{1}{s} \sum_{n=1}^\infty \left(\frac{1}{n-s} - \frac{1}{n+s} \right) \qquad (0 < s < 1).$$

Again, this is initially true for s rational, but we argue as before that it remains valid for all s with $0 < s < 1$.

Euler's evaluation of the integrals. In [13] Euler evaluated the integrals above by a direct method. We shall now discuss his derivation.

The starting point of Euler's derivation is the calculation of the indefinite integral (primitive) of

$$\frac{x^{m-1}}{1 + x^{2n}} dx \qquad (m, n \text{ integers}, 2m > n > 0)$$

which is a beautiful generalization of the fact that $\arctan x$ is the primitive of $(1 + x^2)^{-1} dx$. Writing $\theta = \frac{\pi}{2n}$, Euler begins with the formula

$$\frac{x^{m-1}}{1 + x^{2n}} = \frac{(-1)^{m-1}}{n} \sum_{k=1}^n \frac{x \cos(2k-1)m\theta + \cos(2k-1)(m-1)\theta}{1 + 2x \cos(2k-1)\theta + x^2}.$$

To prove this we recall that if R is a polynomial with distinct roots $a_1, a_2, \ldots, a_N$ (which can be complex) and g is any polynomial of degree $< N$,

$$\frac{g(x)}{R(x)} = \sum_{j=1}^N \frac{g(a_j) R'(a_j)^{-1}}{x - a_j}.$$

In fact, if we assume a representation $\sum_j c_j (x - a_j)^{-1}$ for the right side, we get

$$g(a_j) = c_j \prod_{k \neq j} (a_j - a_k) = c_j R'(a_j).$$

If g, P are real, then the complex roots are even in number and can be grouped in conjugate pairs. Let $R(x) = 1 + x^{2n}, g(x) = x^{m-1}$ and let us write the roots of R as

$$-e^{\pm i(2k-1)\theta} \qquad \left(k = 1, 2, \ldots, n, \theta = \frac{\pi}{2n} \right).$$

Combining the fractions corresponding to the conjugate pairs we get the partial fraction expansion above for

$$\frac{x^{m-1}}{1 + x^{2n}}.$$

Now we know that

$$\int \frac{Ax + B}{1 + 2x \cos \gamma + x^2} dx = \frac{A}{2} \log(1 + 2x \cos \gamma + x^2)$$

$$+ \frac{B - A \cos \gamma}{\sin \gamma} \arctan \frac{x \sin \gamma}{1 + x \cos \gamma}.$$

Substituting the appropriate values for A, B, γ, we then obtain for

$$\int_0^x \frac{x^{m-1}}{1 + x^{2n}} dx$$

the formula

$$\frac{(-1)^{m-1}}{2n}\sum_{k=1}^{n}\cos(2k-1)m\theta\log\left(1+2x\cos(2k-1)\theta+x^2\right)$$

$$+\frac{(-1)^{m-1}}{n}\sum_{k=1}^{n}\sin(2k-1)m\theta\arctan\frac{x\sin(2k-1)\theta}{1+x\cos(2k-1)\theta}$$

where $\theta=\frac{\pi}{2n}$. We now let $x\to\infty$ in this formula. The logarithmic terms become

$$\frac{(-1)^{m-1}}{n}\sum_{k=1}^{n}\cos\frac{(2k-1)m\pi}{2n}\log x+o(1)$$

while the arctan terms converge to

$$\frac{(-1)^{m-1}\pi}{2n^2}\sum_{k=1}^{n}(2k-1)\sin\frac{(2k-1)m\pi}{2n}.$$

We now have

(T)
$$\sum_{k=1}^{n}\cos\frac{(2k-1)m\pi}{2n}=0$$

$$\sum_{k=1}^{n}(2k-1)\sin\frac{(2k-1)m\pi}{2n}=\frac{(-1)^{m-1}n}{\sin\frac{m\pi}{2n}}.$$

Assuming these for the moment we get

$$\int_0^\infty\frac{x^{m-1}}{1+x^{2n}}dx=\frac{\pi}{2n\sin\frac{m\pi}{2n}}.$$

We put $p=m,q=2n$ and rewrite this as

$$\int_0^\infty\frac{x^{p-1}}{1+x^q}dx=\frac{\pi}{q\sin\frac{p\pi}{q}}\qquad(q>p>0).$$

To be sure, we assumed that q is even; but if q is odd, the substitution $x=y^2$ changes the integral to one with the even integer $2q$, and we obtain the above formula for odd q also.

It remains only to prove the trigonometric formulae (T). In [13] Euler evaluates a whole class of such sums. It is not difficult to prove (T) using complex variables. Write $u=e^{\frac{im\pi}{2n}}$. Then $u^{2n}=(-1)^m$. The sum of cosines is then

$$\Re(u+u^3+\cdots+u^{2n-1})=\Re\frac{u(u^{2n}-1)}{u^2-1}=((-1)^m-1)\Re\frac{u}{u^2-1}.$$

As $\bar{u}=u^{-1}$ an easy calculation shows that $u(u^2-1)^{-1}$ changes into its negative under conjugation and so is purely imaginary, thus having zero real part. This proves the first of the relations in (T). The sum of sines is

$$\Im(u+3u^3+\cdots+(2n-1)u^{2n-1})=\Im\left(z\left(\frac{z^{2n+1}-z}{z^2-1}\right)'\bigg|_{z=u}\right).$$

Now

$$\frac{z^{2n+1}-z}{z^2-1}=\frac{1}{2}\left(\frac{z^{2n}}{z+1}+\frac{z^{2n}}{z-1}\right)-\frac{1}{2}\left(\frac{1}{z+1}+\frac{1}{z-1}\right).$$

If we remember that $u^{2n} = (-1)^m$ and note that $z(z \pm 1)^{-2}$ is invariant under conjugation and hence real, we find that the required imaginary part is

$$(-1)^m n \Im \left(\frac{1}{u+1} + \frac{1}{u-1} \right) = \frac{(-1)^{m-1} n}{\sin \frac{m\pi}{2n}}.$$

We have thus proved (T) and hence finished the evaluation of our integral.

Since

$$\int_1^\infty \frac{x^{p-1}}{1+x^q} dx = \int_0^1 \frac{y^{q-p-1}}{1+y^q} dy$$

we then obtain

$$\int_0^1 \frac{x^{p-1} + x^{q-p-1}}{1+x^q} dx = \frac{\pi}{q \sin \frac{p}{\pi q}}.$$

This finishes the evaluation of the original integral of the first type.

Before we go on to the second integral we remark that Euler does not stop with the above calculation but goes on to study the more general integrals

$$J_k = \int_0^\infty \frac{x^{p-1}}{(1+x^q)^k} dx.$$

Indeed, an integration by parts gives the relation

$$\int_0^x \frac{x^{p-1}}{(1+x^q)^k} dx = \frac{x^p}{(k-1)q(1+x^q)^{k-1}} + \frac{(k-1)q-p}{(k-1)q} \int_0^x \frac{x^{p-1}}{(1+x^q)^{k-1}} dx.$$

(One starts with the integral on the right.) Letting $x \to \infty$ we get

$$J_k = \frac{(k-1)q-p}{(k-1)q} J_{k-1}$$

leading to

$$\int_0^\infty \frac{x^{p-1}}{(1+x^q)^k} dx = \frac{(q-p)(2q-p)\ldots((k-1)q-p)}{q \cdot 2q \cdot \ldots \cdot (k-1)q} \frac{\pi}{q \sin \frac{p\pi}{q}}.$$

Following the same method he also finds that

$$\int_0^\infty \frac{x^{m-1}}{1 - 2x^n \cos \omega + x^{2n}} dx = \frac{\pi \sin \frac{n-m}{n}(\pi - \omega)}{n \sin \omega \sin \frac{(n-m)\pi}{n}}.$$

For $\omega = \frac{\pi}{2}$ this reduces to the previous formula.

We now come to the second type of integrals. The first step is to obtain an expression for the indefinite integral

$$\int_0^x \frac{x^{m-1}}{1 - x^{2n}} dx.$$

The method is exactly the same as the previous one. The roots of $x^{2n} - 1 = 0$ are written as

$$\pm 1, -e^{\pm \frac{ik\pi}{n}} \qquad (k = 1, 2, \ldots, n-1).$$

The partial fraction expansion for

$$\frac{x^{m-1}}{1 - x^{2n}}$$

is now, with $\tau = \frac{\pi}{n}$,

$$\frac{1}{2n(1-x)} + \frac{(-1)^{m-1}}{2n(1+x)} + \frac{(-1)^{m-1}}{n} \sum_{k=1}^{n-1} \frac{x \cos km\tau + \cos k(m-1)\tau}{1 + 2x \cos k\tau + x^2}.$$

We integrate as before to get,* for $0 < x < 1$,

$$\int_0^x \frac{x^{m-1}}{1-x^{2n}} dx = \frac{-1}{2n} \log(1-x) + \frac{(-1)^{m-1}}{2n} \log(1+x)$$

$$+ \frac{(-1)^{m-1}}{2n} \sum_{k=1}^{n-1} \cos km\tau \log\left(1 + 2x \cos\tau + x^2\right)$$

$$+ \frac{(-1)^{m-1}}{n} \sum_{k=1}^{n-1} \sin km\tau \arctan \frac{x \sin k\tau}{1 + x \cos k\tau}.$$

The $\log(1-x)$ term will diverge as $x \to 1-$, but we write a similar formula changing m to $2n-m$ and subtract; the logarithmic terms cancel and we get an expression that is convergent when $x \to 1-$, namely

$$\int_0^x \frac{x^{m-1} - x^{2n-m-1}}{1-x^{2n}} dx = \frac{2(-1)^{m-1}}{n} \sum_{k=1}^{n-1} \sin \frac{km\pi}{n} \arctan \frac{x \sin \frac{k\pi}{n}}{1 + x \cos \frac{k\pi}{n}}.$$

We now let $x \to 1-$ and use the formula $\frac{\sin\varphi}{1+\cos\varphi} = \tan\frac{\varphi}{2}$ to get

$$\int_0^1 \frac{x^{m-1} - x^{2n-m-1}}{1-x^{2n}} dx = \frac{(-1)^{m-1}\pi}{n^2} \sum_{k=1}^{n-1} k \sin \frac{km\pi}{n}.$$

We now have

$$\sum_{k=1}^{n-1} k \sin \frac{km\pi}{n} = \frac{(-1)^{m-1}n}{2} \cot \frac{m\pi}{2n}$$

which is proved by exactly the same methods we used earlier. Indeed, with $u = e^{\frac{im\pi}{n}}$, the left side of the above formula is

$$\Im\left(z\left(\frac{z^n-1}{z-1}\right)' \Big|_{z=u}\right) = \Im\left(\frac{-u^n \cdot u}{(u-1)^2} + \frac{nu^{n-1}\cdot u}{u-1} + \frac{u}{(u-1)^2}\right).$$

But $u^n = (-1)^m$ and $\frac{u}{(u-1)^2}$ is unchanged under conjugation so that it has zero imaginary part. Hence the above reduces to

$$(-1)^m n \Im\left(\frac{1}{u-1}\right) = \frac{(-1)^{m-1}n}{2} \cot \frac{m\pi}{2n}.$$

Thus

$$\int_0^1 \frac{x^{m-1} - x^{2n-m-1}}{1-x^{2n}} dx = \frac{\pi}{2n} \cot \frac{m\pi}{2n}.$$

We write $q = 2n, p = m$ and remove the restriction that q be even by using the substitution $x = y^2$ when q is odd. We thus finally obtain

$$\int_0^1 \frac{x^{p-1} - x^{q-p-1}}{1-x^q} dx = \frac{\pi}{q} \cot \frac{p\pi}{q} \qquad (q > p > 0).$$

---

*There is a misprint in §43 of [13] in this formula.

Eulerian integrals. Remarks on the gamma and beta functions. We shall now give a different method for evaluating the integrals that is based on the theory of the beta and gamma functions. These were first introduced by Euler (who else!) and later called *Eulerian integrals of the first and second kind* by Legendre. The discussion below is thus an opportunity to go into this part of analysis that Euler started. This derivation uses nothing that Euler did not know.

What we know now as the *gamma function* is the integral

$$\Gamma(s) = \int_0^\infty e^{-x} x^{s-1} dx \qquad (0 < s < 1).$$

The condition on s is needed for convergence of the integral at 0. Euler would write this as

$$\Gamma(s) = \int_0^1 (-\log x)^{s-1} dx.$$

It was Legendre who wrote the integral in the infinite range and who introduced the notation $\Gamma(s)$ for it. In either representation it is trivial to check, using integration by parts, that

$$\Gamma(s+1) = s\Gamma(s), \qquad \Gamma(1) = 1$$

and hence

$$\Gamma(n+1) = \int_0^1 (-\log x)^n dx = n!$$

Euler viewed the function $\Gamma(s)$ as an extension of the function $n!$, which is defined only over the natural numbers, to a function defined for all real values $s > 0$. Legendre called this the *Eulerian integral of the second kind*.

The *Eulerian integral of the first kind* is the integral

$$B(a, b) = \int_0^1 x^{a-1}(1-x)^{b-1} dx \qquad (a > 0. \, b > 0).$$

The limits on a and b are again needed for convergence. This function is nowadays called the *beta function*. There is also an alternative form of it, as an infinite integral, namely

$$\int_0^\infty \frac{x^{a-1}}{(1+x)^{a+b}} dx = B(a, b);$$

the substitution $x \to \frac{1}{1+x}$ changes this to the standard form. We also have

$$\int_0^\infty \frac{x^{a-1}}{1+x^b} dx = \frac{1}{b} B\left(\frac{a}{b}, 1 - \frac{a}{b}\right)$$

as can be seen by the substitution $x \to x^b$. The relation between the beta and the gamma functions is given by the formula

$$B(a, b) = \frac{\Gamma(a)\Gamma(b)}{\Gamma(a+b)}.$$

In addition to the formula

$$\Gamma(n+1) = n! \qquad (n = 0, 1, 2, \dots)$$

we have

$$\Gamma(1/2) = \sqrt{\pi}$$

and, more generally, the reflection formula

$$B(s, 1-s) = \Gamma(s)\Gamma(1-s) = \frac{\pi}{\sin \pi s} \qquad (0 < s < 1),$$

both of which were known to Euler. We also need the formula obtained by logarithmic differentiation of the reflection formula. We have, with primes denoting derivatives with respect to s,

$$-\frac{\Gamma'(s)}{\Gamma(s)} + \frac{\Gamma'(1-s)}{\Gamma(1-s)} = \pi \cot s\pi.$$

It is in 1729, in a letter to Goldbach* that the gamma function first makes its appearance. Euler's definition, in modern notation, is for positive real m:

(*) $$\Gamma(1+m) = \lim_{n\to\infty} \frac{1.2.3.\ldots n}{(m+1)\ldots(m+n)}(n+1)^m,$$

or, in a form closer to what he preferred to write,

$$\Gamma(1+m) = \lim_{n\to\infty} \frac{1.2^m}{m+1} \frac{2^{1-m}.3^m}{m+2} \cdots \frac{n^{1-m}.(n+1)^m}{m+n}.$$

Subsequently he gave more details in his paper [14] of 1730/1 based on this definition. He knew all the basic properties of this function and used them in an essential manner in his work on the functional equation of the zeta function. We shall very briefly sketch the principal properties of Γ, all of which were known to Euler. For a very complete treatment see [15]; we follow this treatment in our brief discussion. Let

$$\gamma_n = 1 + \frac{1}{2} + \frac{1}{3} + \cdots + \frac{1}{n} - \log(n+1)$$

so that

$$\lim_{n\to\infty} \gamma_n = \gamma$$

is Euler's constant. Then

$$\frac{z(z+1)\ldots(z+n)}{n!}(n+1)^{-z} = ze^{z\gamma_n}\prod_{k=1}^{n-1}\left(1+\frac{z}{k}\right)e^{-z/k}$$

so that we obtain Weierstrass's definition, now valid for all *complex* z,

$$\frac{1}{\Gamma(z)} = ze^{z\gamma}\prod_{n=1}^{\infty}\left(1+\frac{z}{n}\right)e^{-z/n}.$$

This is a classic example of a canonical product of Weierstrass and defines an entire function with simple zeros precisely at $z = 0, -1, -2, \ldots$. So the limit (*) exists for all $z \neq 0, -1, -2, \ldots$ and defines Γ as a meromorphic function, vanishing nowhere and having simple poles precisely at $0, -1, -2, \ldots$. It is immediate from the definition that

$$\Gamma(z+1) = z\Gamma(z).$$

If we start with the integral

$$\mathrm{II}(z,n) = \int_0^n \left(1 - \frac{t}{n}\right)^n t^{z-1}dt$$

and change t to ns and integrate by parts repeatedly, we get

$$\mathrm{II}(z,n) = \frac{1.2.\ldots n}{z(z+1)\ldots(z+n)}n^z$$

*See [1], Ch. 2, Vol. I, 3-6 (Euler to Goldbach).

so that

$$\Gamma(z) = \lim_{n \to \infty} \int_0^n \left(1 - \frac{t}{n}\right)^n t^{z-1} dt.$$

For $0 < x < 1$ we have $-\log(1 - x) = x + x^2/2 + \cdots > x$ so that $1 - x < e^{-x}$. Hence

$$0 \le \left(1 - \frac{t}{n}\right)^n |t^{z-1}| \le e^{-t} t^{\Re(z)-1} \qquad (\Re(z) > 0)$$

and so there is normal convergence and we can pass to the limit under the integral sign. Thus

$$\Gamma(z) = \int_0^\infty e^{-t} t^{z-1} dt = \int_0^1 (-\log u)^{z-1} du \qquad (\Re(z) > 0).$$

Euler always used the second form of the integral. It is clear from this that $\Gamma(z) > 0$ if z is real and > 0 and that

$$\Gamma(1 + z) = z! \qquad (z = 1, 2, 3, \dots).$$

Thus $\Gamma(1 + z)$ *interpolates* $z!$ when z is not a nonnegative integer, a point of view that was emphasized by Euler. From Euler's infinite product for $\sin z$ it follows at once that

$$\Gamma(z)\Gamma(1 - z) = \frac{\pi}{\sin z\pi}.$$

In particular, taking $z = 1/2$ we have

$$\Gamma\left(\frac{1}{2}\right) = \sqrt{\pi}.$$

Reverting back to Euler's definition with $m = \frac{1}{2}$, we can write this as

$$\frac{1}{2}\sqrt{\pi} = \sqrt{\frac{2.4}{3.3} \frac{4.6}{5.5} \cdots}$$

or

$$\frac{\pi}{4} = \frac{2.4}{3.3} \frac{4.6}{5.5} \cdots,$$

which is a famous result of Wallis [16]. It would seem that this result of Wallis acted as a guide to Euler when he set about his theory of the gamma function. From this point of view, the Eulerian infinite products for $\sin x$ are far-reaching generalizations of the Wallis product for $\frac{\pi}{4}$.

The beta function is defined as

$$B(a, b) = \int_0^1 x^{a-1}(1 - x)^{b-1} dx \qquad (\Re(a), \Re(b) > 0).$$

The fundamental formula

$$B(a, b) = \frac{\Gamma(a)\Gamma(b)}{\Gamma(a + b)}$$

may be proved, in post-Eulerian fashion, as follows. We have

$$\Gamma(a)\Gamma(b) = \int_0^\infty \int_0^\infty e^{-t} e^{-u} t^{a-1} u^{b-1} dt du.$$

Change to new variables $t = x^2, u = y^2$ and use polar coordinates in the xy-plane to get

$$\Gamma(a)\Gamma(b) = 2\int_0^\infty e^{-r^2}(r^2)^{a+b-1} d(r^2) \int_0^{\pi/2} (\cos\theta)^{2a-1}(\sin\theta)^{2b-1} d\theta.$$

Changing to $w = \cos^2 \theta$ we get in the end

$$\Gamma(a)\Gamma(b) = \Gamma(a+b)B(a,b).$$

Finally

$$\Gamma'(1) = -\gamma.$$

To prove this we logarithmically differentiate Weierstrass's infinite product to get

$$-\Gamma'(1) = 1 + \gamma + \sum_{n=1}^{\infty} \left(\frac{1}{(n+1)} - \frac{1}{n} \right) = \gamma.$$

It is very interesting to look at Euler's derivation of the formula

$$n! = \int_0^1 (-\log x)^{n-1} dx.$$

Indeed, once he obtained this, he would use this as the definition of the gamma function. To derive this Euler first starts out from the formula, established by expanding $(1-x)^n$ by the binomial theorem,

$$\int_0^1 x^e (1-x)^n dx = \frac{1.2.\ldots.n}{(e+1)(e+2)\ldots(e+n+1)}.$$

He then writes $e = \frac{f}{g}$ and rewrites this as

$$\frac{1.2.3,\ldots n}{(f+g)(f+2g)\ldots(f+ng)} = \frac{f+(n+1)g}{g^{n+1}} \int_0^1 x^{\frac{f}{g}} (1-x)^n dx.$$

He now wants to take $f \to 1$ and let $g \to 0$; the left side will go to $n!$, but the behavior of the right side is not clear. Euler writes the limiting form of the right side as

$$\int_0^1 \frac{x^{\frac{1}{0}}(1-x)^n}{0^{n+1}} dx$$

and proceeds to determine it! He makes the substitution

$$x = y^{\frac{g}{1+g}}$$

and obtains the expression

$$\frac{f+(n+1)g}{f+g} \int_0^1 \frac{(1-x^{\frac{g}{f+g}})^n}{g^n} dx.$$

He now evaluates the limit of the expression inside the integral for $g \to 0, f \to 1$ by a "known" method, namely l'Hopital's rule, to get

$$\lim_{g \to 0} \left(\frac{1 - x^{\frac{g}{1+g}}}{g} \right)^n = (-\log x)^n$$

and hence finally

$$n! = \int_0^1 (-\log x)^n dx$$

(see [17] for a fuller account).

After this brief introduction let us return to our integrals. We wish to prove that

$$\int_0^1 \frac{x^{p-1} + x^{q-p-1}}{1 + x^q} dx = \frac{\pi}{q \sin(\frac{p}{q}\pi)} \qquad (0 < p < q, \quad p, q \text{ integers}).$$

Now

$$\int_0^1 \frac{x^{q-p-1}}{1+x^q} dx = \int_1^\infty \frac{x^{p-1}}{1+x^q} dx$$

as can be seen by the substitution $x \to \frac{1}{x}$. Hence

$$\int_0^1 \frac{x^{p-1} + x^{q-p-1}}{1+x^q} dx = \int_0^\infty \frac{x^{p-1}}{1+x^q} dx = \frac{1}{q} B\left(\frac{p}{q}, 1 - \frac{p}{q}\right).$$

Thus

$$\int_0^1 \frac{x^{p-1} + x^{q-p-1}}{1+x^q} dx = \frac{\pi}{q \sin(\frac{p}{q})\pi} \qquad (0 < p < q, \quad p, q \text{ integers})$$

as we wanted to prove.

The treatment of the second integral is a little more delicate since we cannot separate the two terms. We make the substitution $y = 1 - x^q$ to get, with $s = p/q$ as before,

$$\int_0^1 \frac{x^{p-1} - x^{q-p-1}}{1-x^q} dx = \frac{1}{q} \int_0^1 \frac{(1-y)^{s-1} - (1-y)^{-s}}{y} dy.$$

To separate the two terms we introduce a factor y^h. We have

$$\int_0^1 \frac{y^{s-1} + y^{-s}}{y^{1-h}} dy = B(s,h) - B(1-s,h)$$

$$= \frac{\Gamma(s)\Gamma(h)}{\Gamma(s+h)} - \frac{\Gamma(1-s)\Gamma(h)}{\Gamma(1-s+h)}$$

$$= \frac{1}{h}\left(\frac{\Gamma(s)\Gamma(1+h)}{\Gamma(s+h)} - \frac{\Gamma(1-s)\Gamma(1+h)}{\Gamma(1-s+h)}\right)$$

$$= \frac{1}{h}(a(s,h) - a(1-s,h))$$

where

$$a(s,h) = \frac{\Gamma(s)\Gamma(1+h)}{\Gamma(s+h)}.$$

But

$$a(s,0) = a(1-s,0) = 1$$

so that

$$\int_0^1 \frac{x^{p-1} - x^{q-p-1}}{1-x^q} dx = \lim_{h \to 0} \int_0^1 \frac{y^{s-1} + y^{-s}}{y^{1-h}} dy = a'(s,0) - a'(1-s,0)$$

where the primes denote derivatives with respect to h. As

$$a'(s,0) = \Gamma'(1) - \frac{\Gamma'(s)}{\Gamma(s)}$$

we have, for $a'(s,0) - a'(1-s,0)$, the expression

$$-\frac{\Gamma'(s)}{\Gamma(s)} + \frac{\Gamma'(1-s)}{\Gamma(1-s)} = -\frac{d}{ds}\log(\Gamma(s)\Gamma(1-s)) = \pi \cot s\pi$$

by the reflection formula. Hence we have finally shown that

$$\int_0^1 \frac{x^{p-1} - x^{q-p-1}}{1-x^q} dx = \frac{\pi \cot(p/q)\pi}{q}.$$

As mentioned earlier we may replace p/q by a real parameter s with $0 < s < 1$ in the final formulae.

3.7. Multizeta values

More than 30 years after his great discoveries on the zeta values, Euler introduced a new theme, one whose richness is far from being completely exhausted even today. In a remarkable paper [9] published in 1775 Euler introduced the *double zeta series*

$$\zeta(s_1, s_2) = \sum_{n_1 > n_2 > 0} \frac{1}{n_1^{s_1} n_2^{s_2}} \qquad (s_1, s_2 \in \mathbf{Z}, s_1 \geq 2, s_2 \geq 1).$$

The condition $s_1 \geq 2$ is needed for convergence, although Euler operated boldly with the case when $s_1 = 1$ also. The above definition, which is the one currently used, is a slight modification of the one Euler used. Euler allowed n_1 to be equal to n_2 so that

$$\zeta_{\text{Euler}}(s_1, s_2) = \zeta(s_1, s_2) + \zeta(s_1 + s_2).$$

In the paper [9] Euler obtained remarkable identities involving these, such as

$$\zeta(2, 1) = \zeta(3)$$

as well as the more general

$$\zeta(p, 1) + \zeta(p - 1, 2) + \cdots + \zeta(2, p - 1) = \zeta(p + 1).$$

From this he derived

$$2\zeta(p - 1, 1) = (p - 1)\zeta(p) - \sum_{2 \leq q \leq p-2} \zeta(q)\zeta(p - q).$$

In his paper cited above, Euler's aim was to express any $\zeta(a, b)$ with $a + b = r$ as a linear combination with rational coefficients of $\zeta(r)$ and the $\zeta(p)\zeta(r - p)$, generalizing the above formula. He succeeded in doing this for r odd but not for r even. We shall comment on this later.

Euler's identities for the double zeta values. The starting point to prove the identities described above is the obvious relation

$$\zeta(p)\zeta(q) = \zeta(p, q) + \zeta(q, p) + \zeta(p + q) \quad (p, q > 1).$$

This is immediate from the formula

$$\zeta(p)\zeta(q) = \sum_{m,n} \frac{1}{m^p} \frac{1}{n^q} = \sum_{m>n} + \sum_{n>m} + \sum_{m=n}.$$

The idea now is to look for what happens when p or q is 1 in view of the general principle that at the boundary of the regions where identities hold, interesting things happen. Now both $\zeta(1, q)$ and $\zeta(1)\zeta(q)$ are given by logarithmically diverging series and so one might expect that by examining more closely this divergence we may obtain identities at the points when one of the two values equals 1. The method used below is Euler's.

We start with the partial fraction expansion

$$(x + a)^{-p} x^{-q} = \sum_{0 \leq i \leq q-1} \binom{-p}{i} a^{-(p+i)} x^{-(q-i)}$$

$$+ \sum_{0 \leq j \leq p-1} \binom{-q}{j} (-a)^{-(q+j)} (x + a)^{-(p-j)}.$$

To see this, write

$$(x+a)^{-p}x^{-q} = \sum_{0 \le i \le q-1} A_i x^{-(q-i)} + \sum_{0 \le j \le p-1} B_j(x+a)^{-(p-j)}.$$

Then

$$(x+a)^{-p} = \sum_i A_i x^i + \left((x+a)^{-p}\sum_j B_j(x+a)^j\right)x^q.$$

Then $x = 0$ gives $A_0 = a^{-p}$, and differentiating i times at $x = 0$ gives $A_i = \binom{-p}{i}a^{-(p+i)} = A_i$. We do the same thing at $x = -a$ to get the B_j. This expansion is valid when $p, q \ge 1$. For $p = 2, q = 1$ this is

$$\frac{1}{xa^2} - \frac{1}{(x+a)a^2} = \frac{1}{(x+a)^2 x} + \frac{1}{(x+a)^2 a}.$$

Let us sum over $a, x = 1, 2, \ldots, N$. Write

$$\zeta_N(r) = \sum_{1 \le n \le N} \frac{1}{n^r}, \qquad \zeta_N(r,s) = \sum_{N \ge m > n \ge 1} \frac{1}{m^r n^s}.$$

Then we get

$$\zeta_N(2)\zeta_N(1) - \sum_{1 \le a,x \le N} \frac{1}{x+a}\frac{1}{a^2} = 2\sum_{1 \le a,x \le N} \frac{1}{(x+a)^2}\frac{1}{a}.$$

To estimate the right side we use

$$\zeta_N(2,1) < \sum_{1 \le a,x \le N} \frac{1}{(x+a)^2}\frac{1}{a} < \zeta(2,1)$$

so that the right side $\to 2\zeta(2,1)$. We shall now look at the left side. It is

$$\zeta_N(2)\zeta_N(1) - \sum_{1 \le a \le N} \frac{1}{a^2} \sum_{a+1 \le b \le a+N} \frac{1}{b}$$

$$= \zeta_N(2)\zeta_N(1) - \zeta_N(1,2) - \sum_{1 \le a \le N} \frac{1}{a^2} \sum_{N+1 \le j \le N+a} \frac{1}{j}.$$

We claim that the last term on the right side of this relation goes to 0. Indeed,

$$\sum_{N+1 \le j \le N+a} \frac{1}{j} < \int_N^{N+a} \frac{dx}{x} = \log\frac{N+a}{N} = \log\left(1 + \frac{a}{N}\right) < \frac{a}{N}$$

so that the last term in question is majorized by

$$\sum_{1 \le a \le N} \frac{1}{a^2}\frac{a}{N} = \frac{1}{N}\sum_{1 \le a \le N} \frac{1}{a} = O\left(\frac{\log N}{N}\right) \to 0.$$

We thus have

$$\zeta_N(2)\zeta_N(1) - \zeta_N(1,2) \to 2\zeta(2,1).$$

On the other hand,

$$\zeta_N(2)\zeta_N(1) = \zeta_N(1,2) + \zeta_N(2,1) + \zeta_N(3)$$

so that

$$\zeta_N(2)\zeta_N(1) - \zeta_N(1,2) \to \zeta(2,1) + \zeta(3)$$

also. So

$$\zeta(2,1) + \zeta(3) = 2\zeta(2,1)$$

which is the required identity. The same method works for $p \geq 2, q = 1$. The partial fraction expansion becomes

$$\frac{1}{a^p}\frac{1}{x} - \frac{1}{(x+a)}\frac{1}{a^p} = \frac{1}{(x+a)^p}\frac{1}{x} + \sum_{0 \leq j \leq p-2} \frac{1}{a^{j+1}}\frac{1}{(x+a)^{p-j}}.$$

Once again we sum over $a, x = 1, 2, \ldots, N$. For $r \geq 2$ we have

$$\zeta_N(r, j+1) < \sum_{1 \leq a,x \leq N} \frac{1}{a^{j+1}}\frac{1}{(x+a)^r} < \zeta(r, j+1)$$

so that the right side tends to

$$2\zeta(p, 1) + \zeta(p-1, 2) + \cdots + \zeta(2, p-1).$$

The left side is

$$\zeta_N(p)\zeta_N(1) - \zeta_N(1, p) - \sum_{1 \leq a \leq N} \frac{1}{a^p} \sum_{N+1 \leq j \leq N+a} \frac{1}{j}$$

and the last term of this relation, as before, is $O\left(\frac{\zeta_N(p-1)}{N}\right) = o(1)$. Hence

$$\zeta_N(p)\zeta_N(1) - \zeta_N(1, p) \to 2\zeta(p, 1) + \zeta(p-1, 2) + \cdots + \zeta(2, p-1).$$

But

$$\zeta_N(p)\zeta_N(1) = \zeta_N(p, 1) + \zeta_N(1, p) + \zeta_N(p+1)$$

so that

$$\zeta_N(p)\zeta_N(1) - \zeta_N(1, p) \to \zeta(p, 1) + \zeta(p+1)$$

also. Hence

$$\zeta(p, 1) + \zeta(p-1, 2) + \cdots + \zeta(2, p-1) = \zeta(p+1).$$

This leads to one of Euler's many formulae. We have

$$2\zeta(p+1) = 2\zeta(p, 1) + \sum_{2 \leq q \leq p-1} [\zeta(p-q+1, q) + \zeta(q, p-q+1)]$$

so that, writing $p = \lambda - 1$ as Euler does, we get

$$2\zeta(\lambda-1, 1) = (\lambda-1)\zeta(\lambda) - \sum_{2 \leq q \leq \lambda-2} \zeta(\lambda-q)\zeta(q).$$

For $\lambda = 3$ we get the formula $\zeta(2, 1) = \zeta(3)$. This formula suggests that for each λ, perhaps each $\zeta(m, n)$ with $m + n = \lambda$ is a linear combination of $\zeta(\lambda)$ and the $\zeta(p)\zeta(q)$ for $p + q = \lambda$. This is true in many cases as Euler himself found out, but not always [18a].

Multizeta values. The double zeta series can obviously be generalized further. We then obtain the *multizeta values* defined by

$$\zeta(s_1, s_2, \ldots, s_r) = \sum_{n_1 > n_2 > \cdots > n_r} \frac{1}{n_1^{s_1} n_2^{s_2} \ldots n_r^{s_r}} \qquad (s_i \in \mathbf{Z}, s_i \geq 1, s_1 \geq 1).$$

The identities of Euler discussed above can be generalized to these multizeta values as we shall see below. In recent years the MZV have generated tremendous interest because of their relationship to some fundamental objects in number theory, topology, and algebraic geometry. Some basic conjectures have been formulated, and this area remains an extremely hot one today.

The shuffle identities between the MZV. I follow Deligne [19]. Let $[1, r]$
denote the set $\{1, 2, \ldots, r\}$. A *shuffle* of $[1, r]$ and $[1, s]$ is a bijection σ from the
disjoint sum $I(r, s)$ of $[1, r]$ and $[1, s]$, identified with $[1, r] \times \{1\} \cup [1, s] \times \{2\}$, onto
$[1, r + s]$, such that σ is strictly increasing on each of $[1, r] \times \{1\}$ and $[1, s] \times \{2\}$. A
shuffle is thus a way of splicing together two ordered sequences into one sequence
such that the order within each sequence is preserved. Thus if $r = 3, s = 2$, then
there are 10 shuffles; a shuffle in this case is determined as soon as we decide the
images of $(1, 2)$ and $(2, 2)$. If $\sigma(1, 2) = a$ where $a = 1, 2, 3, 4$, then $\sigma(2, 2)$ has to be
in $\{a + 1, \ldots 5\}$, so that the total number is $\sum_{1 \le a \le 4}(5 - a) = 10$. A *shuffle with
equalities* is heuristically just like a shuffle, but the two sequences are allowed to
overlap when we splice them together. Thus a shuffle with equalities of $[1, r]$ and
$[1, s]$ is a map σ of $I(r, s)$ onto $[1, t]$ where $t \le r + s$ which is strictly increasing on
each of $[1, r] \times \{1\}$ and $[1, s] \times \{2\}$. Let $\mathrm{Sh}(r, s)$ and $\mathrm{Sh}^+(r, s)$ be the set of shuffles
and shuffles with equalities of $[1, r]$ and $[1, s]$.

If f, g are two functions on positive integers, then

$$\sum_{0 < n_1 < N} f(n_1) \sum_{0 < n_2 < N} g(n_2) = \sum_{n_1 > n_2} + \sum_{n_2 > n_1} + \sum_{n_1 = n_2}.$$

The terms on the right correspond to the 3 shuffles with equalities $\sigma \in \mathrm{Sh}^+(1, 1)$,
according to whether $\sigma(1, 1) >, <, = \sigma(1, 2)$. This can be generalized immediately
and leads us to the following identities:

$$\zeta(a_1, \ldots, a_r)\zeta(b_1, \ldots b_s) = \sum_{\sigma \in \mathrm{Sh}^+(r,s)} \zeta(c_1, \ldots, c_t)$$

where $[1, t]$ is the range of σ and

$$c_k = \begin{cases} a_u & \text{if } \sigma(u, 1) = k \\ b_v & \text{if } \sigma(v, 2) = k \\ a_u + b_v & \text{if } \sigma(u, 1) = \sigma(v, 2) = k. \end{cases}$$

For instance,

$$\zeta(a_1, a_2)\zeta(b) = \zeta(b, a_1, a_2) + \zeta(a_1, b, a_2)$$
$$+ \zeta(a_1, a_a, b) + \zeta(a_1 + b, a_2) + \zeta(a_1, a_2 + b).$$

MZV as iterated integrals. We have already seen that Euler had an integral
representation of $\zeta(2)$, namely,

$$\zeta(2) = \int_0^1 \frac{-\log(1 - t)}{t} dt.$$

We can write this as

$$\zeta(2) = \int_{1 > t > s > 0} \frac{dt}{t} \frac{ds}{1 - s}.$$

More generally, let us define the *dilogarithm* by

$$\mathrm{Li}_2(z) = \int_{z > t > s > 0} \frac{dt}{t} \frac{ds}{1 - s}.$$

Then

$$\mathrm{Li}_2(z) = \sum_{n > 0} \frac{z^n}{n^2}, \qquad \zeta(2) = \mathrm{Li}_2(1).$$

There is an obvious generalization of this to multizeta values. We define

$$\omega_0 = \frac{dt}{t}, \qquad \omega_1 = \frac{dt}{1-t},$$

and for any sequence $\mathbf{i} = (i(1), i(2), \ldots, i(r))$,

$$\omega(\mathbf{i}) = \omega_{i(1)}\omega_{i(2)} \cdots \omega_{i(r)}.$$

We define the *iterated integral*

$$\mathrm{It} \int_0^z \omega(\mathbf{i}) := \int_{z>t_1>t_2>\cdots>t_r>0} \omega_{i(1)}\omega_{i(2)} \cdots \omega_{i(r)} dt_1 dt_2 \ldots dt_r.$$

It is then an easy calculation, by expanding

$$\frac{1}{1-t} = \sum_{n \geq 0} t^n,$$

that

$$\mathrm{It} \int_0^z \omega(0,1) = \int_{z>t_1>t_2>0} \frac{dt_1}{t_1} \frac{dt_2}{1-t_2}.$$

More generally, let $s_1, s_2, \ldots s_r$ be integers ≥ 1, and let $\mathbf{i} = (i(1), \ldots, i(r))$ be any sequence of 0's and 1's such that $i(m) = 1$ precisely for $m = s_1, s_1 + s_2, \ldots s_1 + s_2 + \cdots + s_r$. Write

$$\omega_{s_1,s_2,\ldots,s_r} = \omega(\mathbf{i}).$$

Then

$$\mathrm{It} \int_0^z \omega(\mathbf{i}) = \mathrm{Li}_r(z)$$

where $\mathrm{Li}_r(z)$ is the *polylogarithm* defined by

$$\mathrm{Li}_r(z) = \frac{z^{n_1}}{n_1^{s_1} n_2^{s_2} \ldots n_r^{s_r}}.$$

If $s_1 \geq 2$ we can take $z = 1$ and get

$$\zeta(s_1, s_2, \ldots, s_r) = \mathrm{Li}_r(1) = \mathrm{It} \int_0^z \omega_{s_1,s_2,\ldots,s_r}.$$

The representation of the MZV as iterated integrals gives rise to a second family of shuffle identities among the MZV and, more generally, among the polylogarithms. To see this, note that

$$\{t > 0\} \times \{s > 0\} \approx \{t > s > 0\} \sqcup \{s > t > 0\}$$

($\sqcup$ means that the union is disjoint) where $\approx$ means that the two sides are equal up to a set of measure 0, namely the set where $t = s$. This can be generalized immediately to the following:

$$\{t_1 > t_2 > \cdots > t_a > 0\} \times \{s_1 > s_2 > \cdots > s_b > 0\}$$

$$\approx \bigsqcup_{\sigma \in \mathrm{Sh}(a,b)} \{u_{\sigma 1} > u_{\sigma 2} > \cdots > u_{\sigma a+b} > 0\}$$

where

$$u_{\sigma k} = \begin{cases} t_i & \text{if } \sigma(i) = k \\ s_j & \text{if } \sigma(j) = k. \end{cases}$$

Integration and Fubini's theorem now lead at once to the shuffle identities

$$\text{It} \int_0^z \omega_{s_1,s_2,\ldots,s_a} \cdot \text{It} \int_0^z \omega_{t_1,t_2,\ldots,t_b}$$

$$= \sum_{\sigma \in \text{Sh}(a,b)} \text{It} \int_0^z \sigma[\omega_{s_1,s_2,\ldots,s_a}, \omega_{t_1,t_2,\ldots,t_b}].$$

Here

$$\sigma[\omega_{s_1,s_2,\ldots,s_a}, \omega_{t_1,t_2,\ldots,t_b}]$$

means the form obtained by splicing the two forms according to σ. If $s_1, t_1 \geq 2$, we can take $z = 1$ and get the shuffle identity between the MZV's:

$$\zeta(s_1, s_2, \ldots, s_a)\zeta(t_1, t_2, \ldots, t_b) = \sum_{\sigma \in \text{Sh}(a,b)} \sigma[(s_1, \ldots, s_a), (t_1, \ldots, t_b)].$$

Here

$$\sigma[(s_1, \ldots, s_a), (t_1, \ldots, t_b)]$$

means the sequence obtained by splicing the two sequences $(s_i), (t_j)$ according to σ.

The reason why the shuffles with equalities do not enter in this identity is that the regions where two of the variables are equal have measure zero and so do not contribute to the integrals. These identities are very different from the earlier ones. For instance, the first set of identities gives

$$\zeta(2)\zeta(3) = \zeta(2,3) + \zeta(3,2) + \zeta(5)$$

while the second set gives

$$\zeta(2)\zeta(3) = \zeta(2,3) + 4\zeta(3,2) + 5\zeta(4,1).$$

There is a third set of identities due to Ecalle [20], and it is an open question whether all the identities between the MZV's are consequences of these three sets. For more detailed results and discussions of open questions as well as the connections of the MZV to other branches of mathematics, the reader should consult the literature. There is an interpretation of the various integrals as periods; see [18b].

Notes and references

[1] Raymond Ayoub, *Euler and the zeta function*, Amer. Math. Monthly, 81(1974), 1067-1087.

[2] L. Euler, *Opera Omnia. De summatione innumerabilium progressionum*, I-14 (1730/1), 25-41.

[3a] L. Euler, *Opera Omnia. Methodus generalis summandi progressiones*, I-14 (1732/3), 42-72.

[3b] L. Euler, *Opera Omnia. Invention summae cuiusque seriei ex dato termino generali*, I-14 (1736), 108-123.

[3c] L. Euler, *Opera Omnia. Institutiones calculi differentialis*, I-10 (1755), 309-336.

[3d] L. Euler, *Opera Omnia. Introductio in analysin infinitorum*, I-8 (1748), Ch. 11, 201-204. *Introduction to Analysis of the Infinite*, Book I, translated by John D. Blanton, Springer-Verlag, 1988.

[4] L. Euler, *Opera Omnia. De summis serierum reciprocarum*, I-14 (1734/5), 73-86.

[5] A. Weil, *Number Theory: An Approach through History from Hammurapi to Legendre*, Birkhäuser, 1984.

[6] *De summis serierum reciprocarum ex potestatibus numerorum naturalium ortarum dissertatio altera in qua eaedem summationes ex fonte maxime diverso derivantur,* I-14 (1743), 138-155.

[7] L. Euler, *Opera Omnia. De seriebud quibusdam considerationes,* I-14 (1740), 407-462.

[8] L. Euler, *Opera Omnia. Exercitationes analyticae,* I-15 (1772), 131-167.

[9] L. Euler, *Opera Omnia. Meditationes circa singulare serierum genus,* I-15 (1775), 217-267.

[10] K. Knopp, *Theory and Application of Infinite Series,* Blackie, 1928.

[11] P. Cartier, *Fonctions polylogarithmes, nombres polyzetas et groupes prounipotent,* Sem. Bourbaki, 2000–2001, n° 885, 1-37.

[12] P. Stäckel, *Eine vergessene abhandlung Leonhard Eulers Über die summe der reziproken quadrate der natüralichen zahlen,* Bibliotheca Mathematica, 8 (1907-1908), 37-54, in I-14, 156-176.

(1) This article is a review of the entire history of Euler's discovery of the zeta values.

(2) The forgotten article of Euler was published in a Swiss journal and includes a list of the exact values of $\zeta(n)$ for even $n \leq 26$; see I-14, 177-186.

[13] L. Euler, *Opera Omnia. De inventione integralium si post integrationem variabli quantitati determinatus valor tribuatur,* I-17, 35-69.

[14] L. Euler, *Opera Omnia. De progressionibus transcendentibus sen quarum termini generales algebraice dari nequeunt,* I-14, 1-24.

[15] E. T. Whittaker and G. N. Watson, *A Course of Modern Analysis,* Cambridge University Press, 1965.

[16] J. F. Scott, *The Mathematical Work of John Wallis,* Chelsea, 1981.

[17] P. J. Davis, *Leonhard Euler's integral: a historical profile of the gamma function,* Amer. Math. Monthly, 66 (1959), 849-869.

[18a] D. Zagier, *Values of zeta functions and their applications,* in First European Congress of Mathematics, Vol. II, Progress in Math., 120, Birkhäuser, 1994, 497-512.

[18b] M. Kontsevich and D. Zagier, *Periods,* Preprint IHES/M/01/22.

[19] P. Deligne, Lectures at UCLA, Spring 2005.

CHAPTER 4

Euler-Maclaurin Sum Formula

4.1. Formal derivation

After his initial evaluation of $\zeta(2)$ to a large number of decimal places, Euler began studying the general question of finding the approximate values of an expression of the form

$$f(0) + f(1) + \cdots + f(n)$$

where f is a general function. He introduced the function $S(x)$ by requiring that

$$S(x) - S(x-1) = f(x).$$

Of course S is not uniquely determined since one can add any function of period 1 to it without changing this relation. Now there are many instances where

$$\int_0^n f(x)dx$$

is a good approximation to

$$f(0) + f(1) + \cdots + f(n),$$

and so, with remarkable insight, Euler sought to find $S(x)$ in the form

$$S(x) = \int f(x)dx + \alpha f(x) + \beta f'(x) + \gamma f^{(3)}(x) + \delta f^{(4)}(x) + \dots$$

where $\alpha, \beta, \gamma, \delta, \dots$ are constants. In his first paper on this subject he showed how these constants can be obtained recursively. For this he first expanded $S(x-1)$ by Taylor's series as

$$S(x-1) = S(x) - \frac{S'(x)}{1!} + \frac{S''(x)}{2!} - \frac{S^{(3)}(x)}{3!} + \dots.$$

From the relation $S(x) - S(x-1) = f(x)$ we get, omitting the argument x,

$$f = S' - \frac{S''}{2!} + \frac{S^{(3)}}{3!} - \frac{S^{(4)}}{4!} + \dots.$$

One now has

$$S' = f + \alpha f' + \beta f'' + \gamma f^{(3)} + \dots$$
$$S'' = f' + \alpha f'' + \beta f^{(3)} + \gamma f^{(4)} + \dots$$
$$S^{(3)} = f'' + \alpha f^{(3)} + \beta f^{(4)} + \gamma f^{(5)} + \dots.$$

We substitute these in the previous formula and equate to 0 the respective coefficients of $f', f'', f^{(3)}, \ldots$, to get the equations

$$\alpha = \frac{1}{2}$$

$$\beta = \frac{\alpha}{2!} - \frac{1}{3!}$$

$$\gamma = \frac{\beta}{2!} - \frac{\alpha}{3!} + \frac{1}{4!}$$

$$\delta = \frac{\gamma}{2} - \frac{\beta}{3!} + \frac{\alpha}{4!} - \frac{1}{5!}$$

$$\cdots\cdots\cdots$$

which can be solved recursively. Hence we have

$$\alpha = \frac{1}{2}$$

$$\beta = \frac{1}{12}$$

$$\gamma = 0$$

$$\delta = -\frac{1}{720}$$

$$\cdots\cdots\cdots$$

He carried the determination of these constants to many more than indicated above and found, for example, that the odd numbered ones are always 0.

Euler later realized while writing a subsequent paper that the coefficients $\alpha, \beta, \gamma, \ldots$ may be calculated in a more streamlined fashion. In fact, Euler's procedure for determining $S(x)$ may be understood very simply by using the symbolic method. Let

$$D = \frac{d}{dx}.$$

Then the equation

$$f(x) = S(x) - S(x-1) = \frac{S'(x)}{1!} - \frac{S''(x)}{2!} + \frac{S^{(3)}(x)}{3!} - \cdots$$

can be written as

$$f = \left(D - \frac{D^2}{2!} + \frac{D^3}{3!} - \frac{D^4}{4!} + \ldots \right) S$$

or

$$(1 - e^{-D})S = f$$

which leads to the equation

$$S = (1 - e^{-D})^{-1} f.$$

Now

$$\frac{1}{1 - e^{-z}} = 1 + \frac{1}{z}\frac{z}{e^z - 1} = \frac{1}{z} + \frac{1}{2} + \sum_{k=1}^{\infty} \frac{B_{2k}}{(2k)!} z^{2k-1}$$

where the B_k are the Bernoulli numbers that arose in Euler's work on the zeta values, with the convention that

$$B_0 = 1, \ B_1 = \frac{1}{2}, \ B_k = 0 \qquad (k \geq 2, \quad k \text{ an odd number}).$$

Hence, interpreting $D^{-1}f$ as $\int f dx$ we have

$$S(x) = \int^x f(t)dt + \frac{1}{2}f(x) + \sum_{k=2}^{\infty} \frac{B_k}{k!} f^{(k-1)}(x).$$

There is an arbitrary constant involved in the term $\int^x f(t)dt$ that can be determined formally by evaluation for some value of x, say $x = r$. Hence we get:

$$S(x) - S(r) = \int_r^x f(t)dt + \frac{1}{2}[f(x) - f(r)]$$

$$+ \sum_{k=2}^{\infty} \frac{B_k}{k!}[f^{(k-1)}(n) - f^{(k-1)}(r)].$$

Taking $r = 0$ we get the celebrated Euler-Maclaurin formula:

(EM)
$$\sum_{r=0}^{n} f(r) = \int_0^x f(t)dt + \frac{1}{2}[f(0) + f(n)]$$

$$+ \sum_{k=2}^{\infty} \frac{B_k}{k!}[f^{(k-1)}(n) - f^{(k-1)}(0)].$$

The above derivation has been purely formal. In particular, we have not discussed the convergence of the infinite series on the right. Now the Bernoulli numbers grow rather rapidly. In fact, since

$$1 < \zeta(2k) = 1 + \frac{1}{2^{2k}} + \frac{1}{3^{2k}} + \cdots < 1 + \int_1^{\infty} \frac{dx}{x^{2k}} = 1 + \frac{1}{2k-1} \leq 2$$

we have, from Euler's formula,

$$1 < \frac{2^{2k-1}\pi^{2k}|B_{2k}|}{(2k)!} < 2 \qquad (k > 1)$$

from which it follows that

$$\frac{|B_{2k+2}|}{|B_{2k}|} > \frac{k^2}{2\pi^2}$$

showing that

$$|B_{2k}| \to \infty$$

rather rapidly. So the series on the right side of the Euler-Maclaurin formula will seldom converge. Euler was aware of this fact. Nevertheless one can use this formula for very accurate summations of various series, and Euler used it this way to calculate $\zeta(n)$ up to 16 or more decimal places for $2 \leq n \leq 16$.

For the case

$$f(x) = x^p$$

the summation formula will yield the explicit expressions for

$$1^p + 2^p + 3^p + \cdots + n^p$$

in which the Bernoulli numbers will figure decisively, as we shall see presently. Jacob Bernoulli had already worked out the expressions for these sums in his *Ars Conjectandi*. In view of this and following a suggestion of De Moivre, Euler proposed that the numbers B_n be called *Bernoulli numbers*. Around 1740, Maclaurin had also independently obtained this formula, but with characteristic serenity, Euler did not assert his undoubted priority. In fact for quite some time the summation formula was known as *Maclaurin's formula*, and it was only much later that Euler's indisputable priority was discovered. This was typical of Euler; he never got involved in priority disputes and was equally happy at the discoveries of others as he was with his own.

Clearly the series occurring in the Euler-Maclaurin formula is analogous to a Taylor series, which is often divergent but can be used for numerical computations. Euler used it often in computing zeta values; in determining Euler's constant to many decimal places; in studying numerically the functional equation of the zeta function; in summing divergent series, especially the factorial series; and so on. The sureness of his touch in handling the formula which is at best asymptotic is quite impressive. To use the summation formula effectively one must have remainder terms, and this was done by Euler's successors. Poisson was the first to discuss remainder terms, and the complete formula was treated rigorously by Jacobi. Much later, in modern times, Hardy established the summability of the Euler-Maclaurin formula under some fairly general conditions. The formula was extensively used as a summation procedure by Ramanujan. We shall treat this aspect of the summation formula briefly here and treat some applications, for instance, the derivation of *Euler's constant* and *Stirling's approximation for n!*. See [2], [3] for more details.

4.2. The case when the function is a polynomial

The simplest case of the summation formula arises when f above is a *polynomial*. Then the above derivation is *rigorous as all derivatives of f order $>$ the degree of f are* 0. Euler was aware of this. If we take for example

$$f(x) = x^p, \qquad S_p(n) = 1^p + 2^p + \cdots + n^p,$$

then

$$f^{(r)}(x) = 0 \ (r > p), \qquad f^{(p)}(x) = p!.$$

Hence, by the summation formula,

$$S_p(x) = \frac{x^{p+1}}{p+1} + \frac{x^p}{2} + \sum_{2 \leq k \leq p} \frac{p(p-1)\ldots(p-k+2)}{k!} B_k x^{p-k+1}$$

which can be rewritten as

$$S_p(x) = \frac{1}{p+1} \left(x^{p+1} + (p+1)B_1 x^p + \sum_{2 \leq k \leq p} \binom{p+1}{k} B_k x^{p+1-k} \right).$$

We can write this symbolically as

$$S_p(x) = \frac{(x+B)^{p+1} - B_{p+1}}{p+1}$$

with the convention that when we expand the binomial expression, we write B_r for B^r. The term B_{p+1} appears above because in the formula before that there is no constant term in the summation appearing there. In particular

$$1^p + 2^p + \cdots + n^p = \frac{(n+B)^{p+1} - B_{p+1}}{p+1}$$

which is essentially what Jacob Bernoulli had obtained. For instance, taking $p = 2, 3$ we have

$$1^2 + 2^2 + \cdots + n^2 = \frac{1}{3}(n^3 + 3(1/2)n^2 + 3(1/6)n) = \frac{n(n+1)(2n+1)}{6}$$

$$1^3 + 2^3 + \cdots + n^3 = \frac{1}{4}(n^4 + 4n^3(1/2) + 6n^2(1/6)) = \frac{n^2(n+1)^2}{4}.$$

4.3. Summation formula with remainder terms

We now take up the question of *exact summation formulae*. We shall proceed in the spirit of the Taylor expansions. To motivate this approach let us review briefly the Taylor expansions.

Let f be a smooth function and let x be a fixed real number which we may assume to be > 0. Then we first have

$$f(x) - f(0) = \int_0^x f'(t)dt.$$

We integrate by parts the right side and use $-(x - t)$ as the antiderivative of 1. We get

$$f(x) = f(0) + \left[-(x - t)f'(t) \right]_0^x + \int_0^x (x - t)f''(t)dt$$

which becomes

$$f(x) = f(0) + f'(0)x + \int_0^x (x - t)f''(t)dt.$$

The next step is to integrate by parts again to get the last term on the right side as

$$\left[\frac{-(x - t)^2}{2} f''(t) \right]_0^x + \int_0^x \frac{(x - t)^2}{2} f^{(3)}(t)dt$$

which reduces to

$$\frac{x^2}{2} f''(0) + \int_0^x \frac{(x - t)^2}{2} f^{(3)}(t)dt.$$

This obviously can be continued, and we get the Taylor series to an arbitrary *finite* number of terms with remainder term:

$$f(x) = f(0) + f'(0)x + \frac{f''(0)}{2!}x^2 + \cdots + \frac{f^{(n)}(0)}{n!}x^n + \int_0^x \frac{(x - t)^n}{n!} f^{(n+1)}(t)dt.$$

Of course estimates for the remainder integral can be obtained if we know bounds for $f^{(n+1)}$ in the interval in which we vary x.

We use a similar technique based on integration by parts in our treatment of the summation formula. First we have

$$(*) \qquad \int_0^n f(t)dt = nf(n) - \int_0^n tf'(t)dt$$

by integration by parts. Similarly we can use *summation by parts* to rewrite the sum $\sum_{0 \le k \le n} f(k)$. In fact,

$$f(0) + f(1) + \cdots + f(n) = \sum_{k=0}^{n} ((k+1) - k)f(k)$$

$$= (n+1)f(n) + \sum_{k=0}^{n-1} (k+1)f(k) - \sum_{k=1}^{n} kf(k)$$

$$= (n+1)f(n) + \sum_{k=1}^{n} kf(k-1) - \sum_{k=1}^{n} kf(k)$$

$$= (n+1)f(n) - \sum_{k=1}^{n} k(f(k) - f(k-1))$$

in exact analogy with $(*)$. So we get, in view of $(*)$,

$$f(0) + \cdots + f(n) = \int_0^n f(t)dt + f(n) + \int_0^n tf'(t)dt - \sum_{k=1}^{n} k(f(k) - f(k-1)).$$

Now

$$k(f(k) - f(k-1)) = k \int_{k-1}^{k} f'(t)dt = \int_{k-1}^{k} (1 + [t])f'(t)dt$$

where

$$[t] = \text{largest integer } \le t.$$

Hence

$$f(0) + \cdots + f(n) = \int_0^n f(t)dt + f(n) + \int_0^n (t - [t] - 1)f'(t)dt.$$

We can rewrite this as follows. Let

$$P_1(t) = t - [t] - \frac{1}{2}.$$

Then

$$(\text{EM}_1) \qquad f(0) + \cdots + f(n) = \int_0^n f(t)dt + \frac{1}{2}[f(n) + f(0)] + \int_0^n P_1(t)f'(t)dt.$$

A comparison of this with (EM) shows that it is reasonable to call this the Euler-Maclaurin summation formula with remainder term of order 1. We can now integrate by parts the last term repeatedly to get additional terms. For this we need the following lemma.

LEMMA. *Let $g(t)$ be a periodic function of period 1, assumed bounded and Riemann integrable, such that $\int_0^1 g(t)dt = 0$. Then there is a unique absolutely continuous periodic function $g_1(t)$ of period 1 such that*

$$\frac{dg_1}{dt} = g, \qquad \int_0^1 g_1(t)dt = 0.$$

In particular, there is a unique sequence of absolutely continuous periodic functions P_n $(n \ge 2)$ of period 1 such that

$$\frac{dP_n}{dt} = P_{n-1}, \qquad \int_0^1 P_n(t)dt = 0 \qquad (n \ge 2)$$

where $P_1(t) = t - [t] - 1/2$ as defined earlier.

PROOF. It is clear that

$$g_1(t) = \int_0^t g(u)du + C$$

where C is a constant which can be uniquely determined by the condition that

$$\int_0^1 g_1(t)dt = 0.$$

So we need only prove that g_1 has period 1. Now as g is periodic of period 1, we have $\int_t^{t+1} g(u)du = \int_0^1 g(u)du = 0$ for any t. Hence

$$g_1(t+1) - g_1(t) = \int_t^{t+1} g(u)du = 0$$

proving that g_1 has period 1. The existence and uniqueness of the P_n is now clear by a repeated application of the above result starting from $g = P_1$. Of course $\int_0^1 P_1(t)dt = 0$, so we can do this. This proves the lemma.

Now

$$\int_0^n P_1(t)f'(t)dt = \Big[P_2(t)f'(t)\Big]_0^n - \int_0^n P_2(t)f''(t)dt.$$

If we observe that $P_2(n) = P_2(0)$ we get the summation formula with remainder of order 2:

$$f(0) + \cdots + f(n) = \int_0^n f(t)dt + \frac{1}{2}[f(n) + f(0)] + P_2(0)[f'(n) - f'(0)]$$

(EM$_2$)

$$- \int_0^n P_2(t)f''(t)dt.$$

Clearly we can continue this to any order. The final formula is

$$f(0) + \cdots + f(n) = \int_0^n f(t)dt + \frac{1}{2}[f(n) + f(0)]$$

(EM$_N$)

$$+ \sum_{k=2}^N (-1)^k P_k(0)[f^{(k-1)}(n) - f^{(k-1)}(0)]$$

$$+ (-1)^{N+1} \int_0^n P_N(t)f^{(N)}(t)dt.$$

Let us calculate $P_2(0)$. We first determine $P_2(t)$ for $0 \le t < 1$ since it is determined everywhere else by periodicity. For $0 \le t < 1$ we have

$$P_2(t) = \int_0^t (u - 1/2)du + C = \frac{1}{2}(t^2 - t) + C$$

where C is chosen so that $\int_0^1 P_2(t)dt = 0$. This gives $C = 1/12$. Hence

$$P_2(t) = \frac{1}{2}(t^2 - t) + \frac{1}{12}$$

so that

$$P_2(0) = \frac{1}{12}$$

and the summation formula to order 2 becomes

$$f(0) + \cdots + f(n) = \int_0^n f(t)dt + \frac{1}{2}[f(n) + f(0)] + \frac{B_2}{2}[f'(n) - f'(0)]$$

$$(\text{EM}_2) \qquad\qquad\qquad - \int_0^n P_2(t)f''(t)dt.$$

To identify the further expansions with the original formula we must show that

$$(-1)^k P_k(0) = \frac{B_k}{k!} \qquad (k \geq 2)$$

This is obvious for $k = 1$, and we have already seen this for $k = 2$. But the general case is very easy. Write

$$(-1)^k P_k(0) = \frac{B_k'}{k!} \qquad (k \geq 2).$$

Then for the function $f(x) = x^p$ the formula derived now yields

$$1^p + \cdots + n^p = \frac{1}{p+1}\left(n^{p+1} + \sum_{1 \leq k \leq p} \binom{p+1}{k} B_k' n^{p+1-k}\right).$$

Thus

$$1^p + \cdots + n^p = \frac{(n + B')^{p+1} - B_{p+1}'}{p+1}$$

with the same convention as in our earlier derivation of the value of this sum. Hence

$$(n + B)^{p+1} - B_{p+1} = (n + B')^{p+1} - B_{p+1}'$$

from which it is immediate that

$$B_k = B_k' \qquad (k \geq 2).$$

We thus have the following theorem.

THEOREM (Euler-Maclaurin formula with remainders). *Let f be a smooth function. Then*

$$f(0) + \cdots + f(n) = \int_0^n f(t)dt + \frac{1}{2}[f(n) + f(0)]$$

$$(\text{EM}_N) \qquad\qquad + \sum_{k=2}^N \frac{B_k}{k!}[f^{(k-1)}(n) - f^{(k-1)}(0)]$$

$$+ (-1)^{N+1} \int_0^n P_N(t)f^{(N)}(t)dt.$$

Now, in the range $0 \leq t < 1$, we have $P_1(t) = t - 1/2$ and so $P_k(t)$ is a polynomial of degree k in the same range. Let P_k^* be this polynomial. Notice that $P_k^* \neq P_k$ as P_k is periodic but P_k^* is not. For instance, $P_1^*(t) = t - 1/2$, but $P_1(t) = t - [t] - 1/2$, its periodic version, so to speak. They coincide in $0 \leq t < 1$ of course, and so for any calculation within this interval we can use either. By Taylor's series, for $0 \leq t < 1$,

$$P_k(t) = \sum_{r=0}^k \frac{P_k^{(r)}(0)}{r!}t^r.$$

But, from the relation $P'_k = P_{k-1}$ we get (with $P_0 = 1$)

$$P_k^{(r)} = P_{k-r} \qquad (0 \le r \le k)$$

and so

$$P_k^{(r)}(0) = P_{k-r}(0) = (-1)^{k-r} \frac{B_{k-r}}{(k-r)!}.$$

Hence

$$P_k^*(t) = \sum_{r=0}^{k} (-1)^{k-r} \frac{B_{k-r}}{r!(k-r)!} t^r.$$

The P_k^* are called the *Bernoulli polynomials*, while the P_k may be called the *Bernoulli functions*. We have, symbolically,

$$P_k^*(t) = \frac{1}{k!} (t - B)^k.$$

4.4. Applications

We shall now give two applications of the summation formula with remainders. The first is to *Euler's constant*, which was already due to Euler; the second to Stirling's approximation for $n!$.

Euler's constant. The result is the following. The limit

$$C = \lim_{n \to \infty} \left(1 + \frac{1}{2} + \frac{1}{3} + \cdots + \frac{1}{n} - \log n \right)$$

exists and

$$0 < C < 1$$

where C is *Euler's constant*. Nowadays it is denoted by γ. We shall follow Euler and denote it by C.

We get this by using the formula (EM$_1$) for the function

$$f(x) = \frac{1}{1+x}.$$

Then

$$1 + \frac{1}{2} + \cdots + \frac{1}{n} = \int_0^{n-1} \frac{dx}{1+x} + (1/2)[(1/n) + 1] - \int_0^{n-1} \frac{P_1(u)}{(1+u)^2} du.$$

Since P_1 has period 1 this can be written as

$$1 + \frac{1}{2} + \cdots + \frac{1}{n} - \log n = (1/2)[(1/n) + 1] - \int_1^n \frac{P_1(u)}{u^2} du.$$

But, as

$$|P_1(u)| \le \frac{1}{2}$$

the integral

$$\int_0^\infty \frac{P_1(u)}{u^2} du$$

exists. Hence the right side of the above formula has the limit

$$C = \frac{1}{2} - \int_0^\infty \frac{P_1(u)}{u^2} du.$$

This proves the result with the additional explicit representation for C. Since

$$|P_1(u)| < \frac{1}{2}$$

except when u is an integer, we have

$$\left| \frac{P_1(u)}{u^2} du \right| < \frac{1}{2} \int_0^\infty \frac{du}{u^2} = \frac{1}{2}.$$

Hence

$$0 < C < 1.$$

Euler calculated C to 16 decimals using his summation formula and obtained

$$C = 0.5772156649015329\ldots.$$

Perhaps a word of explanation is in order about the way Euler used his summation formula. He was aware that the full series is almost always divergent, and so, as he explained, *one should continue the summation till the terms begin to diverge.* In this special case, if we use the summation formula with a higher order of remainder we get

$$C = \frac{1}{2} + \frac{B_2}{2} + \frac{B_4}{4} + \ldots \frac{B_{2k}}{2k} - (2k+1)! \int_1^\infty \frac{P_{2k+1}(u)}{u^{2k+2}} du.$$

Euler of course simply wrote

$$C = \frac{1}{2} + \frac{B_2}{2} + \frac{B_4}{4} + \ldots.$$

Stirling's formula for $n!$. Here we take

$$f(x) = \log(1+x).$$

Then the summation formula gives

$$\log n! = \int_1^n \log x\, dx + (1/2) \log n + (1/12)[(1/n) - 1] + \int_0^n \frac{P_2(u)}{u^2} du$$

$$= (n + 1/2) \log n - n + (1/12)n - 1/12 + \int_0^n \frac{P_2(u)}{u^2} du.$$

As before

$$\int_0^\infty \frac{P_2(u)}{u^2} du$$

exists, and so we have the following result: the limit

$$\lim_{n \to \infty} \log n! - [(n + 1/2) \log n - n] = C_1$$

exists and

$$C_1 = -\frac{1}{12} + \int_0^\infty \frac{P_2(u)}{u^2} du.$$

This is Stirling's formula. It is usually written in the exponentiated form

$$n! \sim K e^{-n} n^{n+1/2} \qquad (n \to \infty)$$

where $K = e^{C_1}$ is a constant > 0 and, as usual,

$$a_n \sim b_n$$

means a_n *is asymptotic to* b_n, i.e.,

$$\lim_{n \to \infty} \frac{a_n}{b_n} = 1.$$

To evaluate K we can use Euler's infinite product for $\sin x/x$ when $x = \pi/2$. We have

$$\frac{2}{\pi} = \prod_{n=1}^{\infty}\left(1 - \frac{1}{2n}\right)\left(1 + \frac{1}{2n}\right)$$

so that

$$\frac{2}{\pi} = \lim_{N\to\infty}\prod_{n=1}^{N}\left(1 - \frac{1}{2n}\right)\left(1 + \frac{1}{2n}\right).$$

The product on the right side can be written as

$$\frac{1.3.\ldots(2N-1).3.5.\ldots(2N+1)}{(2.4.\ldots(2N))^2}$$

which simplifies to

$$\frac{((2N)!)^2(2N+1)}{2^{4N}(N!)^4}.$$

Hence

$$\frac{\pi}{2} = \lim_{N\to\infty}\frac{2^{4N}(N!)^4}{((2N)!)^2(2N+1)}.$$

By Stirling's formula the expression on the right side is asymptotic to

$$K^2\frac{2^{4N}N^{4N+2}e^{-4N}}{e^{-4N}(2N)^{4N+1}(2N+1)} = K^2\frac{N}{2(2N+1)} \sim \frac{K^2}{4}.$$

Hence

$$K^2 = 2\pi$$

or

$$K = \sqrt{2\pi}.$$

Thus

$$n! \sim \sqrt{2\pi}e^{-n}n^{n+1/2}.$$

Although we used Euler's formula to get the infinite product for $\pi/2$, that product goes back earlier and is due to Wallis. Here again, one can get more accurate formulae by taking additional terms in the summation formula.

The summation formula can be discussed along these lines for a large class of functions. The treatment is not difficult and depends on a close study of the Bernoulli functions $P_k(t)$. The final result is as follows.

THEOREM. Let f be of constant sign for $x \geq 0$ and let f and its derivatives go to 0 monotonically as $x \to \infty$. Then

$$f(0) + \cdots + f(n) = \int_0^n f(x)dx + (1/2)[f(n) + f(0)]$$

$$+ \sum_{k=1}^{N}\frac{B_{2k}}{(2k)!}[f^{(2k-1)}(n) - f^{(2k-1)}(0)] + R_N$$

where the series is alternating and the remainder satisfies

$$\left|R_N\right| < \frac{|B_{2k+2}|}{(2k+2)!}|f^{(2k+1)}(n) - f^{(2k+1)}(0)|.$$

In other words, the error committed in stopping the evaluation at the term $B_{2N}/(2N)!$ is *at most the first term omitted.* This explains to some extent Euler's description of the calculative process as *stopping when the terms begin to diverge.*

A sketch of the proof is as follows. Since f is positive and f' is monotonic and goes to 0 at infinity, f' is negative and increases to 0 as $x \to \infty$. Then f'' is positive and decreases to 0 as $x \to \infty$. This argument can be continued and leads to the conclusion that all odd derivatives of f are negative and increase to 0 as $x \to \infty$, while all the even derivatives of f are positive and decrease to 0 as $x \to \infty$. Since the Bernoulli numbers alternate in sign this already proves that the Euler-Maclaurin series is alternating. It remains only to estimate the remainder. The remainder after the term $(B_{2k}/k!)[f^{(2k-1)}(n) - f^{(2k-1)}(0)]$ is

$$R_k = -(2k+1)! \int_0^n P_{2k+1}(u) f^{(2k+1)}(u) du.$$

It turns out from a close study of the P_k that R_k and R_{k+1} have opposite signs and hence that

$$|R_k| < |R_k - R_{k+1}|.$$

Since

$$R_k - R_{k+1} = \frac{B_{2k+2}}{(2k+2)!}[f^{(2k+1)}(n) - f^{(2k+1)}(0)]$$

the result follows.

Notes and references

[1] L. Euler, *Opera Omnia*, I-10, I-14.

Euler first presented the summation formula in 1732. Three years later, in 1735, he gave a more complete treatment. He gave another exposition in his treatise I-10 on differential calculus published in 1755.

[2] G. H. Hardy, *Divergent Series*, Oxford, 1973.

[3] K. Knopp, *Theory and Application of Infinite Series*, Blackie, 1928.

CHAPTER 5

Divergent Series and Integrals

5.1. Divergent series and Euler's ideas about summing them

Infinite series like the geometric series

$$\frac{1}{1-r} = 1 + r + r^2 + \cdots + r^n + \ldots$$

and the harmonic series

$$1 + \frac{1}{2} + \frac{1}{3} + \cdots + \frac{1}{n} + \ldots$$

have been known from very old times. The series

$$\frac{4}{3} = 1 + \frac{1}{4} + \frac{1}{4^2} + \cdots$$

goes back to Archimedes and the quadrature of the parabola. Archimedes used the *method of exhaustion*. It was typical of this method that the sum cannot be reached by stopping at any finite number of terms but can be approximated arbitrarily closely by taking more and more terms. The harmonic series as well as its divergence appear to have been known as early as the 14th century [1]. However, statements about convergence and divergence of series did not have in those days the precise meaning they have today; in fact, even in the 18th century, most of the assertions about convergence were based on a mixture of intuition, heuristic reasoning, extensive numerical calculation, and, occasionally, rigorous argumentation. The divergence of the harmonic series, in spite of the fact that its terms tend to 0, must have been one of the striking facts about infinite series in those early times. Later on, infinite series such as the binomial series and other power series occurred in the works of many mathematicians like Newton, Gregory, Leibniz, the Bernoullis,

and others, and ideas about the convergence and divergence of infinite series gradually became more precise. Leibniz already considered series whose terms were both
positive and negative, and his criterion for the convergence of the alternating series

$$a_0 - a_1 + a_2 - \cdots + (-1)^n a_n + \ldots \qquad (a_n > 0, a_n \geq a_{n+1}),$$

namely, that this series is convergent if (and only if) $a_n \to 0$ as $n \to \infty$, is perhaps
one of the earliest general criteria for deciding the convergence or divergence of a
series. Examples of such series were

$$\frac{\pi}{4} = 1 - \frac{1}{3} + \frac{1}{5} - \frac{1}{7} + \ldots$$

$$\log 2 = 1 - \frac{1}{2} + \frac{1}{3} - \frac{1}{4} + \frac{1}{5} - \ldots.$$

Perhaps intellectual curiosity was one of the reasons that mathematicians
started to look at series that do not converge. However, they knew, from experience if not strict logical reasoning, that a series of positive terms, if it does not
converge, diverges to ∞, and so there was nothing further that could presumably
be done for such a series. Their attention was therefore directed to series whose
terms are alternating but which do not converge. In the late 17th century there was
controversy about what sum, if any, should be assigned to the series

(1) $1 - 1 + 1 - 1 + 1 - 1 + 1 - 1 + \ldots.$

The discussions involved many people, including Leibniz. There was a strong suggestion, especially from Leibniz, that the sum of (1) should be $1/2$. Leibniz thought
that since the partial sums of the series (1) are alternately 1 and 0, the sum should
be the average, namely $1/2$, as is suggested by the theory of probability. Actually, if
we interpret the probabilistic statement in terms of the frequencies, this is just the
statement that the *sequences of the averages of the partial sums have the limit* $1/2$.
Thus one could say that Leibniz came quite close to one of the modern definitions
of summability, namely the one associated to Cesàro. We shall of course have more
to say about this later.

The first person who treated divergent series in a serious and systematic manner
was Euler. Euler agreed with Leibniz that the series (1) should be assigned the sum
$1/2$, but for a different reason, namely, since it arises from the expansion

$$\frac{1}{1+x} = 1 - x + x^2 - x^3 + \cdots + (-1)^n x^n + \ldots$$

when we take $x = 1$, the sum should be the value of $1/(1 + x)$ at $x = 1$, namely,
$1/2$. Similarly, the expansion

$$\frac{1}{(1+x)^2} = 1 - 2x + 3x^2 - 4x^3 + \cdots + (-1)^n (n+1) x^n + \ldots$$

led to the sum $1/4$ being assigned to the series

$$1 - 2 + 3 - 4 + \cdots + (-1)^n (n+1) + \ldots$$

and so on. Euler then went on from such examples to his general principle of
assigning to any series its sum as the value at 1 of its generating function.

There was some attempt to point out the difficulties with this reasoning by
considering the example

$$\frac{1 - x^m}{1 - x^n} = 1 - x^m + x^n - x^{m+n} + x^{2n} + \ldots \qquad (m, n \text{ integers } \geq 1, m < n).$$

The argument was made that if one takes $x = 1$ in the above expansion, one gets

$$\frac{m}{n} = 1 - 1 + 1 - 1 + \dots,$$

which is absurd since the left side can then be an arbitrary rational number between 0 and 1. Lagrange already disposed of this example by remarking that one should write

$$\frac{1 - x^m}{1 - x^n} = 1 + 0x + 0x^2 + \dots - x^m + 0x^{m+1} + \dots + x^n + 0x^{n+1} + \dots$$

which would lead to

$$(2) \qquad \frac{m}{n} = 1 + 0 + \dots + 0 - 1 + 0 + \dots + 1 + \dots$$

where the terms repeat with period n and $a_0 = 1, a_m = -1$, a series that is different from Leibniz's. Thus, when $m = 2, n = 3$ for example, we have

$$\frac{2}{3} = 1 + 0 - 1 + 1 + 0 - 1 + \dots.$$

Lagrange also treated more general examples of series

$$\sum_{k \geq 0} a_k x^k, \qquad a_k = a_{k+n}.$$

If

$$P(x) = a_0 + a_1 x + \dots + a_{n-1} x^{n-1},$$

then

$$\sum_{k \geq 0} a_k x^k = \frac{P(x)}{1 - x^n}.$$

For the rational function on the right side of this equation to be well defined at $x = 1$ we must have

$$P(1) = a_0 + a_1 + \dots + a_{n-1} = 0.$$

Under this condition, taking the value at $x = 1$ gives the result

$$\sum_{k \geq 0} a_k = -\frac{P'(1)}{n} \qquad \left(P' = \frac{dP}{dx} \right).$$

The partial sums of the series are periodic of period n and

$$-\frac{P'(1)}{n} = \frac{s_0 + s_1 + \dots + s_{n-2}}{n}$$

so that the Euler procedure leads to

$$\sum_{k \geq 0} a_k = \frac{s_0 + s_1 + \dots + s_{n-2}}{n}.$$

We can use the same probabilistic reasoning as in the case of the series (1) and verify that the right side is indeed the limit of the averages of the sequence of partial sums.

It is not surprising that there was much confusion and controversy about divergent series in the 17th century. We must remember that the foundations of analysis had not been laid properly and that rigor, as we understand it today, came only after the work of Abel and Cauchy. It was not possible, even for the greatest mathematicians of those days, to make definitions and axioms that appear almost obvious to us today. Here is what Hardy, a major figure of the 20th century in the theory of divergent series and analysis in general, has to say about these controversies [2].

> ... It does not occur to a modern mathematician that a collection of mathematical symbols should have a 'meaning' until one has been assigned to it by definition. It was not a triviality even to the greatest mathematicians of the 18th century. They had not the habit of definition: it was not natural to them to say that 'by X we mean Y'.... It is broadly true to say that mathematicians before Cauchy asked not 'How shall we define $1 - 1 + 1 - 1 + 1 - 1 + \ldots$ ' but 'What is $1 - 1 + 1 - 1 + 1 - 1 + \ldots$?'

In many ways the great exception to this was Euler. He took divergent series seriously, used them to obtain beautiful results, and had definite views about what is meant by the sum of a divergent series and how they should be used. Most of the time he used divergent series only in the context of expansions of functions. In correspondence with Nicolaus Bernoulli I, he insisted that assigning sums to divergent series in the context of power series expansions would be a consistent procedure, in the sense that the same series cannot arise from two different expressions. We now know that Euler was correct because a function has at most one power series expansion. However he did not always restrict his use of divergent series to convergent power series. He was always on the lookout for including as many divergent series as possible in his collection for which he could assign a sum by some method of analysis. For instance, he was able to assign a sum to the series

$$(3) \qquad 1 - 1! + 2! - 3! + \cdots + (-1)^n n! + \ldots.$$

which he called the "divergent series par excellence". The associated power series

$$(4) \qquad 1 - 1!x + 2!x^2 - 3!x^3 + \cdots + (-1)^n n! x^n + \ldots$$

converges for no x other than 0, and so there is no function whose expansion is given by the series (4). So the method of power series that Euler relied on most of the time fails here; nevertheless he was able to develop a reasonable way to sum this series.

In the paper entitled *De seriebus divergentibus* communicated in 1755 and published in 1760 [3], Euler formally introduced for the first time the idea that one should try to associate sums to series that are divergent. His reasons for this were quite practical but also based on his feeling that every series had a natural "sum" and it was necessary to find this "sum". He pointed out that this will not do any harm provided one is systematic in the way sums are associated to divergent series, since the procedure does not alter the sums of convergent series.

Here are some extended quotations from his 1760 paper, given in the translation [4]:

> Notable enough, however, are the controversies over the series $1 - 1 + 1 - 1 + 1 - 1 + \ldots$ whose sum was given by Leibniz as $1/2$, although others disagree. No one has yet assigned another value to that sum, and so the controversy turns on the question whether the series of this type have a certain sum. Understanding of this question is to be sought in the word "sum"; this idea, if thus conceived—namely the sum of a series is said to be that quantity to which it is brought closer as more terms of the series are

taken—has relevance only for convergent series, and we should in general give up this idea of sum for divergent series. Wherefore, those who thus define a sum cannot be blamed if they claim they are unable to assign a sum to a series. On the other hand, as series in analysis arise from the expansion of fractions or irrational quantities or even transcendentals, it will in turn be permissible in calculation to substitute in place of such a series that quantity out of whose development it is produced. For this reason, if we employ this definition of sum, that is to say, the sum of a series is that quantity which generates the series, all doubts with respect to divergent series vanish and no further controversy remains on this score, inasmuch as this definition is applicable equally to convergent or divergent series. Accordingly, Leibniz, without any hesitation, accepted for the series $1 - 1 + 1 - 1 + 1 - 1 + \ldots$, the sum $1/2$, which arises out of the expansion of the fraction $1/1 + 1$, and for the series $1 - 2 + 3 - 4 + 5 - 6 + \ldots$, the sum $1/4$, which arises out of the expansion of the formula $1/(1+1)^2$. In a similar way a decision for all divergent series will be reached, where always a closed formula from whose expansion the series arises should be investigated. However, it can happen very often that this formula itself is difficult to find, as here where the author treats an exceptional example, that divergent series par excellence $1 - 1 + 2 - 6 + 24 - 120 + 720 - 5040 + \ldots$, which is Wallis' hypergeometric series, set out with alternating signs; this series, in whatever formula it finds its origin and however much this formula is valid, is seen to be determinable by only the deepest study of higher Analysis. Finally, after various attempts, the author by a wholly singular method using continued fractions found that the sum of this series is about 0.596347362123, and in this decimal fraction the error does not affect even the last digit. Then he proceeds to other similar series of wider application and he explains how to assign them a sum in the same way, where the word "sum" has that meaning which he has here established and by which all controversies are cut off.

Actually Euler had discussed his ideas about divergent series much earlier, in correspondence with Goldbach and Nicolaus Bernoulli I. In a letter to Goldbach [5] written on August 7, 1745, discussing his attempts to sum the divergent series

$$1 = 1! + 2! - 3! + \ldots$$

Euler has this to say (free translation from German):

... I believe that every series should be assigned a certain value. However, to account for all the difficulties that have been pointed out in this connection, this value should not be denoted by the name sum, because usually this word is connected with the notion that a sum has been obtained by a real summation: this idea however is not applicable to "seriebus divergentibus". ...

Earlier in the same year he had written a letter to Nicolaus Bernoulli I [6] on July 17, 1745, in which he had discussed in some detail his method of summing the series

$$1 = 1! + 2! - 3! + \cdots + (-1)^n n! + \ldots.$$

The glimpse into Euler's views on divergent series provided by these and other letters is quite remarkable. Indeed, Hardy, while discussing Euler's remarks in his letter to Goldbach mentioned above, says that

> ... this is language which might almost have been used by Cesàro or Borel.

> ... it is a mistake to think of Euler as a 'loose' mathematician, though his language may sometimes seem loose to modern ears; and even his language sometimes suggests a point of view far in advance of the general ideas of his time.

It is therefore very clear that Euler had an understanding of the issues involving divergent series that was very much ahead of his time. He knew not only that assigning a sum for a divergent series was *a matter of convention* but also that such sums have to be assigned in a systematic manner so that one could use the notion in a flexible manner in analysis. Although it would appear that different conventions could assign different sums to the same series, he had the conviction that any divergent series would be found to have a unique "sum", irrespective of the method used to sum it. For the most part, except in the study of singular examples like (3), he used the method of power series expansions: for the series

$$(5) \qquad\qquad a_0 + a_1 + \cdots + a_n + \ldots$$

one would determine the generating function

$$f(x) = a_0 + a_1 x + a_2 x^2 + \cdots + a_n x^n + \ldots$$

and define

$$(6) \qquad\qquad \sum_{n=0}^{\infty} a_n = f(1).$$

More than a century after Euler wrote his paper this method came to be known as the *Abel summation* when the theory of summation of divergent series was developed systematically.

In this chapter I shall start with Euler's ideas on divergent series and follow the trail of these ideas till I reach the current mathematical scene. In §2 I shall give a glimpse of some of Euler's work on divergent series by discussing his discovery, by the methods of divergent series, of the functional equation of the zeta function, anticipating by more than a hundred years the work of Riemann. In §3 I shall discuss Euler's treatment of the series (3) by the method of differential equations which anticipates the theory, due to Borel, of what we now call Borel summation and its relation to solutions of differential equations and their asymptotic expansions, as well as the work of Stieltjes on continued fractions. I shall then develop briefly in §4 the theory of summability of divergent series along modern lines mostly following the book of Hardy [2]; the reader can also consult Knopp's classic and wonderful book on infinite series [7]. In §5 I treat Borel summation and discuss the results of Borel which grew out of Euler's work on the factorial series and became a vast generalization of it. In §6 I shall discuss what are called Tauberian theorems; these are theorems that assert that if a series is summable by some method and its coefficients satisfy a supplementary (Tauberian) condition appropriate to this method, then the series is convergent. These theorems are very interesting because they assert that every summation method will *fail* to sum series which diverge *slowly*. In §7 I give some applications of the theory of divergent series to problems of classical analysis.

In §8 I return to Tauberian theorems and treat the work of Wiener, who discovered that most summation methods are described as the actions of convolution operators of L^1-functions and so can be studied by the methods of *Fourier integral theory* on **R**. This point of view led him to a general Tauberian theorem from which most of the classical Tauberian theorems could be deduced as corollaries. I shall give a proof of the Wiener Tauberian theorem in this section. Miraculously, Gel'fand, a few years later, took Wiener's work and placed it in the framework of the theory of commutative Banach algebras which he created. Gel'fand's work opened a new era in mathematics, in which abstract algebras were represented as algebras of functions on a topological space canonically associated to the algebra. This is a point of view that was taken to its absolute limit by Grothendieck in his revolutionary approach to algebraic geometry, but which is also at the heart of recent ideas in physics regarding the structure of spacetime at small distances.

Then I shall discuss in §9 the method of what may be called *smeared summation,* which is really summation in the weak sense in the language of distributions, and show how it allows us to sum all trigonometric series with coefficients that are of at most polynomial growth. The theory of distributions, with its many ramifications and applications, is at the basis of everything in modern analysis. I shall conclude the chapter with a treatment in §10 of analytic continuation as a method of summation of divergent Gaussian integrals, which will point the way to themes that are encountered in quantum mechanics and quantum field theory under the name of the Feynman path integrals.[†]

5.2. Euler's derivation of the functional equation of the zeta function

As an illustration of Euler's ideas let me discuss here his derivation of what we now call the functional equation of the zeta function. As usual, let

$$\zeta(s) = 1 + \frac{1}{2^s} + \frac{1}{3^s} + \ldots \qquad (s > 1).$$

Euler considered $\zeta(s)$ only for integral and some rational values of s, although in his proof of the infinitude of primes he had implicitly considered the behavior of $\zeta(s)$ as $s \to 1+$. It was only after the work of Dirichlet that the necessity and usefulness of viewing $\zeta(s)$ and other series of the form

$$\sum_{n \geq 1} \frac{a_n}{n^s}$$

(nowadays called Dirichlet series) as functions of a real variable s became clear. Of course, after Riemann, ζ and other Dirichlet series became functions of a *complex* variable. Riemann showed in 1859 [9] that $\zeta(s)$ can be continued to the whole

---

[†]It may seem to be stretching a point to start with Euler and end with spacetime and Feynman. The reader who may be skeptical about such long chains of ruminations may like to read the chapter entitled *From Fermat's Last Theorem to the Abolition of Capital Punishment* in the delightful book of Littlewood (J. E. Littlewood, *A Mathematician's Miscellany*, 1960). Coincidentally, a very interesting comparison of Euler and Feynman has been made in the beautiful essay of P. Cartier [8].

s-plane as an analytic function except for a simple pole at $s = 1$ and that the extended function, still denoted by ζ, satisfies the functional equation

(FE)
$$\zeta(1 - s) = 2(2\pi)^{-s} \cos \frac{s\pi}{2} \Gamma(s)\zeta(s)$$

where $\Gamma(s)$ is Euler's gamma function. He then proceeded to show how this function plays a fundamental role in the theory of the *distribution* of primes. From the functional equation it is clear that the only zeros outside the *critical strip* defined by the inequality $0 \leq \Re(s) \leq 1$ are at the even negative integers (these are called the *trivial* zeros). It was in the course of this work that Riemann formulated his famous *Riemann hypothesis*, namely the conjecture that apart from the trivial zeros, all other zeros of $\zeta(s)$ are located on the line $\Re s = 1/2$. This problem remains the most famous unsolved problem in mathematics [9].

For a long time this was thought to be a problem in analysis and only analysts were working on the problem. The work of Hardy, Selberg, Levinson and many others showed that a positive fraction of all the zeros in the critical strip actually lie on the line $\Re(s) = 1/2$. But progress in the analytic approach has not been significant after these results, and many people believe that ideas outside analysis are perhaps needed to crack this problem. Actually people have been able to use computers to determine the zeros of the zeta function in the critical strip, and all of the computed ones (millions of them in fact) are on the critical line. Meanwhile, the algebraic geometers and algebraic number theorists had begun to associate zetalike functions to algebraic and geometric objects like algebraic number fields and algebraic curves. In the years 1930-1940 Weil was able to prove that the zeta function of a smooth projective algebraic curve defined over a *finite* field satisfies the analogue of the Riemann Hypothesis. Furthermore he conjectured in 1949 that such results should be true for the zeta functions of smooth projective varieties of any dimension defined over finite fields. In 1973 this was proved by Deligne, building on the foundations of algebraic geometry erected by Grothendieck. This is one of the greatest achievements of the mathematics of the 20th century. Ever since these discoveries were made, the connection between zeta functions and geometry has been very close. It is widely believed that the solution to the original Riemann Hypothesis will have a strong algebraic geometric component [10].

But, in 1749 [11], over a hundred years before Riemann, Euler discovered and proved the relation **(FE)** for all integer values of s as well as $s = 1/2$, verified it numerically for $s = 3/2$, and *conjectured* that it is true for all values of s! Euler preferred to work with the function $\eta(s)$ where

$$\eta(s) = 1 - \frac{1}{2^s} + \frac{1}{3^s} - \frac{1}{4^s} + \dots \qquad (s > 0)$$

which is convergent for $s > 0$ by the Leibniz test for convergence of alternating series. The alternating nature of the series for $\eta(s)$ was also valuable to Euler in numerical calculations, since in such series the error committed in truncating the series is less than the first term omitted. On the other hand, we have

$$\eta(s) = (1 - 2^{1-s})\zeta(s)$$

and so the functional equation for ζ becomes the following equation for η:

(FE)
$$\frac{\eta(1-s)}{\eta(s)} = -\frac{2^s - 1}{2^{s-1} - 1} \frac{\cos \frac{s\pi}{2}}{\pi^s} \Gamma(s).$$

Euler wrote m in place of s and so wrote this as

$$\frac{1 - 2^{m-1} + 3^{m-1} - \text{etc.}}{1 - 2^{-m} + 3^{-m} - \text{etc.}} = -\frac{1.2.3.\dots(m-1)(2^m - 1)}{(2^{m-1} - 1)\pi^m \cos\frac{m\pi}{2}}.$$

Euler wanted to prove this relation even when m was *not* a positive integer. The presence of the factor $(m-1)!$ makes it unclear how to make sense of the formula when m is not a positive integer. But this was not a problem for him. He had already discovered (see Chapter 3) that the function

$$1 \cdot 2 \cdot \dots \cdot \lambda = \lambda! \qquad (\lambda = 1, 2, \dots)$$

can be extended to nonintegral values of the variable k and wrote it as

$$1 \cdot 2 \cdot \dots \cdot \lambda = [\lambda]$$

with the characteristic property

$$[\lambda] = \lambda[\lambda - 1].$$

He had also worked out many properties of this function, including the identity

$$[\lambda][-\lambda] = \frac{\lambda\pi}{\sin\lambda\pi}.$$

Euler's theory was further continued by Legendre, who introduced the modern notation

$$[\lambda] = \Gamma(\lambda + 1).$$

The identity just mentioned above translates into

$$\Gamma(s)\Gamma(1 - s) = \frac{\pi}{\sin s\pi}.$$

It was with the understanding that $(m-1)!$ should be replaced by $[m-1]$ or $\Gamma(m)$ that Euler wanted to prove (**FE**) for values of m that were not positive integers. He verified (**FE**) for all integer values of s and also for $s = 1/2$. He also considered $s = 3/2$. In this case he could not evaluate the values of $\eta(3/2)$ and $\eta(-1/2)$ but verified it numerically with high accuracy.

For $s > 0$ there is no difficulty in defining

$$\eta(s) = 1 - 2^{-s} + 3^{-s} - \dots \qquad (s > 0)$$

as the series is alternating and so converges by the Leibniz criterion. But if $s \leq 0$ the series does not converge and Euler's calculations amounted to the definition

$$1 - 2^{-s} + 3^{-s} + \dots = \lim_{x \to 1-0} \sum_{n=1}^{\infty} (-1)^{n-1} n^{-s} x^n \qquad (s \leq 0).$$

If $s = m$ is a positive integer, he verified that the power series above is a rational function with $(1 + x)^{m+1}$ as its denominator and so its limit as $x \to 1$ exists and could be explicitly computed.

The first step in Euler's treatment of this question is to compute

$$\eta(-m) = 1^m - 2^m + 3^m - 4^m \dots \qquad (m = 0, 1, 2, 3, \dots).$$

The series of course diverges, and after remarking that its sum cannot be found by the usual methods and referring to his 1760 paper, Euler proceeds to evaluate the sum for $m = 2k - 1$ by his generating function method (nowadays called *Abel summation*; see §4). So, for him,

$$1^{2k-1} - 2^{2k-1} + 3^{2k-1} - \dots = \lim_{x \to 1-0} (1^{2k-1} x - 2^{2k-1} x^2 + 3^{2k-1} x^3 - \dots).$$

This method inevitably leads to the Bernoulli numbers, which figured in his evaluations of the series

$$1^{-2k} - 2^{-2k} + 3^{-2k} - \dots.$$

This must have been the signal to Euler that $\zeta(2k)$ and $\zeta(1-2k)$ were related. I shall follow Hardy's discussion in his book [2] (pp. 23-26). Instead of x we write

$$x = e^{-y} \qquad (y > 0)$$

so that one has

$$a_1 - a_2 + a_3 - \dots = \lim_{y \to 0+} \left(a_1 e^{-y} - a_2 e^{-y} + a_3 e^{-3y} - \dots \right).$$

To start with we have

$$e^{-y} - e^{-2y} + e^{-3y} - \dots = \frac{1}{e^y + 1} \qquad (y > 0).$$

Then, for any integer $m \geq 0$, we have

$$1^m e^{-y} - 2^m e^{-2y} + 3^m e^{-3y} - \dots = (-1)^m \frac{d^m}{dy^m} \left(\frac{1}{e^y + 1} \right) \qquad (y > 0).$$

We have already seen that

$$\frac{1}{e^y + 1} = \frac{1}{2} + \sum_{k=1}^{\infty} \frac{(1 - 2^{2k}) B_{2k}}{(2k)!} y^{2k-1}.$$

Hence

$$(-1)^m \frac{d^m}{dy^m} \left(\frac{1}{e^y + 1} \right) \bigg|_{y=0} = \begin{cases} \frac{1}{2} & \text{if } m = 0 \\ 0 & \text{if } m = 2, 4, 6, \dots \\ \frac{2^{2k}-1}{2k} B_{2k} & \text{if } m = 2k - 1 \end{cases}.$$

Thus Euler obtained his remarkable formulae

$$1^m - 2^m + 3^m - \dots = \begin{cases} \frac{1}{2} & \text{if } m = 0 \\ 0 & \text{if } m = 2, 4, 6, \dots \\ \frac{2^{2k}-1}{2k} B_{2k} & \text{if } m = 2k - 1 \end{cases}.$$

In particular

$$\eta(1 - 2k) = \frac{2^{2k} - 1}{2k} B_{2k}.$$

On the other hand we have the formulae discovered earlier by him:

$$\eta(2k) = 1 - \frac{1}{2^{2k}} + \frac{1}{3^{2k}} - \dots = (-1)^{k-1} \frac{2^{2k-1} - 1}{(2k)!} \pi^{2k} B_{2k} \qquad (k = 1, 2, \dots).$$

It follows from this that

$$\frac{\eta(1 - 2k)}{\eta(2k)} = -\frac{2^{2k} - 1}{2^{2k-1} - 1} \frac{(-1)^k}{\pi^{2k}} (2k - 1)!$$

which is just (**FE**) for $s = 2, 4, 6, 8, \dots$. Since

$$\eta(-2k) = 1^{2k} - 2^{2k} + 3^{2k} - \dots = 0 \qquad (k = 1, 2, 3, \dots)$$

by our earlier calculation, the equation (**FE**) is true for $s = 3, 5, 7, \dots$, both sides being 0. This is in fact the reason why Euler substituted $\cos k\pi$ for $(-1)^k$ in the previous calculation, a substitution that clearly suggests that he was thinking of the functional equation for nonintegral values of s. Euler then proceeded to write that "... *I shall hazard the conjecture that the relation* (**FE**) *is true for all* $s \dots$."

He then proceeded to verify this conjecture, as he called it, for other values of s. For $s = 1$ we have

$$\eta(1) = \log 2, \qquad \eta(0) = \frac{1}{2}.$$

In fact,

$$\eta(0) = 1 - 1 + 1 - 1 + 1 - 1 + \ldots,$$

the Leibniz series for which long ago Euler had assigned the sum $1/2$. The equation (**FE**) is valid provided we interpret the right side as its limit when $s \to 1$, for, by the l'Hôpital rule,

$$\lim_{s \to 1} \frac{\cos s\pi/2}{2^{s-1} - 1} = -\frac{\pi}{2 \log 2}.$$

For $s = 0$ the calculations are the same since the roles of s and $1 - s$ are reversed. In fact, as Euler notes explicitly, the transformation $s \mapsto 1 - s$ leaves the functional equation invariant, a fact that is easily verified and depends on the relation

$$\Gamma(s)\Gamma(1-s) = \frac{\pi}{\sin s\pi}$$

which Euler had proved earlier. Thus the functional equation is verified whenever s is any integer, positive, negative, or zero.

Euler now takes up the verification for fractional values of s. For $s = 1/2$ we have $\eta(s) = \eta(1-s) = \eta(1/2)$ and the verification of (**FE**) depends on the relation

$$\Gamma\left(\frac{1}{2}\right) = \sqrt{\pi}$$

which Euler knew. It is however clear that for nonintegral values of s other than $s = 1/2$ the method of Euler, which is just verification, will not succeed because for these values no exact determination of the sums was available to him. Euler treated one particular value, $s = 3/2$, although it is clear that he had verified by this technique the functional equation for a large number of values of s, in particular, for $s = (2i + 1)/2$ $(i = 1, 2, 3, 4, 5, 6, \ldots)$.

When $s = 3/2$, we have

$$\eta\left(-\frac{1}{2}\right) = 1 - \sqrt{2} + \sqrt{3} - \ldots$$

$$\eta\left(\frac{3}{2}\right) = 1 - \frac{1}{2\sqrt{2}} + \frac{1}{3\sqrt{3}} - \ldots.$$

The first series is divergent, while the second one is an alternating convergent series. For both of them Euler used the Euler-Maclaurin sum formula to compute the sum. This of course is a departure from the assumption that divergent series are supposed to be summed by the Abel or the generating function method, but Euler could not have been disturbed by this since he believed that all methods of summation lead to the same value. His technique was to sum the first nine terms by hand accurately and then for the rest use the sum formula applied to the function $x^{1/2}$ in the first case and $x^{-3/2}$ in the second. The reason for this is that the sum formula is a linear combination of the successive derivatives of the function at the values starting from 10 and so the denominators will involve higher and higher powers of 10. Here are his results:

$$1 - \sqrt{2} + \sqrt{3} - \sqrt{4} + \cdots + \sqrt{9} = 1.9217396632\ldots,$$

while for the remainder

$$\sqrt{10} - \sqrt{11} + \ldots$$

he obtains

$$1.541610\dots.$$

Subtracting this from the previous number, he gets

$$\eta\left(-\frac{1}{2}\right) = 1 - \sqrt{2} + \sqrt{3} - \dots = 0.380129\dots.$$

The same method yields

$$\eta\left(\frac{3}{2}\right) = 0.765158\dots$$

so that

$$\frac{\eta\left(\frac{3}{2}\right)}{\eta\left(\frac{1}{2}\right)} = 0.4967738\dots.$$

The right side of the functional equation leads to

$$\frac{2\sqrt{2}-1}{2(2-\sqrt{2})\pi} = \frac{3+\sqrt{2}}{2\pi\sqrt{2}} = 0.4967736\dots,$$

and Euler remarks that the result is verified to six decimal places and proceeds to say that as the result is verified to a very high degree of accuracy there cannot be any doubt about the result in the case where the variable s is fractional. One must admire Euler's sureness of judgment and incredible powers of divination that he was led to this over one hundred years before Riemann.

We have already seen that Euler attempted throughout his life to determine the values of the series

$$1 - \frac{1}{2^{2\lambda+1}} + \frac{1}{3^{2\lambda+1}} - \dots \qquad (\lambda = 1, 2, \dots)$$

without success. He made one last attempt to see if the functional equation would lead to any clue. He wrote the functional equation as

$$1 - \frac{1}{2^{2\lambda+1}} + \frac{1}{3^{2\lambda+1}} - \dots = -\frac{2^{2\lambda}-1}{(2\lambda)!\,(2^{2\lambda+1}-1)} \frac{1 - 2^{2\lambda} + 3^{2\lambda} - \dots}{\cos\frac{2\lambda+1}{2}\pi}.$$

The fraction

$$\frac{1 - 2^{2\lambda} + 3^{2\lambda} - \dots}{\cos\frac{2\lambda+1}{2}\pi}$$

is an indeterminate $0/0$, and Euler evaluated it by l'Hôpital's rule to get the value

$$\frac{2(1^{2\lambda}\log 1 - 2^{2\lambda}\log 2 + 3^{2\lambda}\log 3 - \dots)}{-\pi\cos\lambda\pi}$$

and obtained the formula

$$1 - \frac{1}{2^{2\lambda+1}} + \frac{1}{3^{2\lambda+1}} - \dots = (-1)^{\lambda}\frac{2(2^{2\lambda}-1)\pi^{2\lambda}}{(2\lambda)!\,(2^{2\lambda+1}-1)}(1^{2\lambda}\log 1 - 2^{2\lambda}\log 2 + \dots).$$

He must have been disappointed, because the sum on the right appeared to be more complicated than the sum on the left. This of course is one of the many formulae involving the series on the left that are scattered throughout his work.

Euler concludes his beautiful paper with a discussion of the series

$$L(s) = 1 - \frac{1}{3^s} + \frac{1}{5^s} - \dots$$

and mentions that the same methods as the ones for $\eta(s)$ have led him to the conjecture

$$\frac{L(1-s)}{L(s)} = \frac{2^s \Gamma(s)}{\pi^s} \sin \frac{s\pi}{2}.$$

He mentions also that further results of this type should be sought after. He was thus a true pioneer in the theory of arithmetically defined Dirichlet series and their functional equations.

As a final historical note let me observe that Landau [12] discussed Euler's paper in detail. Recall that the Euler-Maclaurin sum formula that Euler used for the series

$$1 - \sqrt{2} + \sqrt{3} - \ldots$$

is divergent and so from a rigorous point of view it has to be replaced by Abel's method. However, it is not obvious at all that the series for $\eta(s)$ for fractional values of s is Abel summable. In his exposition [12] Landau comments that Euler's calculations may be transformed to yield a proof that the Abel sum of the series for $\eta(-1/2)$, i.e.,

$$\lim_{x \to 1-0} \left(x - \sqrt{2}x^2 + \sqrt{3}x^3 - \ldots \right),$$

exists and computes it, thus making the Euler procedure consistent. He gives a proof of the existence of the Abel sum for all values of s. Let us consider the function $\zeta(s, w)$,

$$\zeta(s, w) = \sum_{n=0}^{\infty} \frac{1}{(w+n)^s} \qquad (w > 0),$$

first considered by Hurwitz in Zeit. für Math. und Phys., **27** (1882), 86. For $w = 1$ we get $\zeta(s)$. Then it can be shown (see for example the book *A Course of Modern Analysis*, by E. T. Whittaker and G. N. Watson, 1965, p. 265) that $\zeta(s, w)$ has an analytic continuation to the entire s-plane except for a simple pole at $s = 1$ with residue 1. Landau's proof starts with the result attributed by Landau to Mellin (*Über eine Verallgemeinerung der Riemannschen Function $\zeta(s)$*, Acta societatis scientiarum Fennicae, **24** (1898), 40) which asserts that when s is not equal to a positive integer,

$$\lim_{x \to 1-0} \left(\sum_{n=0}^{\infty} \frac{x^{w+n}}{(w+n)^s} - \Gamma(1-s) \left(\log \frac{1}{s} \right)^{s-1} \right) = \zeta(s, w).$$

Hence for $w = 1$ we get

$$\lim_{x \to 1-0} \left(\sum_{n=1}^{\infty} \frac{x^n}{n^s} - \Gamma(1-s) \left(\log \frac{1}{s} \right)^{s-1} \right) = \zeta(s, w).$$

Let

$$P_s(x) = \sum_{n=1}^{\infty} \frac{x^{n-1}}{n^s}, \qquad Q_s(x) = \sum_{n=1}^{\infty} (-1)^{n-1} \frac{x^{n-1}}{n^s}.$$

Then, parallel to the relation between ζ and η we have

$$Q_s(x) = P_s(x) - 2^{1-s} x P_s(x^2).$$

So we get from the limit formula above the formulae

$$\lim_{x \to 1-0} \left(x P_s(x) - \Gamma(1-s) \left(\log \frac{1}{s} \right)^{s-1} \right) = \zeta(s)$$

$$\lim_{x \to 1-0} \left(x^2 P_s(x) - \Gamma(1-s) \left(2 \log \frac{1}{s} \right)^{s-1} \right) = \zeta(s)$$

so that

$$\lim_{x \to 1-0} \left(x P_s(x) - 2^{1-s} x^2 P_s(x^2) \right) = (1 - 2^{1-s}) \zeta(s) = \eta(s).$$

Omitting a factor x which goes to 1 this is just

$$\lim_{x \to 1-0} Q_s(x) = \eta(s),$$

which is the desired result. If one takes the values $w = 1/4$, $w = 3/4$, the two limit formulae of Mellin yield, by subtraction, the result

$$\lim_{x \to 1-0} \left(\sum_{n=0}^{\infty} \frac{x^{\frac{1}{4}+n}}{(1+4n)^s} - \sum_{n=0}^{\infty} \frac{x^{\frac{3}{4}+n}}{(3+4n)^s} \right) = \frac{1}{4^s} \Big(\zeta(s, 1/4) - \zeta(s, 3/4) \Big).$$

Replacing x by x^2, one then obtains

$$\lim_{x \to 1-0} \left(\sum_{n=1}^{\infty} (-1)^{n+1} \frac{x^{n-1}}{(2n-1)^s} \right) = \frac{1}{4^s} \Big(\zeta(s, 1/4) - \zeta(s, 3/4) \Big).$$

This shows that

$$L(s) = \frac{1}{4^s} \Big(\zeta(s, 1/4) - \zeta(s, 3/4) \Big)$$

has an analytic continuation over the entire s-plane and further that for s not equal to a positive integer the series

$$1 - \frac{1}{3^s} + \frac{1}{5^s} - \ldots$$

is Abel summable for all s different from a positive integer to the sum

$$\frac{1}{4^s} \Big(\zeta(s, 1/4) - \zeta(s, 3/4) \Big).$$

This rounds out the picture regarding the Euler paper.

Nowadays analytic continuation is the method used to define $\zeta(s)$ for all s and to establish the functional equation. This is what Riemann did. I shall discuss in the next chapter the significance of Riemann's method from the point of view of the theory of modular functions and modular forms. Here I shall merely point out that analytic continuation is often viewed as a summation procedure. In particular, this point of view has become pervasive in quantum mechanics and quantum field theory. I shall touch upon this aspect briefly later on in this chapter.

5.3.   Euler's summation of the factorial series

The divergent series that intrigued Euler the most, the one that he called the *divergent series par excellence* in his 1760 paper, is the *factorial series* which he wrote always as

$$1 - 1 + 2 - 6 + 24 - 120 + 720 - \ldots .$$

In more concise notation this is the series

$$1 - 1! + 2! - 3! + 4! - 5! + 6! - \cdots + (-1)^n n! + \ldots .$$

The method of generating function cannot be applied here because the power series

$$\sum_{n \geq 0} (-1)^n n! x^n = 1 - 1! x + 2! x^2 - 3! x^3 + 4! x^4 - \cdots + (-1)^n n! x^n + \ldots$$

converges nowhere in the complex plane except at the trivial point $x = 0$. Nevertheless Euler applied an ingenious method that even allowed him to get a very accurate numerical value. He had already discussed this method in his letters to Nicolaus Bernoulli I and Goldbach that I mentioned earlier. He then gave a detailed discussion of it in his 1760 paper on divergent series. In my discussion below I follow Euler, as well as Barbeau and Leah [4] (pp. 149-153).

In the first few sections of the paper Euler employs his method of successive differences to numerically sum the series. We shall discuss this as a general method of summation in the next section but point out here that all the series that he gets from the factorial series are divergent but have the property that the error committed is less than the first omitted term. So he obtains a value ≈ 0.6 by this method which has other questionable aspects. In any case he abandons this line of work quickly and goes on to an analytical treatment which is of the greatest interest.

Euler's idea is very simple. If

$$f(x) = 1 - 1! x + 2! x^2 - 3! x^3 - \ldots, \qquad g(x) = x f(x) = x - 1! x^2 + 2! x^3 - \ldots,$$

then formal differentiation gives the differential equation

$$x^2 g' + g = x.$$

Euler then proceeds to solve this equation. The equation can be written in the standard form

$$g' + Pg = Q, \qquad P = \frac{1}{x^2}, \qquad Q = \frac{1}{x}$$

for which the solution is

$$g = e^{-\int P\, dx} \int Q e^{\int P\, dx}\, dx + c e^{-\int P\, dx}$$

where c is a constant. So, we get, formally,

$$g \sim e^{1/x} \int \frac{1}{x} e^{-1/x}\, dx + c e^{1/x}$$

where we have replaced equality by $\sim$ to keep in mind the formal relation between the series and the integral. The function $e^{-1/x}$ is rapidly decreasing as $x \to 0+$ (it is the Cauchy function, all of whose derivatives vanish at the origin) and so by integration by parts we have, for $x > 0$,

$$\int_0^x \frac{1}{t} e^{-1/t}\, dt = x e^{-1/x} - \int_0^x e^{-1/t}\, dt \leq x e^{-1/x}$$

from which we get

$$e^{1/x} \int_0^x \frac{1}{t} e^{-1/t}\, dt = O(x) \qquad (x \to 0+).$$

So, if we try to match the fact that the formal power series for g starts with the x term by saying that the analytical solution vanishes as $x \to 0+$, we should choose $c = 0$. This we shall do and write

$$g(x) \sim e^{1/x} \int_0^x \frac{1}{t} e^{-1/t}\, dt$$

so that

$$f(x) \sim \frac{1}{x} e^{1/x} \int_0^x \frac{1}{t} e^{-1/t} \, dt.$$

Then evaluating the integral at $x = 1$ Euler finds for the sum of the series the value

$$\int_0^1 \frac{e^{1-(1/t)}}{t} \, dt.$$

Euler evaluates this by a Riemann sum (modified at the end points)

$$\int_0^1 h(x) \, dx \approx \frac{1}{10} [(1/2)h(0) + h(1/10) + h(2/10) + \cdots + (1/2)h(9/10)]$$

which gives

$$\int_0^1 \frac{e^{1-(1/t)}}{t} \, dt \approx 0.59637255\ldots.$$

Thus he gets his final result

$$1 - 1! + 2! - 3! + 4! - \cdots = \int_0^1 \frac{e^{1-(1/t)}}{t} \, dt \approx 0.59637255\ldots.$$

He now makes another transformation of the integral. Let

$$I = \int_0^\infty \frac{e^{-w}}{1+w} \, dw.$$

Euler makes the substitution $z = e^{1-(1/t)}$ to get

$$I = \int_0^1 \frac{du}{1 - \log u} = \int_0^1 \frac{dt}{1 - \log(1-t)} = \int_0^1 \frac{dt}{1 + t + \frac{t^2}{2} + \frac{t^3}{3} + \ldots}.$$

The integrand can also be evaluated numerically as a Riemann sum as the one before was.

Euler was still not done with the series! It is clear that he must have felt that the method of numerical integration was not adequate for ensuring accuracy to very high orders because the evaluation of the integrands was becoming a matter of extreme effort. He then proceeded to obtain a beautiful *continued fraction* for the formal series. His result was that

$$f(x) = \sum_{n \geq 0} (-1)^n n! x^n = \frac{1}{1+} \frac{x}{1+} \frac{x}{1+} \frac{2x}{1+} \frac{2x}{1+} \frac{3x}{1+} \frac{3x}{1+} \text{ etc.}$$

This is not a standard type of continued fraction, but Euler must have been aware of this from his own work on infinite continued fractions as well as extensive numerical calculations [13]. In any case he takes the convergence for granted and finds the result by evaluating this continued fraction at $x = 1$. Euler employs several ingenious shortcuts in his evaluation of the value of the continued fraction (for a more detailed discussion of some of the aspects of this numerical evaluation of the continued fraction, see [4]). Thus he is finally able to report the value

$$I = 0.596347362123\ldots (I = 0.5963473623194 \text{ by MATHEMATICA })$$

which he mentions is accurate to the last digit! The value by MATHEMATICA is the same for both the continued fraction and numerical integration methods. His oft-repeated assertion that different methods of summing divergent series must always lead to the same value must have been based on such remarkable numerical calculations as well as a feeling for the general structure of analytical transformations.

In his letter to Nicolaus Bernoulli I he mentions that similar methods can be applied to many other series related to corresponding continued fractions. Here he discusses some of these in greater detail. He considers for instance the class of series defined by

$$g = g_{mpq} = x^m - px^{m+q} + p(p+q)x^{m+2q} - p(p+q)(p+2q)x^{m+3q} - \dots.$$

For $m = p = q = 1$ one obtains the factorial series $f = g_{111}$. He finds for g the formal differential equation

$$x^{q+1}g' + [(p-m)x^q + 1]g = x^m.$$

He then integrates this to find the solution

$$g \sim e^{\frac{1}{qx^q}} x^{m-p} \int_0^x e^{-1/qt^q} t^{p-q-1} \, dt + cx^{m-p} e^{1/qx^q}$$

where c is a constant. As before we find by an integration by parts that

$$x^{m-p} \int_0^x e^{-1/qt^q} t^{p-q-1} \, dt = O(x^m) \qquad (x \to 1+0)$$

so that we take $c = 0$ to find

$$g \sim e^{\frac{1}{qx^q}} x^{m-p} \int_0^x e^{-1/qt^q} t^{p-q-1} \, dt.$$

Euler finds for g the continued fraction

$$g_{mn} = \cfrac{x^m}{1+} \cfrac{px^q}{1+} \cfrac{qx^q}{1+} \cfrac{(p+q)x^q}{1+} \cfrac{2qx^q}{1+} \cfrac{(p+2q)x^q}{1+} \cfrac{3qx^q}{1+} \cfrac{(p+3qx^q)}{1+} \quad \text{etc.}$$

If we take $p = m = 1$ and $q = 2$ we find that

$$g = x - 1x^3 + 1.3x^5 - 1.3.5x^7 + 1.3.5.7x^9 - \dots.$$

Setting $x = 1$ gives the divergent series

$$1 - 1 + 1.3 - 1.3.5 + 1.3.5.7 - \dots.$$

For g the integral is given by

$$g \sim e^{\frac{1}{x^2}} \int_0^x e^{-1/2t^2} \frac{1}{t^2} \, dt.$$

Thus, for the sum of the divergent series he finds

$$1 - 1 + 1.3 - 1.3.5 + 1.3.5.7 - \dots = \int_0^1 e^{[1-(1/t^2)]/2} \frac{1}{t^2} \, dt$$

as well as the convergent continued fraction

$$\cfrac{1}{1+} \cfrac{1}{1+} \cfrac{2}{1+} \cfrac{3}{1+} \cfrac{4}{1+} \cfrac{5}{1+} \quad \text{etc.}$$

Using the same methods as before he finds for this series the value

$$\approx 0.65568 \qquad (\approx 0.655679542418798 \text{ by MATHEMATICA}).$$

I shall now follow Hardy [2] (pp. 26-29). Then

$$f(x) \sim \int_0^\infty \frac{e^{-w}}{1+xw} \, dw \qquad (x > 0)$$

so that

$$f(x) \sim \int_0^\infty \frac{e^{-w}}{1+xw} dw \qquad (x > 0).$$

If we expand the term $1/(1+xw)$ as a geometric series in xw and formally integrate we get the factorial series. We could have gotten to this point directly by writing

$$f(x) \sim \sum_{n \geq 0} (-1)^n x^n \int_0^\infty e^{-w} w^n \, dw = \int_0^\infty e^{-w} \left(\sum_{n \geq 0} (-1)^n x^n w^n \right) dw.$$

Let us now examine the relation between the integral and the series a little more closely. We change x to z to emphasize that we consider it as a complex variable. Now $|1 + zw| = \sqrt{1 + 2rw\cos\theta + r^2 w^2}$ where $z = re^{i\theta}$; if $a \geq 0$ and c is a real constant with $|c| \leq 1$, then

$$1 + 2ac + a^2 \geq \begin{cases} 1 & \text{if } c \geq 0 \\ 1 - c^2 & \text{if } c \leq 0 \end{cases}.$$

Hence

$$|1 + zw| \geq \begin{cases} 1 & \text{if } |\theta| \leq \pi/2 \\ |\sin\theta| & \text{if } \pi/2 \leq |\theta| < \pi \end{cases}.$$

So the integral is convergent for all z in the sector $|\theta| < \pi$ which is the complex plane from which the negative real axis (and 0) has been removed. If we truncate the geometric series at any given stage we get the result that the formal series is the *asymptotic expansion* of the integral. More precisely, since

$$\frac{1}{1+xw} = 1 - xw + x^2 w^2 - \cdots + (-1)n x^n w^n + (-1)^{n+1} x^{n+1} \frac{w^{n+1}}{1+xw}$$

we get

$$\int_0^\infty \frac{e^{-w}}{1+zw} \, dw = 1 - 1!z + 2!z^2 - \cdots + (-1)^n n! z^n + R_n(z)$$

where

$$R_n(z) = (-1)^{n+1} z^{n+1} \int_0^\infty \frac{e^{-w} w^{n+1}}{1+zw} \, dw.$$

By our estimate for $|1 + zw|$ above we get

$$|R_n(z)| \leq \begin{cases} (n+1)! r^{n+1} & \text{if } |\theta| \leq \pi/2 \\ (n+1)! r^{n+1} |\sin\theta|^{-1} & \text{if } \pi/2 \leq |\theta| < \pi. \end{cases}$$

Hence, *in the sector* $|\theta| \leq \pi - \delta$ *with* $0 < \delta < \pi/2$ *and for all* $n \geq 0$ we have

$$|R_n(z)| = K_\delta (n+1)! \left(|z|^{n+1} \right).$$

Thus we can write

$$\int_0^\infty \frac{e^{-w}}{1+zw} \, dw \sim 1 - 1!z + 2!z^2 - 3!z^3 + \ldots \qquad (|\theta| < \pi)$$

where the notation $\sim$ now has a precise meaning: the difference between the function on the left side and the sum of n terms of the right side is $O\left((n+1)! |z|^{n+1} \right)$ on any sector $\Gamma_\delta := |\theta| < \pi - \delta$, where the O is uniform with respect to n and $z \in \Gamma_\delta$.

If V is an open sector with vertex at the origin of the complex plane and V_ρ is the part of the sector where $|z| < \rho$, and if g is a function defined and analytic on V_ρ, let us write

$$g \sim a_0 + a_1 z^2 + a_2 z^2 + \dots (V)$$

if for any n,

$$g(z) - (a_0 + a_1 z + \dots + a_n z^n) = O\left(|z|^{n+1}\right)$$

uniformly on any open subsector $V'_{\rho'} << V_\rho$ ($<<$ means that the closure of V' is still contained in V). We refer to the formal series

$$a_0 + a_1 z^2 + a_2 z^2 + \dots$$

as the *asymptotic expansion of g at* 0. This notion of *asymptotic series* was first introduced by Poincaré [14]. However in our present case we have a much sharper situation where the estimate for the difference between g and the sum of the first n terms of its asymptotic expansion has a majorant that is *uniform in n*. Such sharper asymptotics arise naturally in the summation of series like the factorial series. Eventually, as we shall see later on in this chapter, the Euler theory was subsumed under the theory of summation developed by Borel, and so we shall refer to the sharper asymptotics encountered above as *strong asymptotics*. We shall write

$$g \sim a_0 + a_1 z + a_2 z^2 + \dots \qquad (strong)$$

if

$$a_n = O(n!\sigma^n)$$

and

$$|g(z) - (a_0 + a_1 z + a_2 z^2 + \dots + a_n z^n)| = O\left((n+1)!\sigma^{n+1}|z|^{n+1}\right)$$

where the O is uniform for all n and all z in any fixed sector $V'_{\rho'} << V_\rho$, ρ' being a positive number $< \rho$.

The reader must be warned that the formal series does not determine the analytic function on the sector that is asymptotic to it. This is because for any sector there are *flat functions* defined on the sector, where flat means that the function is $O(|z|^n)$ as $z \to 0$ in the sector, the O being uniform on proper subsectors. The classic example is $e^{-1/z}$ which is flat on the sector $|\theta| < \pi/2$. If $z^{1/2}$ denotes the square root that is analytic on $V = \{|\theta| < \pi\}$ and is 1 at $z = 1$, then $e^{-1/z^{1/2}}$ is flat on V. So Poincaré asymptotics *does not possess uniqueness*. However strong asymptotics *does possess uniqueness if the angle of the sectors is strictly greater than $\pi/2$*; this is a famous theorem of Watson [15]. If we assume this, we can say that the analytic functions discovered by Euler, which in our terminology are strongly asymptotic to the factorial series and their generalizations on the sector $|\theta| < \pi\}$, are *uniquely determined by this requirement*. Hence the sums associated by Euler to these series as the values of these analytic functions at $z = 1$ are *canonical*. We shall return to this point when we discuss Borel summation in §5 below. One final point: The flat functions lead to *sheaves* on the unit circle with nontrivial cohomology, and so *efforts to get around the nonuniqueness of Poincaré asymptotics in general have cohomological obstructions*.

Let us return to the factorial series and write $F(x)$ for the integral asymptotic to the factorial series so that

$$F(x) = \frac{1}{x} e^{1/x} \int_0^{1/x} \frac{e^{-1/t}}{t} \, dt.$$

But

$$\int_y^\infty \frac{e^{-u}}{u}\, du = \int_1^\infty \frac{e^{-u}}{u}\, du - \int_0^1 \frac{1-e^{-u}}{u}\, du - \int_1^u \frac{du}{u} + \int_0^y \frac{1-e^{-u}}{u}\, du.$$

Using the formula

$$\int_1^\infty \frac{e^{-u}}{u}\, du - \int_0^1 \frac{1-e^{-u}}{u}\, du = \gamma$$

where γ is Euler's constant, we get

$$\int_y^\infty \frac{e^{-u}}{u}\, du = -\gamma - \log y + y - \frac{y^2}{2.2!} + \frac{y^3}{3.3!} - \cdots + (-1)^{n-1}\frac{y^n}{n.n!} + \cdots.$$

Hence

$$F(x) = \frac{1}{x}e^{1/x}\int_{1/x}^\infty \frac{e^{-u}}{u}\, du = -\frac{1}{x}e^{1/x}\log\frac{1}{x} + S\left(\frac{1}{x}\right)$$

where

$$S(y) = -ye^y\left(\gamma - y + \frac{y^2}{2.2!} - \frac{y^3}{3.3!} + \cdots\right).$$

The function S is an entire function on the whole complex plane $\mathbf{C}$. The formula for f is valid in the sector obtained by removing from $\mathbf{C}$ the half-line of real numbers ≤ 0. This function of course has an analytic continuation over the entire plane but is multi-valued. But the asymptotic relation

$$F(z) \sim f(z) = 1 - 1!z + 2!z^2 - 3!z^3 + \cdots$$

is valid only in the sector described above. In particular, for $z = 1$, we have

$$1 - 1! + 2! - 3! + 4! - \cdots = -e\left(\gamma - 1 + \frac{1}{2.2!} - \frac{1}{3.3!} + \frac{1}{4.4!} - \cdots\right).$$

This series expression for $F(1)$ can also be used to determine $F(1)$, which was Euler's answer to the problem of summing the factorial series. Euler's constant γ can be evaluated by numerical quadrature from its integral representation, and the series which is rapidly convergent can be evaluated by taking a few terms. We have

$$1 - 1! + 2! - 3! + 4! - \cdots = 0.5963\ldots.$$

The series is alternating, its terms decrease rapidly, and so it is beautifully suited for numerical evaluations with error bounds.

Two things stand out from all of this discussion. First, that although the power series with which we start may converge nowhere except at $z = 0$, there is a sector on which it is asymptotic to an analytic function expressed by an explicit integral, and one can evaluate this analytic function at $z = 1$ to obtain his "sum" of the divergent series. This suggests a definite method of summation applicable not only to this series but to a whole class of them. Eventually this suggestion was justified when Borel developed his *Borel summation* and clarified the relation between divergent series and certain classes of analytic functions defined from them. We shall take this up in §5.

5.4.   The general theory of summation of divergent series

In spite of the many remarkable successes obtained by using divergent series, especially by Euler, their use declined and died altogether after him. In fact they were not resurrected for over a hundred years after Euler's death. Mathematicians, guided by Abel and Cauchy, realized that working carelessly with divergent series was likely to lead to incorrect results. Abel, in fact, made the remark in a letter [16] to his former teacher Holmboë that

> ... *Divergent series are in general the work of the devil and it is shameful to base any demonstration whatever on them.... The most essential part of mathematics is without basis. For the most part the results are valid, it is true, but it is a curious thing. I am looking for the reason, and it is a very interesting problem....*

Gradually, the foundations of analysis laid down by Abel and Cauchy became more secure, and so it was only a question of time before a systematic investigation of divergent series began in earnest. Still, even in 1948, in his preface to Hardy's book[2] (Hardy had died before its appearance in print), Littlewood, Hardy's lifelong friend and collaborator, could say

> ... *The title holds curious echoes of the past, and of Hardy's past. Abel wrote in 1828: "Divergent series are the invention of the devil, and it is shameful to base any demonstration on them whatsoever." In the ensuing period of critical revision they were simply rejected. Then came a time when it was found that something after all could be done about them. This is now a matter of course, but in the early years of the century the subject, while in no way mystical or unrigorous, was regarded as sensational, and about the present title, now colourless, there hung an aroma of paradox and audacity.*

The modern theory of divergent series began in 1880 with a short but beautiful paper of Frobenius [17]. In this paper Frobenius studied the questions arising out of the series (1), (2), and their generalizations by Lagrange, in complete generality. The theorem of Frobenius asserts that if $s_n = \sum_{0 \le k \le n} a_k$, and if

$$\lim_{n \to \infty} \frac{s_0 + s_1 + s_2 + \cdots + s_n}{n+1}$$

exists, then $\sum_{n \ge 0} a_n x^n$ converges for $|x| < 1$,

$$\lim_{x \to 1-0} \sum a_n x^n$$

exists, and

$$\lim_{x \to 1-0} \sum a_n x^n = \lim_{n \to \infty} \frac{s_0 + s_1 + s_2 + \cdots + s_n}{n+1}.$$

This result makes precise the sense in which the value $1/2$ for Leibniz's series and the value m/n for the series (2), as well as the values for Lagrange's generalizations of (2), are to be interpreted. Clearly, the theorem of Frobenius is a beautiful generalization of Abel's. Within a few years of the paper of Frobenius, Hölder and Cesàro published papers [18] in which they formulated the notions of what we now call *Hölder* and *Cesàro summations*. These papers and those of Borel heralded the rapid growth of the theory of divergent series. The monograph [19a] of Borel which appeared in 1901 began an intensive development of the theory of summability of divergent series in which Hardy, Littlewood and many others played a major role. This period ended with the path-breaking work of Wiener on Tauberian theorems in which Wiener showed how the whole subject is a chapter in the theory of the Fourier integral. The book [2] of Hardy gives a systematic treatment of the whole subject. I have made extensive use of the books of Borel, Hardy and Knopp [7] in my account below, and the reader should look into these for further clarification of most topics on divergent series that I discuss.

Abel, Cesàro, and Hölder methods. We begin with *Abel summation,* which was the preferred method of summing divergent series for Euler, although historically the first modern definitions involved Cesàro and Hölder summations. Let $\sum_{n\geq 0} a_n$ be a series which has a radius of convergence at least 1. Then the series is summable (A) to the sum s,

$$\sum_{n=0}^{\infty} a_n = s \qquad (A)$$

in symbols, if

$$\lim_{x\to 1-0} \sum_{n=0}^{\infty} a_n x^n = s.$$

The justification for naming this after Abel is Abel's theorem. A sequence $(s_n)_{n\geq 0}$ is said to have the limit s in the $(C,1)$ or $(H,1)$ sense,

$$\lim_{n\to\infty} s_n = s \ (C,1) \quad \text{or} \ (H,1)$$

in symbols, if

$$\lim_{n\to\infty} \frac{s_0 + s_1 + s_2 + \cdots + s_n}{n+1} = s$$

in the usual sense. A series $\sum_{n\geq 0} a_n$ is said to have the sum s in the $(C,1)$ or $(H,1)$ sense,

$$\sum_{n=0}^{\infty} a_n = s \qquad (C,1) \text{ or } (H,1),$$

if its partial sums $s_n = a_0 + a_1 + \cdots + a_n$ have the limit s in the $(C,1)$ or $(H,1)$ sense. The series is then said to be *Cesàro or Hölder summable* to the sum s. Thus we have

$$1 - 1 + 1 - 1 + \cdots = \frac{1}{2} \qquad (C,1)$$

and, for integers $1 < m < n$,

$$1 + 0 + 0 + \cdots - 1 + 0 + 0 + \cdots + 1 + 0 + \cdots = \frac{m}{n} \qquad (C,1)$$

where the series on the left is Lagrange's series (2). For a series $\sum_{n\geq 0} a_n$,

$$\frac{s_0 + s_1 + \cdots + s_{n-1}}{n} = \sum_{r=0}^{n-1} \left(1 - \frac{r}{n}\right) a_r$$

so that

$$\sum_n a_n = s \ (C,1) \iff \lim_n \sum_{r=0}^{n-1} \left(1 - \frac{r}{n}\right) a_r = s.$$

We shall come back to this formula when we deal with $(C,1)$ summability of integrals.

The $(C, 1)$ or $(H, 1)$ summation procedure may be extended in two ways that are equivalent but differ slightly from each other. The method of *Hölder means* is to form, given a sequence $(s_n)_{n \geq 0}$ or a series $\sum_{n \geq 0} a_n$ with partial sums $s_n = a_0 + a_1 + \cdots + a_n$, the *Hölder means* $h_n^{(k)}$ defined inductively by

$$h_n^{(k)} = \frac{h_0^{(k-1)} + h_1^{(k-1)} + \cdots + h_n^{(k-1)}}{n+1}, \qquad h_n^{(0)} = s_n.$$

The sequence (s_n) or the series $\sum_n a_n$ is summable (H, k) if

$$\lim_{n \to \infty} h_n^{(k)} = s$$

in the usual sense. We write

$$\lim s_n = s \quad (H, k) \qquad \text{or} \qquad \sum_n a_n = s \quad (H, k).$$

We have, writing C for usual convergence,

$$C \implies (H, 1) \implies (H, 2) \implies \ldots \implies (H, k) \implies \ldots$$

where the implication carries with it the assertion that the limit or the sum remains the same. The implications follow from the following easily proved lemma.

LEMMA. *If (s_n) is a sequence with limit s and*

$$\sigma_n = \frac{s_0 + s_1 + \cdots + s_n}{n+1},$$

then

$$\lim_{n \to \infty} s_n = \lim_{n \to \infty} \sigma_n.$$

The following lemma gives the limitation of the summation procedure by Hölder means.

THEOREM. *If $\sum_n a_n$ is (H, k)-summable, then $a_n = o(n^k)$, i.e.,*

$$\lim \frac{a_n}{n^k} = 0.$$

PROOF. We have $h_n^{(k)} = s + o(1)$ and so

$$h_n^{(k-1)} = (n+1)h_n^{(k)} - n h_{n-1}^{(k)} = s + o(n) = o(n).$$

Changing k to $k - 1$ we conclude that $h_n^{(k-2)} = o(n^2)$ and so on, till we come to $h_n^{(0)} = s_n = o(n^k)$. But then, $a_n = s_n - s_{n-1} = o(n^k)$ also.

It follows from this that the series

$$1^r - 2^r + 3^r - \cdots + (-1)^n (n+1)^r + \ldots$$

is not summable (H, r).

The second way to extend $(C, 1)$ summation is to define, for any given sequence (s_n), the sums

$$s_n^{(k)} = s_0^{(k-1)} + s_1^{(k-1)} + \cdots + s_n^{(k-1)}, \qquad s_n^{(0)} = s_n.$$

We can rewrite these formulae in terms of generating functions as

$$\sum_n s_n^{(k)} z^n = (1 - z)^{-1} \sum_n s_n^{(k-1)} z^n$$

so that

$$\sum_n s_n^{(k)} z^n = (1-z)^{-k} \sum_n s_n z^n.$$

Hence

$$s_n^{(k)} = \binom{n+k-1}{k-1} s_0 + \binom{n+k-2}{k-2} s_1 + \cdots + s_n.$$

Taking $s_n = 1$ for all n and writing $\gamma_n^{(k)}$ for the corresponding $s_n^{(k)}$, we see that $\gamma_n^{(k)}$ is the coefficient of z^n in $(1-z)^{-(k+1)}$, i.e.,

$$\gamma_n^{(k)} = \binom{n+k}{k}.$$

For $k = 1$ we have $\gamma_n^{(1)} = n+1$ and

$$c_n^{(1)} = \frac{s_0 + s_1 + \cdots + s_n}{n+1}.$$

The Cesàro means of order k of a sequence (s_n) are now defined by

$$c_n^{(k)} = \frac{s_n^{(k)}}{\binom{n+k}{k}} = \frac{1}{\binom{n+k}{k}} \sum_{r=0}^n \binom{n-r+k-1}{k-1} s_r.$$

Since

$$\binom{n+k}{k} = \frac{(n+1)\ldots(n+k)}{k!} \sim \frac{n^k}{k!} \qquad (k \to \infty)$$

we have

$$s_n \to s \quad (C,k) \iff \frac{s_n^{(k)}}{n^k} \to \frac{s}{k!}.$$

A series $\sum_n a_n$ or a sequence (s_n) is summable (C,k) to s if the limit

$$\lim c_n^{(k)} = s$$

exists in the usual sense, and we write

$$\sum_n a_n = s \qquad (C,k) \qquad \text{or} \qquad \lim s_n = s \qquad (C,k).$$

It turns out that the Cesàro and Hölder summation procedures are entirely equivalent, i.e.,

$$(C,k) \iff (H,k) \qquad (k \geq 1).$$

This fact and the fact that the $c_n^{(k)}$ are given explicitly for all k in terms of the s_n make the (C,k) summation a little easier to work with than the (H,k) summation. There is however one additional point in favor of the (C,k) summation. The binomial coefficients

$$\binom{n+r}{r} = \frac{(n+r)!}{n!\,r!} = \frac{\Gamma(n+r+1)}{\Gamma(r+1)\Gamma(n+1)}$$

make sense for all *real* $r > -1$, and so we can define $c_n^{(k)}$ for all *real* $k > -1$. Hence the Cesàro means are defined for all $k > -1$ and give a family of summation procedures indexed by a *continuous parameter*.

It is true that

$$(C,k) \implies (A).$$

For $k = 1$ this is just the theorem of Frobenius quoted above. But there are series which are not summable (H, r), hence not summable (C, r) for *any* r, but Abel summable. The classic example is obtained by defining the a_n by

$$f(z) = e^{1/1+z} = \sum_{n \geq 0} a_n z^n.$$

The function f is defined on the whole complex plane except for $z = -1$ and is analytic on $\mathbf{C} \setminus \{-1\}$. So the power series converges in $|z| < 1$, and its limit as $z \to 1 - 0$ through real values is $e^{1/2}$. But the a_n are not $O(n^r)$ for any r. For, if $a_n = O(n^r)$, then for some constant $K > 0$, we have, *uniformly for all $|z| < 1$,*

$$|f(z)| \leq K \sum_{n \geq 0} n^r |z|^n \leq Kr! \sum_{n \geq 0} \binom{n + r}{r} |z|^n \leq Kr!(1 - |z|)^{-(r+1)}.$$

Taking $z = x$ to be real with $-1 < x < 0$, and writing $y = 1 + x$, we see from the above estimate that

$$f(x) = e^{1/y} = O(y^{-(r+1)}) \qquad (y \to 0+),$$

a contradiction. Thus this series $\sum_n a_n$ is not summable (C, k) for any k.

General methods. The summation procedures I have discussed are all based on averages or means, and it is possible to generalize them and unify them at the same time under what are called *Toeplitz means.* To introduce them we need some preliminaries which explain the point of view taken. We consider the space $\mathcal{S}$ of all sequences $(s_n)_{n \geq 0}$ and the subset $\mathcal{S}_c$ of convergent sequences; for any element $s \in \mathcal{S}_c$ we write $\ell(s)$ for its limit. Both $\mathcal{S}$ and $\mathcal{S}_c$ are linear spaces. We consider a summation procedure as a *linear operator*, say T, from a linear subspace of $\mathcal{S}$ into $\mathcal{S}$; i.e., it is a linear operation that converts sequences into sequences. All the summation processes that we have considered so far are of this type; this is obvious for the Cesàro and Hölder means, as these are defined by certain linear combinations of the s_n. In the case of the Abel summation, the operator converts the sequence (s_n) into the *function $f(z)$,* which is technically not a sequence; but by restricting z to a sequence of values tending to 1, we can think of it as an operator converting sequences to sequences. T is called a *summation operator* if it has the following property: if s is a convergent sequence, $T(s)$ is defined and convergent, and $\ell(s) = \ell(T(s))$. In this case we define $D(T)$ to be the set of all sequences s for which $T(s)$ is defined and is convergent. Then the elements of $D(T)$ are called *T-summable*; for any element $s \in D(T)$, if $A = \ell(T(s))$, we say that s is *T-summable* to the limit A and write

$$\lim s_n = A \qquad (T) \qquad (A = \ell(T(s))).$$

Finally, if $\sum_{n \geq 0} a_n$ is a series, we say it is *T-summable* to the sum A if its partial sum sequence (s_n) is in $D(T)$ and

$$\lim s_n = A \qquad (T).$$

There is a very general result due to Toeplitz which characterizes summation operators among those that are defined by matrices. Let

$$T = (c_{mn})$$

be an infinite matrix. We identify T with the corresponding linear operator it defines, namely

$$T : (s_n) \longmapsto (t_m), \qquad t_m = \sum_n c_{mn} s_n.$$

Since t_m is defined as a possibly infinite sum, the domain of T will consist only of those sequences for which t_m is finite for all m. For instance, if T is *triangular*, i.e., if for each m there is an integer k_m (often $k_m = m$) such that $c_{mn} = 0$ for $n > k_m$, then T is defined on all of $\mathcal{S}$.

THEOREM. *T, defined by the matrix (c_{mn}), is a summation operator if and only if the following conditions are all satisfied:*

    (i) *For each m, $C_m := \sum_n |c_{mn}| < \infty$; and $\sup_m C_m < \infty$.*
    (ii) $\lim_{m \to \infty} c_{mn} = 0$ *for each n.*
    (iii) $d_m := \sum_n c_{mn} \to 1$ *as $m \to \infty$.*

PROOF. See Hardy [2].

REMARK 1. The tricky part of the proof of the necessity is to show that the sequence (C_m) is bounded. If we are willing to use a little Banach space theory, this is easy. The space of all sequences which converge to 0, say $\mathcal{N}$, is a *Banach space* with the norm

$$||s|| = \sup_n |s_n| \qquad (s = (s_n)).$$

The operator T is then a linear operator from $\mathcal{N}$ to $\mathcal{N}$. It is also *closed;* this means that if $s^{(k)} = (s_{kn})$ is a sequence of elements of $\mathcal{N}$ such that $s^{(k)}$ converges to s *and* $T(s^{(k)})$ converges to t, then $T(s) = t$. Let $T(s^{(k)}) = (t_{mk})$ so that

$$t_{mk} = \sum_n c_{mn} s_{kn}.$$

Also let $s = (s_n)$, $t = (t_m)$. Since $s^{(k)} \to s$ in $\mathcal{N}$, $||s^{(k)}||$ is bounded and so $|s_{kn}| \le A$ for all k, n, A being a constant. As $|c_{mn} s_{kn}| \le A|c_{mn}|$ and $\sum_n |c_{mn}| < \infty$, the series $\sum_n c_{mn} s_{kn}$ converges uniformly in k and so

$$t_m = \lim_k t_{mk} = \lim_k \sum_n c_{mn} s_{kn} = \sum_n c_{mn} s_n$$

so that

$$t = T(s).$$

The *closed graph theorem* then implies that the operator T is *bounded*, i.e.,

$$\sup_{s \in \mathcal{N}, ||s|| \le 1} ||T(s)|| < \infty.$$

Let $\mathrm{sgn}(z)$ for complex z be $|z|/z$ for $z \ne 0$ and 0 otherwise. Fix integers $k, M \ge 1$ and take s to be the sequence

$$s_n = \begin{cases} \mathrm{sgn}(c_{kn}) & \text{if } n \le M \\ 0 & \text{if } n > M. \end{cases}$$

Then $s \in \mathcal{N}$ and $||s|| \le 1$, while for $t = T(s)$ we have

$$t_k = \sum_{n \le M} |c_{kn}|.$$

Hence

$$\sum_{n \leq M} |c_{kn}| \leq ||T||.$$

Letting $M \to \infty$ we see that

$$C_k \leq ||T||.$$

This proves the result. It can easily be shown that

$$||T|| = \sup_m C_m$$

so that this method is entirely natural.

REMARK 2. The summation procedures defined by matrices (t_{mn}) satisfying the conditions of the theorem are often called *regular summation methods*.

Euler's method of finite differences. There is also the method of finite differences often used by Euler to improve the speed of convergence of slowly convergent series and to sum divergent series. In his notation let us write a series as

$$a - b + c - d + e - f + \dots.$$

Then Euler forms the differences

$$\alpha = b - a, c - b, d - c, e - d, f - e, \dots,$$

then the second differences

$$\beta = c - 2b + a, d - 2c + b, e - 2d + c, \dots.$$

The transformed series is then

$$\frac{a}{2} - \frac{\alpha}{4} + \frac{\beta}{8} - \frac{\gamma}{16} + \dots.$$

If we take the sequence $(-1)^n a_n$ and write the sequence of successive differences as $\Delta a_n, \Delta^2 a_n, \dots$, then

$$\Delta^n a_0 = \binom{n}{0} a_0 - \binom{n}{1} a_1 + \dots + (-1)^n \binom{n}{n} a_n.$$

If we now write a_n instead of $(-1)^n a_n$, we get the following: the *Euler transform* of the series $\sum_{n \geq 0} a_n$ is

$$\sum_{n \geq 0} b_n, \qquad b_n = \frac{1}{2^{n+1}} \left[\binom{n}{0} a_0 + \binom{n}{1} a_1 + \dots + \binom{n}{n} a_n \right].$$

For the partial sums the corresponding transformation is

$$(s_n) \to (t_n), \qquad t_n = \frac{1}{2^n} \left[\binom{n}{0} s_0 + \binom{n}{1} s_1 + \dots + \binom{n}{n} s_n \right].$$

This is a special case of the general summation method defined by a matrix $T = (t_{mn})$ if we take

$$t_{mn} = \begin{cases} \frac{\binom{m}{n}}{2^m} & \text{if } 0 \leq n \leq m \\ 0 & \text{if } n > m. \end{cases}$$

It is easily verified that the conditions of regularity are satisfied and so we have a regular summation procedure. We call it E-summation and write $\sum_n a_n = s$ (E). If

$$f(z) = \sum_{n=0}^{\infty} a_n z^n, \qquad g(w) = \sum_{n=0}^{\infty} b_n w^n,$$

the change from f to g is given by

$$g(w) = f(\mu(w)), \qquad \mu(w) = \frac{\frac{w}{2}}{1 - \frac{w}{2}}.$$

For instance, the Euler transform of the series

$$\log 2 = 1 - \frac{1}{2} + \frac{1}{3} - \cdots$$

is the series

$$\log 2 = \frac{1}{2} + \frac{1}{2^2 \, 2} + \frac{1}{2^3 \, 3} + \cdots.$$

The sum remains the same because the method is regular. We recognize in this the transformation that Euler used in his first accurate computation of $\zeta(2)$. Typically the Euler transform of a convergent series will be a series with the same sum but converging much more rapidly. Thus it is immensely useful in numerical evaluations.

5.5. Borel summation

The summation method that is closest to Euler's work on the factorial series is that of Borel. In his monograph [19a] Borel was concerned with the following program. Let $\mathbf{C}[[z]]$ be the algebra of *formal* power series in z with complex coefficients. $\mathbf{C}[[z]]$ comes equipped with a derivation, namely formal differentiation d/dz, and so is an example of what is known as a *differential algebra*. Then Borel's idea was to construct *explicitly* a "large" subalgebra $\mathbf{A}$ of $\mathbf{C}[[z]]$ closed under d/dz (a differential subalgebra) together with a map

$$\beta : \mathbf{A} \longrightarrow \mathcal{O}(V)$$

where V is a sector with vertex at 0 in $\mathbf{C}$ and $\mathcal{O}(V)$ is the differential algebra of analytic functions on V satisfying the following conditions:

(i) β is a homomorphism of differential algebras; i.e., it is a homomorphism commuting with d/dz.

(ii) $\beta(f) \sim f$ for all $f \in \mathbf{A}$ where $\sim$ means that the function $\beta(f)$ has the power series f as its *asymptotic expansion* at 0.

It is immediate that (i) and (ii) imply at once that if $f_1, f_2, \ldots, f_m$ are elements of $\mathbf{A}$ satisfying formally a differential equation

$$P(f_1, \ldots, f_m, f_1', \ldots, f_m', \ldots, f_1^{(k)}, \ldots, f_m^{(k)}) = 0$$

where P is a polynomial, then the functions $F_i = \beta(f_i)$ satisfy the differential equation

$$P(F_1, \ldots, F_m, F_1', \ldots, F_m', \ldots, F_1^{(k)}, \ldots, f_m^{(k)}) = 0.$$

Borel's idea was to build the algebra $\mathbf{A}$ and the map β by a summation process. It is this that we shall call the *Borel summation*.

Borel was thus interested in a summation method that will sum formal power series which may not converge anywhere to suitable domains on which they will converge. If one wants to use a summation method to continue analytically an analytic function given on its disk of convergence, the Cesàro method will be of no use, since no power series will converge (C, k) outside its circle of convergence for any k. Indeed, if $\sum_n a_n z_0^n$ is summable (C, k), then $a_n z_0^n = o(n^k)$ and so $\sum a_n z^n$ converges for $|z| < |z_0|$. Borel's idea was to examine whether the choice of a matrix $T = (t_{mn})$ with $t_{mn} \neq 0$ for *infinitely many* n for each m will lead to a more powerful method of summation. For instance if one starts with an entire function

$$J(z) = \sum_n j_n z^n \qquad (j_n > 0 \text{ for all } n)$$

and chooses the matrix of summation by defining the functions

$$t_n(x) = \frac{j_n x^n}{J(x)} \qquad (x > 0),$$

then it is clear that the conditions of regularity are satisfied and so we have a regular method of summation. In fact, $J(z) = J(|z|)$ is increasing with $|z|$ and so must go to ∞, as otherwise J is a constant by Liouville's theorem; hence $t_n(x) \to 0$ for $x \to \infty$, while $\sum_n |t_n(x)| = \sum_n t_n(x) = 1$. The choice of $J(z) = e^z$ leads to B-summation:

$$\sum_n a_n = s \quad (B) \iff \lim_{x \to \infty} e^{-x} \sum_n A_n \frac{x^n}{n!} = s \quad (A_n = a_0 + \cdots + a_n).$$

We shall use instead a closely related method of summation, also due to Borel. For this purpose let us introduce for any $f \in \mathbf{C}[[z]]$ the power series $f^\sim$ defined as follows:

$$f = \sum_{n \geq 0} a_n z^n \qquad f^\sim = \sum_{n \geq 0} a_n \frac{z^n}{n!}.$$

We shall call $f^\sim$ the *Borel transform* of f. If f has a positive radius of convergence, $f^\sim$ is an entire function. But even if f has zero radius of convergence, it is possible that $f^\sim$ may have positive radius of convergence (if $a_n = O((n!)^\alpha)$ where $\alpha < 1$, then $f^\sim$ is entire). For such an f one can make a tentative definition of a summation method by the following:

$$\sum_{n \geq 0} a_n = s \quad (\text{Bo}) \iff \lim_{x \to \infty} \int_0^x e^{-t} f^\sim(t) dt = s.$$

But the assumption that $f^\sim$ is entire is too strong. For example, for the Euler factorial series $\sum_{n \geq 0} (-1)^n n! z^n$, its Borel transform

$$f^\sim(z) = \sum_{n \geq 0} (-1)^n z^n = \frac{1}{1+z}$$

converges only for $|z| < 1$. But its sum $1/1 + z$ *is defined on the nonnegative real axis and so one can apply the method above.* This is the crux of Borel's idea.

Motivated by these remarks, let us return to $\mathbf{C}[[z]]$. The tentative definition of (Bo) suggests that we make the following definition: an element $f \in \mathbf{C}[[z]]$ is *admissible* if $f^\sim$ has a positive radius of convergence and has an analytic continuation to a domain D_f that includes the positive real axis including 0. The domain D_f is of course allowed to depend on f. The set of admissible elements of $\mathbf{C}[[z]]$ is a differential subalgebra of $\mathbf{C}[[z]]$. Thus the Euler factorial series is certainly admissible. For any admissible f we shall continue to write $f^\sim$ for the function which analytically continues the power series to the domain D_f, but in calculations we must remember that $f^\sim$ is given by the power series

$$f^\sim(z) = \sum_{n \geq 0} a_n \frac{z^n}{n!}$$

only for small $|z|$. However this does not cause any serious difficulties. Indeed, as we shall see, in most cases the arguments used by assuming that $f^\sim$ is entire go through even when f is only admissible. We shall formally define the definition of the summation (Bo) for any admissible $f = \sum_{n \geq 0} a_n z^n$:

$$\sum_{n \geq 0} a_n = s \quad \text{(Bo)} \iff \lim_{x \to \infty} \int_0^x e^{-t} f^\sim(t)\, dt = s.$$

It is obvious that this summation is a linear process. But it is not well behaved under multiplication. However for any method to be able to build an *algebra* of summable series it is essential that the method sums the Cauchy products of two summable series to the product of the sums of the individual series. So Borel was forced to strengthen the definition of admissibility to obtain a viable theory of multiplication of series summable by (Bo).

DEFINITION. A power series $f = \sum_n a_n z^n$ is *strongly admissible* if it is admissible and

$$\int_0^\infty e^{-t} |f^{\sim(p)}(t)|\, dt < \infty$$

for all $p \geq 0$.

Then one of Borel's main results is the following. Let $\mathbf{A}_0$ be the set of all strongly admissible elements of $\mathbf{C}[[z]]$.

THEOREM 1. $\mathbf{A}_0$ *is a differential subalgebra of* $\mathbf{C}[[z]]$ *and contains any power series* $\sum_{n \geq 0} a_n z^n$ *for which* $\sum_{n \geq 0} |a_n| < \infty$. *For every element* $f = \sum_n a_n z^n$ *of* $\mathbf{A}_0$, *the series* $\sum_{n \geq 0} a_n$ *is summable* (Bo), *and if we denote its sum by* $\beta_0(f)$, *then the map*

$$\beta_0 : \sum_{n \geq 0} a_n z^n \longmapsto \text{the (Bo)-sum of } \sum_{n \geq 0} a_n$$

is a homomorphism. Finally, if $f \in \mathbf{A}_0$, *then* $zf \in \mathbf{A}_0$ *and*

$$\beta_0(f) = \beta_0(zf).$$

Specialization will take us from this theorem to a summability procedure that will encompass what Euler did with the factorial series and its generalizations. Let us define $\mathbf{A}$ to be the set of all strongly admissible elements $f = \sum_{n \geq 0} a_n z^n$ with the property that for each real number $\xi \geq 0$ the power series

$$f_\xi := \sum_{n \geq 0} a_n \xi^n z^n$$

is again strongly summable. Notice that for any $\xi > 0$ the map $w \longmapsto \xi w$ of $\mathbf{C}$ leaves the nonnegative reals $\mathbf{R}_{\geq 0}$ invariant and so it preserves the set of functions analytic on $\mathbf{R}_{\geq 0}$. This shows that if f is admissible, then so is f_ξ for all $\xi > 0$. Clearly $f_\xi^\sim(t) = f_\xi^\sim(\xi t)$ and so $f \in \mathbf{A}$ if and only if

$$\int_0^\infty e^{-t/\xi}|f^{\sim(p)}(t)|\,dt < \infty$$

for all $p \geq 0$, $\xi > 0$.

THEOREM (Borel). **A** *is a differential algebra. For each* $f \in \mathbf{A}$ *let* Φ_f *be the function on the nonnegative reals defined by*

$$\Phi_f(\xi) = \beta_0(f_\xi);$$

namely, $\Phi(f)$ *is the* (Bo)*-sum of the series* $\sum_n a_n \xi^n$. *Then* Φ_f *extends to an analytic function on the right half plane* $\{\Re(z) > 0\}$, *also denoted by* Φ_f. *The map*

$$\beta : f \longmapsto \Phi_f$$

is an injection of the differential algebra **A** *into the differential algebra of all analytic functions on this half plane. Moreover, for any* $f \in \mathbf{A}$, Φ_f *is asymptotic to* f.

PROOF. We have

$$\Phi_f(\xi) = \int_0^\infty e^{-t} f_\xi^\sim(t)\,dt = \frac{1}{\xi}\int_0^\infty e^{-t/\xi} f^\sim(t)\,dt.$$

Let

$$\Phi_f(\zeta) = \frac{1}{\zeta}\int_0^\infty e^{-t/\zeta} f^\sim(t)\,dt \qquad (\Re(\zeta) > 0).$$

It is clear that the integral converges absolutely and uniformly for ζ in compact sets of the right half plane showing that Φ_f is analytic there. To see that Φ_f is asymptotic to f, fix $N \geq 0$ and expand $\mathbf{F}^\sim$ as a Taylor series with remainder in the integral form. Then

$$\Phi_f(\zeta) = \sum_{r=0}^N a_r \zeta^r + \frac{1}{p!\zeta}\int_0^\infty e^{-t/\zeta} \int_0^t (t-u)^p f^{\sim(p+1)}(u)\,du.$$

If $\xi = \Re(1/\zeta)$, then $\Re(1/\zeta) = \xi/|\zeta|^2 = \eta$ (say), and the last term is majorized by

$$\frac{1}{p!|\zeta|}\int_0^\infty e^{-t\eta}\int_0^t (t-u)^p |f^{\sim(p+1)}(u)|\,du$$

which is

$$\leq \frac{1}{\eta^{p+1}|\zeta|}\int_0^\infty e^{-t\eta}|f^{\sim(p+1)}(u)|\,du.$$

If $\zeta = \rho e^{i\theta}$ and we assume that $\rho \leq R$, $|\theta| \leq \varphi < \pi/2$, then $\xi \geq |\zeta|\cos\varphi$ and the integral above is less than or equal to $K|\zeta|^p$ where K is a constant independent of ζ. This shows that Φ_f is asymptotic to f in the sense of Poincaré. Finally, if $\Phi_f = 0$, then all the a_r have to be 0 and so $f = 0$. This finishes the proof.

As an illustration, let us take Euler's series

$$f = \sum_{n\geq 0}(-1)^n n! z^n.$$

Then for any $\xi \geq 0$

$$f_\xi = \sum_{n\geq 0}(-1)^n n! \xi^n z^n.$$

Its Borel transform is

$$f_\xi^\sim = \sum_{n\geq 0}(-1)^n \xi^n z^n = \frac{1}{1+\xi z}$$

which is analytic on $\{\Re(z) > -1\}$ and so f_ξ is admissible. Moreover

$$f_\xi^{\sim(p)}(t) = (-1)^{p-1}\xi^p p! \frac{1}{(1+\xi t)}^{p+1}$$

and so it is clear that f_ξ is strongly admissible. Hence $f \in \mathbf{A}$. We then have

$$\Phi(f)(\xi) = \int_0^\infty \frac{e^{-t}}{(1+\xi t)}\, dt.$$

If we put

$$\Phi_f(w) = \int_0^\infty \frac{e^{-t}}{(1+wt)}\, dt \qquad (w = re^{i\theta},\ |\theta| < \pi),$$

then we have seen in §3 that the integral defines an analytic function on the sector $\Gamma = |\theta| < \pi$ and so is an analytic extension from the nonnegative reals to the sector Γ. The Borel sum at $w = 1$ is of course

$$\Phi_f(1) = \int_0^\infty \frac{e^{-t}}{(1+t)}\, dt.$$

Thus

$$1 - 1! + 2! - 3! + \cdots = \int_0^\infty \frac{e^{-t}}{(1+t)}\, dt = \int_0^1 \frac{e^{1-(1/t)}}{t}\, dt \qquad (\text{Bo}).$$

Actually the series $\sum_{n\geq 0}(-1)^n n! w^n$ is summable (Bo) on the entire sector $\Gamma = |\theta| < \pi$:

$$\sum_{n\geq 0}(-1)^n n! w^n = \int_0^\infty \frac{e^{-t}}{(1+wt)}\, dt \qquad (\text{Bo}) \qquad (|\theta| < \pi).$$

This is clear from the fact that for $f_w := \sum_{n\geq 0}(-1)^n n! w^n z^n$ the Borel transform is $1/1 + wt$.

For another example let us take the series discussed by Euler in his 1760 paper:

$$1 - pw + p(p+q)w^2 - p(p+q)(p+2q)w^3 + p(p+q)(p+2q)(p+3q)w^4 - \ldots$$

where $p, q > 0$. The formal power series associated to this is

$$f = \sum_{n\geq 0}(-1)^n p(p+q)\ldots(p+(n-1)q)z^n$$

and its Borel transform is

$$f^\sim = \sum_{n\geq 0}(-1)^n \frac{p(p+q)\ldots(p+(n-1)q)}{n!}z^n$$

which is convergent and represents

$$\frac{1}{(1+qz)^{-p/q}}.$$

It follows as before that f_ξ is strongly admissible for every $\xi > 0$ and so $f \in \mathbf{A}$. So, with $w = re^{i\theta}$,

$$\Phi_f(w) = \int_0^\infty \frac{e^{-t}}{(1+qwt)^{-p/q}}\, dt \qquad (|\theta| < \pi).$$

Hence Φ_f is analytic on the sector $|\theta| < \pi$ and

$$\sum_{n\geq 0}(-1)^n p(p+q)\ldots(p+(n-1)q)w^n = \int_0^\infty \frac{e^{-t}}{(1+qwt)^{-p/q}}\, dt \qquad (\text{Bo}).$$

For $p = 1, q = 2, w = 1$ we get the series

$$1 - 1 + 1.3 - 1.3.5 + 1.3.5.7 - \ldots$$

and the sum

$$\int_0^\infty \frac{e^{-t}}{(1 + 2wt)^{-1/2}} \, dt = \int_0^1 e^{[1 - (1/t^2)]/2} \frac{1}{t^2} \, dt.$$

A calculation of this integral by MATHEMATICA shows that

$$1 - 1 + 1.3 - 1.3.5 + \cdots = 0.655679542418798\ldots,$$

confirming what Euler gets. Of course with the numerical integration there is no bound for the error, but as in the original case of the factorial series Euler found a convergent continued fraction for the integral and, using it, appears to have made sure that his result was accurate to any desired level of accuracy.

The two examples point to an unsatisfactory aspect of the Borel theorem described above, namely that the formal series defines analytic functions only on the half plane and also that the asymptotics are only in the sense of Poincaré. However Watson discovered in 1912 that strong asymptotics in a sector of *angles strictly greater than* $\pi/2$ implies uniqueness and that the correspondence between the formal series and the analytic function representing it asymptotically becomes expressible explicitly. We shall give a brief discussion of Watson's theorem [15].

Watson's theorem deals with functions analytic on sectors of the form

$$D = \{z = re^{i\theta} \mid 0 < r \le \alpha, -\pi/2 - \lambda < \theta < \pi/2 + \lambda\}.$$

With Watson we consider first the differential algebra $\mathbf{C}_W[[z]]$ of all $f = \sum_{n \ge 0} a_n z^n$ such that for some $\sigma > 0$ we have

$$a_n = O(n! \sigma^n).$$

In tandem with this we have the differential algebra $\mathcal{O}_W(D)$ of all functions F analytic on D such that F is strongly asymptotic to some $f \in \mathbf{C}_W[[z]]$. Actually it is easy to see that if F is analytic in D and is strongly asymptotic to an $f \in \mathbf{C}[[z]]$, then f is necessarily in $\mathbf{C}_W[[z]]$. If $f \in \mathbf{C}_W[[z]]$, then its Borel transform

$$f^\sim(t) = \sum_n a_n \frac{t^n}{n!}$$

is convergent in $|t| < 1/\sigma$.

THEOREM (Watson). *Let $F \in \mathcal{O}_W(D)$ and let F be strongly asymptotic to $f \in \mathbf{C}_W[[z]]$. Then we have the following:*

(a) *$f^\sim$ has an analytic extension to the sector $\{re^{i\theta} \mid |\theta| < \lambda\}$ and is given there by the integral*

$$\frac{1}{2\pi i} \oint_C (\nu, \beta) F(tu) \frac{e^{1/u}}{u} \, du \qquad (t = \rho e^{i\varphi}, |\varphi| < \lambda)$$

provided ν and β are properly chosen: if $\rho < R$ and $|\theta| < \delta < \lambda$, then we can take $0 < \beta < \alpha/R, \nu = \lambda - \delta$, the integral converging absolutely and uniformly.

(b) *The function F can be obtained from $f^\sim$ by the inversion formula*

$$F(z) = \int_0^\infty e^{-w} f^\sim(wz) \, dw \qquad (z = re^{i\theta}, r < \alpha, |\theta| < \lambda).$$

(c) *The function F can be obtained from $f^\sim$ by the inversion formula*

$$F(z) = \int_0^\infty e^{-w} f^\sim(wz)\, dw \qquad (z = re^{i\theta}, r < \alpha, |\theta| < \lambda).$$

In particular F is uniquely determined by f.

REMARK. Unlike the examples from Euler's paper treated above, this theorem exhibits an interesting phenomenon. In the Euler cases the series is summable (Bo) in the full sector $|\theta| < \pi$. In Watson's theorem the summability is valid in the small sector $|\theta| < \lambda$ while the function lives on the big sector $|\theta| < \frac{\pi}{2} + \lambda$. This cannot be avoided, as the following example shows.

Let us consider the series

$$f = \sum_n (-1)^n \frac{(2n)!}{n!} z^{2n} \in \mathbf{C}[[z]].$$

Then

$$f^\sim(t) = e^{-t^2}$$

is entire, but the integral

$$\int_0^\infty e^{-t - w^2 t^2}\, dt < \infty$$

only when w lies in the sector $|\theta| < \pi/4$.

The relationship of these integrals and divergent series suggested by the work of Euler, Borel, and Watson is extremely interesting. The relation with analytic continued fractions which is revealed in Euler's work, and which was taken up by Stieltjes [14], appears to be worth pursuing in a general context.

It is also obvious that the Borel theory cannot apply to divergent series of the form

$$\sum_{n \geq 0} (-1)^n (n!)^s \qquad (s > 1).$$

Such series occur in many problems of interest in the theory of differential equations with *irregular singularities*. One needs an extension of the Borel theory for such situations. In recent years, mainly due to the efforts of a host of people such as Balser, Ecalle, Ramis, Sibuya, Ilyashenko, and many others, a vast generalization of Borel summability has been created, the so-called *multisummability*. This theory has many profound applications to such diverse subjects as irregular singular differential equations, Galois differential groups, dynamical systems and the Hilbert problem of number of limit cycles of polynomial vector fields in the plane, and so on. For a very nice introductory account of these themes, see the paper of J. P. Ramis [19b] and the references therein.

5.6. Tauberian theorems

No account of divergent series will be complete without a discussion of Tauberian theorems. They formed a natural landmark in the classical development of the subject; moreover, Wiener's epoch-making discoveries in the subject led to ideas and discoveries of great import in mathematics and physics. This section and the next two discuss these matters.

There is a natural question that arises with respect to any summation process. Suppose that a series $\sum_{n \geq 0} a_n$ is summable to s in a given summation process. Is it possible to specify some general conditions on the a_n which imply that $\sum_{n \geq 0} a_n$

is convergent? Such theorems may be called *converse theorems* and one of the simplest for the Cesàro summation is Hardy's theorem [2]:

THEOREM. *If* $\sum_{n\geq 0} a_n$ *is summable* $(C, 1)$ *and* $a_n = O(1/n)$, *then the series* $\sum_{n\geq 0} a_n$ *is convergent.*

For Abel summability the first converse theorem was proved [20] by A. Tauber and so all converse theorems are called *Tauberian*. Tauber's result was that

$$\sum_{n\geq 0} a_n = s \quad (A), \quad a_n = o(1/n) \implies \sum_{n\geq 0} a_n = s.$$

In 1910 Littlewood [21] proved the much deeper result that in the above theorem it is enough to suppose only that

$$a_n = O(1/n).$$

Clearly Littlewood's theorem is the analogue for Abel summation of the converse theorem proved above for Cesàro summation but is considerably more difficult to prove.* It has also a variant in which the above condition on a_n is replaced by one-sided versions

$$a_n > -K\frac{1}{n} \quad \text{or} \quad a_n < K\frac{1}{n}.$$

Littlewood's theorem generated a huge amount of activity in proving such Tauberian theorems. Eventually, the whole subject was completely transformed in 1932 when Wiener placed the subject of Tauberian theorems in the context of Fourier analysis and proved a very general Tauberian theorem from which all earlier Tauberian theorems followed as special cases. The Wiener Tauberian theorem was so powerful that it led to a proof of the prime number theorem, as Wiener himself showed in his great 1932 paper [22].

In order to bring out clearly the ideas of Wiener let me start by discussing the Littlewood Tauberian theorem and give it a formulation that makes the transition to Wiener's theory a little easier. The two variables are n and x where n goes to ∞ and $x \to 1 - 0$. Replacing x by $e^{-1/x}$ allows us to work with a single continuous real variable which goes to ∞. The basic assumption is that

$$f(x) = \sum_{n\geq 0} a_n e^{-n/x} \to \ell \quad (x \to \infty).$$

Let us replace $s_n = a_0 + \cdots + a_n$ by $s(x)$ defined by

$$s(x) = s_{[x]} = \sum_{n\leq x} a_n.$$

Under the assumption that

$$a_n = O\left(\frac{1}{n}\right)$$

it is enough to show that

$$\sum_{n\geq 0} a_n = \ell \quad (C, 1)$$

---

*For some very interesting personal reminiscences regarding his proof of the theorem, see Littlewood's book referred to in the footnote at the end of §5.1, p. 131.

because we can then apply Hardy's converse theorem to conclude that the series $\sum_{n\geq 0} a_n$ is itself convergent. The $(C,1)$ convergence can be obviously rewritten in the integral form

$$\frac{1}{x}\int_0^x s(y)dy \to \ell \qquad (x \to \infty).$$

On the other hand, the Littlewood condition that $a_n = O(1/n)$ is entirely natural, as can be seen from the following lemma.

LEMMA. *If f is bounded as $x \to \infty$ and $a_n = O(1/n)$, then $s(x)$ is bounded as $x \to \infty$.*

PROOF. Let $n|a_n| < K$ where $K > 0$ is a constant. For $u > 0$ we have

$$1 - e^{-u} = \int_0^u e^{-t}dt < \int_0^u dt = u.$$

Moreover $x - 1 \leq [x] \leq x$, and so for $x \geq 1$,

$$|s(x) - f(x)| = \left| \sum_0^{[x]} a_n(1 - e^{-n/x}) - \sum_{[x]+1}^{\infty} a_n e^{-n/x} \right|$$

$$\leq \sum_0^{[x]} |a_n|(n/x) + K \sum_{[x]+1}^{\infty} (1/n)e^{-n/x}$$

$$\leq \frac{[x]+1}{x}K + K \int_{[x]}^{\infty} e^{-t/x}\frac{dt}{t}$$

$$\leq 2K + K \int_1^{\infty} e^{-t}\frac{dt}{t}.$$

This proves the lemma.

So the Littlewood condition is entirely natural. Let

$$k(y) = \frac{1}{y}e^{-1/y} \qquad (y > 0).$$

Then, with $s_{-1} = 0$,

$$\sum_{n\geq 0} a_n e^{-n/x} = \sum_{n\geq 0} (s_n - s_{n-1})e^{-n/x}$$

$$= \sum_{n\geq 0} s_n(e^{-n/x} - e^{-(n+1)/x})$$

$$= \frac{1}{x}\int_0^{\infty} s(y)e^{-y/x}\,dy.$$

Hence the theorem of Littlewood may be formulated as follows, with the understanding that all statements are for $x \to \infty$:

$$\frac{1}{x}\int_0^{\infty} s(y)e^{-y/x}\,dy \to \ell, \quad s(x) = O(1) \Longrightarrow \frac{1}{x}\int_0^x s(y)\,dy \to \ell.$$

Let

$$k(y) = \frac{1}{y}e^{-1/y} \ (y > 0) \quad k_1(y) = \begin{cases} 1/y & \text{if } y > 1 \\ 0 & \text{if } 0 < y \leq 1. \end{cases}$$

Let us also write

$$d^\times y = \frac{dy}{y} \qquad (y > 0).$$

Then $d^\times y$ is the measure invariant under the multiplications $a \longmapsto ay$ of the multiplicative group $\mathbf{R}^\times = \{y|y > 0\}$. We have

$$\int_0^\infty k(y)\, d^\times y = \int_0^\infty k_1(y)\, d^\times y = 1,$$

and the above formulation of Littlewood's theorem can be rewritten as

$$\int_0^\infty k(x/y)s(y)\, d^\times y \to \ell, \quad s(x) = O(1) \implies \frac{1}{x}\int_0^\infty k_1(x/y)s(y)\, d^\times y \to \ell.$$

In other words we have the following situation. For any function K which is integrable with respect to $d^\times y/y$ on $(0, \infty)$ we have a summation procedure $W(K)$ (W for Wiener) that is defined by the statement that

$$f(y) \to \ell \ \ (W(K)) \iff \int_0^\infty K(x/y)f(y)\, d^\times y \to \ell \int_0^\infty K(y)\, d^\times y.$$

The appearance of $\int_0^\infty K(y)\, d^\times y$ is to allow a little flexibility and use kernels for which this integral need not be 1, as it happened to be in the above discussion. Then the theorem of Littlewood becomes

$(W_{\mathbf{R}^\times})$ $\qquad s(x) \to \ell \ \ (W(k)),\ s(x) = O(1) \implies s(x) \to \ell \ \ W((k_1)).$

It is thus very clear that the condition of boundedness of s is the *Tauberian condition* that allows one to show that Wiener summation with respect to k implies Wiener summation with respect to k_1.

Notice that this last formulation can be transferred to $(-\infty, \infty)$ by the logarithmic map $x \mapsto \log x$ which takes multiplication to addition and $d^\times y$ to dy. Over $\mathbf{R}$ with a function K integrable over $(-\infty, \infty)$ the definition of $W(K)$ becomes

$$f(x) \to \ell \ \ (W(K)) \iff \int_{\mathbf{R}} K(x - y)f(y)\, dy \to \ell \int_{\mathbf{R}} K(y)\, dy.$$

Then Wiener formulated his result as follows:

$(W_{\mathbf{R}})$ $\qquad f(x) \to \ell \ \ (W(K)),\ f(x) = O(1) \implies f(x) \to \ell \ \ (W(K_1)).$

Wiener's great discovery was that in order to prove that $W(K) \implies W(K_1)$ it is enough to require that *the Fourier transform of K does not vanish anywhere on* $(-\infty, \infty)$.

The *Fourier transform* of a function g which is integrable on $\mathbf{R}$ is the function $\widehat{g}$ defined by

$$\widehat{g}(\xi) = \frac{1}{\sqrt{2\pi}} \int_{\mathbf{R}} g(y)e^{iy\xi}\, dy.$$

The integrability of g implies that $\widehat{g}$ is well defined on $\mathbf{R}$ and that

$$|\widehat{g}(\xi)| \le ||g|| = \frac{1}{\sqrt{2\pi}} \int_{\mathbf{R}} |g(y)|\, dy.$$

Then the *Wiener Tauberian theorem* takes the following form:

THEOREM (Wiener). *Suppose that K is integrable on $\mathbf{R}$ and $\widehat{K}$ does not vanish anywhere on $\mathbf{R}$. Then for any K_1 which is integrable on $\mathbf{R}$, and any bounded function f,*

$$f(x) \to \ell \quad (W(K)) \; f(x) = O(1) \Longrightarrow f(x) \to \ell \quad (W(K_1)).$$

If we assume this, then to prove the theorem of Littlewood it is enough to transfer k to $\mathbf{R}$ and verify that the Fourier transform of the corresponding function on $\mathbf{R}$ never vanishes. But if $K(t)$ on $\mathbf{R}$ corresponds to $k(y)$ on $(0, \infty)$, we have, writing $u = e^t$ and omitting the factors involving π,

$$\begin{aligned}
\widehat{K}(\xi) &= \int_{\mathbf{R}} e^{i\xi t} k(e^t) \, dt \\
&= \int_0^\infty u^{i\xi} k(u) d^\times u \\
&= \int_0^\infty u^{i\xi} e^{-1/u} \frac{du}{u^2} \\
&= \int_0^\infty y^{-i\xi} e^{-y} \, dy \\
&= \Gamma(1 - i\xi)
\end{aligned}$$

where Γ is the classical gamma function. But we know that Γ has no zero in the complex plane; indeed, $1/\Gamma$ *is an entire function.* Hence Wiener's criterion is met, and we have proved the theorem of Littlewood!

Although we have treated only the Tauberian theorem for convergence, there are Tauberian theorems for other summations also. Thus the proof above already shows that

$$\sum_n a_n = s \quad (A), \quad s_n = O(1) \Longrightarrow \sum_n a_n = s \quad (C, 1).$$

For a discussion of applications of Wiener's theorem to other summations see [29], p. 299.

We shall postpone the proof of the Wiener Tauberian theorem to §8 after we have some applications of the theory developed thus far. The Wiener theorem depends in a deep way on the theory of Fourier integrals, and the discussion of Fourier series in the next section will also be a good introduction to the theory of Fourier integrals. That Fourier transform theory enters into these questions should not be a surprise. Our discussion above shows that the basic structure of any summation method is determined by an operator of the form

$$f \longmapsto K * f, \qquad K * f(x) = \int_{\mathbf{R}} K(x - y) f(y) \, dy.$$

These operators are *convolution operators*. They have the most important property of *commuting with translations*. What is more important, every operator that maps a function space on the real line into another function space and which *commutes with translations* is a convolution operator in some generalized sense. Under Fourier transforms, convolutions become *multiplications* and so convolution operators are naturally studied using Fourier transform theory. This was Wiener's point of view, and it formed the basis of his application of Fourier analysis to Tauberian theorems and to his many profound works in signal theory.

5.7. Some applications

Cesàro's theorem on multiplication of series. When Cesàro introduced his summation process he made a beautiful application to the theory of multiplication of series. Let

$$\sum_{n \geq 0} a_n, \qquad \sum_{n \geq 0} b_n$$

be two series, both of them convergent, to sums A, B respectively. The *Cauchy product* of the two series is the series $\sum_{n \geq 0} c_n$ where

$$c_n = a_0 b_n + a_1 b_{n-1} + \cdots + a_r b_{n-r} + \cdots + a_n b_0.$$

This definition is natural if we associate to each series $\sum_n u_n$ the power series $\sum_n u_n z^n$ and then multiply the two power series $\sum_n a_n z^n$, $\sum_n b_n z^n$ in the usual manner to form the product power series $\sum_n c_n z^n$. The question is whether the product series converges, and if so, is its sum AB? The classical theorems did not find a decisive answer to this. The best result was that if one of the series converges absolutely, then the product series converges to the sum AB. It was also known that when neither series converges absolutely, the product series *need not* converge and that if both series converge absolutely, the product series converges absolutely as well. Cesàro proved that if the series $\sum_n a_n, \sum_n b_n$ both converge to sums A, B respectively, then the product series is *always summable* $(C,1)$ *to the sum AB.* It follows in particular that if the product series converges at all, its sum should be AB. This beautiful and general result, which completely clarifies the classical problem, is a very nice instance of the usefulness of the notion of Cesàro summability.

THEOREM. *Let $\sum_{n \geq 0} a_n, \sum_{n \geq 0} b_n$ be two convergent series with sums A, B respectively. Let*

$$c_n = a_0 b_n + a_1 b_{n-1} + \cdots + a_{n-1} b_1 + a_n b_0.$$

Then

$$\sum_{n \geq 0} c_n = AB \qquad (C,1).$$

In particular, if the product series converges, its sum has to be AB.

PROOF. This comes down to proving that if $u_n, v_n \longrightarrow 0$, then

$$w_n = \frac{u_0 v_n + u_1 v_{n-1} + \cdots + u_n v_0}{n+1} \longrightarrow 0.$$

The proof is by the standard argument of splitting the sum.

REMARK. It can be proved that if $\sum_{n \geq 0} a_n$ and $\sum_{n \geq 0} b_n$ are summable (C,h), (C,k) respectively to A, B, then the product series is summable to AB by $(C, h + k + 1)$. The above result is the special case when $h = k = 0$. However, just as in the case of ordinary convergence, when h, k are arbitrary, the series may converge (C, ℓ) where $\ell < h + k + 1$. For example, the square of Euler's series

$$\frac{1}{2} = 1 - 1 + 1 - 1 + 1 - 1 + \ldots \qquad (C,1)$$

is

$$\frac{1}{4} = 1 - 2 + 3 - 4 + 5 - 6 + \ldots \qquad (C,2)$$

although the general theorem predicts only that it is summable $(C,3)$.

Since the product series is defined through the power series expansions, it is to be expected that the Cauchy product is best behaved with respect to the Abel summation. We have:

THEOREM. *If*

$$\sum_{n\geq 0} a_n = A \quad (A), \qquad \sum_{n\geq 0} b_n = B \quad (A),$$

then

$$\sum_{n\geq 0} c_n = AB \quad (A).$$

Fourier series: some historical remarks. The beginnings of the theory of what are now called Fourier series may be traced to the middle of the 18th century and the investigations of d'Alembert, Daniel Bernoulli, and Euler on the motion of a vibrating string. For small perturbations of the string the equation of the transverse displacement $u(x,t)$ of the point x of the string $(0 \leq x \leq a)$ at time t is

$$\frac{\partial^2 u}{\partial x^2} = \frac{\partial^2 u}{\partial t^2}.$$

The general solution of this equation is of the form

$$u(x,t) = f(x+t) + g(x-t)$$

where f and g are *arbitrary* functions, a result that had been obtained by d'Alembert. Since

$$u(x,0) = f(x) + g(x), \quad u_t(x,0) = f'(x) - g'(x), \quad u'(x,0) = f'(x) + g'(x)$$

it follows that $u(x,0)$ and $u_t(x,0)$ determine $u(x,t)$, a fact that was observed by Euler a year after d'Alembert's discovery. Bernoulli's investigations began a few years later and were concerned with the problem of determining the position of the string at a time $t > 0$ given that the string is fixed at its ends throughout the motion. In view of Euler's observation, this is clearly possible. This is the same as solving the above equation with the initial and boundary conditions

$$u(0,t) = u(a,t) = 0, \quad u(x,0) = f(x), \quad \frac{\partial u}{\partial t}(x,0) = 0 \quad (0 \leq x \leq a).$$

The functions

$$u(x,t) = \cos\frac{k\pi t}{a} \sin\frac{k\pi x}{a} \quad (k = 1,2,3,\dots)$$

satisfy the equation with

$$u(x,0) = f(x) = \sin\frac{k\pi x}{a}.$$

Building on the idea that the general vibration is a superposition of the various "harmonics", Bernoulli sought for a solution in the form

$$u(x,t) = A\cos\frac{\pi t}{a}\sin\frac{\pi x}{a} + B\cos\frac{2\pi t}{a}\sin\frac{2\pi x}{a} + C\cos\frac{3\pi t}{a}\sin\frac{3\pi x}{a} + \dots$$

which satisfies the initial and boundary conditions mentioned provided that $A, B, C, \dots$ are determined by expressing the initial function $f(x)$ as

$$f(x) = A\sin\frac{\pi x}{a} + B\sin\frac{2\pi x}{a} + C\sin\frac{3\pi x}{a} + \dots.$$

Bernoulli was now led to ask *if any function $f(x)$ can be expressed as a series of sines as above.* It appeared to him that such an expansion was no less general than the Taylor expansions which were known at that time and that the fact that one had at one's disposal an infinite number of constants $A, B, C, \ldots$ should allow these to be determined by requiring that the function pass through an infinite number of points. However he was unable to determine the coefficients $A, B, C, \ldots$. They were eventually written down by Euler in the form we know today by simply integrating the expansion of the function after multiplying by the sines and cosines. However these investigations were clouded by controversy, because even the greatest mathematicians of that era, like Euler, Lagrange, and others, could not understand how a *nonperiodic* function such as $x(\ell - x)$ which represents between 0 and ℓ a possible position of the string whose ends are fixed at 0 and ℓ could be expanded in terms of periodic functions [23].

The true beginning of the theory of Fourier series is more appropriately assigned to the year 1822, when the epoch-making treatise *Théorie analytique de la chaleur* of Fourier first appeared in print [24]. Fourier's great discovery was not only that any function of a real variable could be represented *on any finite interval* by a series of sines (and cosines) but also that once the interval and the type of the series (sines, cosines, or containing both) were fixed, *the coefficients were uniquely determined and in fact expressible in a direct and explicit manner in terms of the given function,* a discovery that had eluded Bernoulli almost 80 years previously. These expansions provided Fourier with a very powerful mathematical tool for solving a remarkable variety of problems in arbitrary dimension involving the conduction of heat in material bodies. To do this Fourier had to abandon the somewhat rigid conception of functions prevailing at that time and replace it by a more subtle one whereby a function could be defined by different analytic expressions in different intervals, something that is very natural from the physical point of view but which presented great difficulties to some of the greatest masters of that era. In fact their opposition delayed the publication of Fourier's treatise.

This is not the place to discuss the controversy generated by Fourier's epoch-making ideas and their role in the development of real analysis. I wish only to point out that a full appreciation and acceptance of his work would become possible only after a completely new foundation was laid for the theory of functions of real variables and the theory of integration of such functions; this was the achievement of Dirichlet and Riemann. Dirichlet was Fourier's great protégé, and Riemann studied under Dirichlet. Dirichlet's proof that the Fourier series of a piecewise monotonic function converges to the function at each of its points of continuity appeared in 1829 [25]. Riemann took up the theory of trigonometric series and under its impetus developed what we now know as the theory of Riemann integration and heat conduction summation. Riemann's work on the representability of functions as a trigonometric series was presented in 1854 but appeared only in 1867 [26].

For a function f defined on an interval $[-\ell, \ell]$ the Fourier series takes the form

$$f \sim \frac{1}{2}a_0 + \sum_{n \geq 1} \left(a_n \cos \frac{n\pi x}{\ell} + b_n \sin \frac{n\pi x}{\ell} \right)$$

where the coefficients a_n, b_n are determined by the expressions that Fourier found for them:

$$a_n = \frac{1}{\ell} \int_{-\ell}^{\ell} f(x) \cos \frac{n\pi x}{\ell} dx, \quad b_n = \frac{1}{\ell} \int_{-\ell}^{\ell} f(x) \sin \frac{n\pi x}{\ell} dx.$$

Much later, it became clear that it would be simpler to work with the *complex* version:

$$f \sim \sum_{-\infty < n < \infty} c_n e^{in\pi x/\ell}, \quad c_n = \frac{1}{2\ell} \int_{-\ell}^{\ell} f(x) e^{-in\pi x/\ell} dx.$$

We write $\sim$ instead of $=$ to emphasize the formal nature of the process; no convergence is involved. Also, to keep the classical flavor I have retained the sines and cosines instead of the exponentials which are more usual when we go to Fourier integrals. The fact that the expressions for the coefficients involved integration made it very clear that a proper foundation for the subject cannot be laid *until the theory of integration and that of convergence of series were developed in full rigor;* this was Riemann's starting point. The theory of Fourier series, and later Fourier integrals, would always be dominated by these two aspects. The formulae for the constants a_n, b_n are found by formally integrating the relation obtained by equating f to the series and using the orthogonality of the trigonometric functions. Fourier however did not proceed by this method. He had a much more elaborate but not totally convincing argument to derive these formulae (see the discussion on pp. 30-34 in [2]). In any case it was discovered much later that the formulae were buried in Euler's work.

The period following the publication of Fourier's book was almost exclusively concerned with the understanding and consolidation of his ideas. This required a rethinking, mainly due to Dirichlet and Riemann, of the foundations of real analysis and the theory of functions of real variables and their integrals. The question of pointwise convergence of the Fourier series occupied the central place during this time. In particular, a whole series of tests for the convergence of the Fourier series of a function at a continuity point were developed; however it was known that mere continuity alone would not suffice. The situation was of course similar but more involved for Fourier integrals.

Summation methods applied to Fourier series. We shall now take a closer look at the theory of summation of Fourier series. Let f be an integrable function on an interval. We take the interval to be $[-\pi, \pi]$ instead of $[-\ell, \ell]$, since we can go from one to the other by a scale change. We then extend f to the entire real line by requiring that it is of period 2π. The integrability of f was classically in the sense of Riemann, but eventually the theory had to be enlarged to deal with functions integrable in the sense of Lebesgue. Following classical notation we *define* the Fourier coefficients of f as follows:

$$a_n = \frac{1}{\pi} \int_{-\pi}^{\pi} f(t) \cos nt \, dt \ (n \geq 0) \qquad b_n = \frac{1}{\pi} \int_{-\pi}^{\pi} f(t) \sin nt \, dt \ (n \geq 1).$$

The *Fourier series of f* is then the series

$$\frac{1}{2} a_0 + \sum_{n \geq 1} (a_n \cos nx + b_n \sin nx).$$

There is no consideration of any convergence for this series as yet; it is a series formally associated to f, and we write

$$(*) \qquad\qquad f \sim \frac{1}{2}a_0 + \sum_{n \geq 1}(a_n \cos nx + b_n \sin nx).$$

One of the questions that was investigated intensively in classical analysis was whether this series converges to $f(x)$. In his book [24] Fourier had considered many special instances of these series and so had become aware of the delicacy of the question of convergence. In the examples he worked out, the Fourier series converged to the function at points where it is continuous, but the convergence was generally not absolute. The simplest example is obtained when

$$f(x) = x \qquad (-\pi < x < \pi).$$

Then

$$a_n = 0\ (n \geq 0), \qquad b_n = \frac{2(-1)^{n-1}}{n}\ (n \geq 1)$$

so that

$$x \sim 2\sum_{n=1}^{\infty}(-1)^{n-1}\frac{\sin nx}{n} \qquad (-\pi < x < \pi).$$

The series on the right converges conditionally but not absolutely. It is a delicate question whether its sum is x.

The first rigorous results on the convergence of the right side of $(*)$ to $f(x)$ were due to Dirichlet. Dirichlet proved that if f is a function with a finite number of maxima and minima between which it is monotonic, then we have

$$\frac{f(x+) + f(x-)}{2} = \frac{1}{2}a_0 + \sum_{n \geq 0}(a_n \cos nx + b_n \sin nx)$$

where

$$f(x\pm) = \lim_{\varepsilon \to 0+} f(x \pm \varepsilon).$$

In particular, at any continuity point x, we have

$$f(x) = \frac{1}{2}a_0 + \sum_{n \geq 0}(a_n \cos nx + b_n \sin nx).$$

It was soon realized that Dirichlet's analysis could be pushed to show that if f is of *bounded variation* in a neighborhood of a point x, then its Fourier series converges to

$$\frac{f(x+) + f(x-)}{2}$$

at any point x_0; the existence of this limit is of course a consequence of the bounded variation of f because of the classical result that f is a difference of two monotonic increasing functions.

Two remarks need to be made at this time. In the first place, the condition for convergence at x depends only on the behavior of the function in an arbitrarily small neighborhood of x. It was Riemann who first noticed this, and it is called the *localization principle* for the convergence of Fourier series. The second remark is that considerably more than continuity is needed to ensure the convergence of the Fourier series, namely the requirement of bounded variation. Indeed, although much weaker conditions were developed by a host of classical mathematicians coming after Dirichlet, it became clear that some condition about the variation of f in a neighborhood of x was essential to ensure convergence at x. This point will be discussed briefly later.

However the situation becomes significantly better when we consider the $(C, 1)$ *summability* of the Fourier series. Fejér [27] proved the beautiful result that the Fourier series of f is *always* summable $(C, 1)$ to

$$\frac{f(x+) + f(x-)}{2}$$

provided the limits $f(x\pm)$ exist, in particular that if f is continuous at x, its Fourier series is summable $(C, 1)$ to $f(x)$. Fejér's proof also showed that if f is everywhere continuous and periodic of period 2π, then its Fourier series converges *uniformly* to f on $[-\pi, \pi]$. This result has the famous Weierstrass approximation theorem as an immediate corollary, as we shall see later.

Let

$$s_n(x) = s_n(x; f) = \frac{1}{2}a_0 + \sum_{k=1}^{n}(a_k \cos kx + b_k \sin kx)$$

$$\sigma_n(x) = \frac{s_0(x) + s_1(x) + \cdots + s_{n-1}(x)}{n}.$$

Making use of the identities which go back at least to Euler,

$$1 + 2\cos u + 2\cos 2u + \cdots + 2\cos nu = \frac{\sin(1/2)(2n+1)u}{\sin(1/2)u}$$

$$\sin u + \sin 3u + \cdots + \sin(2n-1)u = \frac{\sin^2 nu}{\sin u},$$

we get

$$s_n(x) = \int_0^{\pi/2} \frac{[f(x+2v) + f(x-2v)]}{2} D_n(v)dv$$

$$\sigma_n(x) = \int_0^{\pi/2} \left[\frac{f(x+2v) + f(x-2v)}{2}\right] F_n(v)dv$$

where

$$D_n(v) = \frac{2}{\pi}\frac{\sin(2n+1)v}{\sin v}$$

$$F_n(v) = \frac{D_0(v) + D_1(v) + \cdots + D_{n-1}(v)}{n} = \frac{2}{n\pi}\left(\frac{\sin nv}{\sin v}\right)^2.$$

The function D_n is called the *Dirichlet kernel*; the function F_n is called the *Fejér kernel*. The main point is to show that the kernels D_n and F_n converge to the delta function when $n \to \infty$. For D_n this is a delicate question, because the function $\sin(2n+1)v$ has the *very small* period $2\pi/2n+1$ and so D_n oscillates rapidly for *large n*. For F_n it is much simpler since it is always positive. This explains why the usual convergence of Fourier series is more delicate than its $(C, 1)$ summation.

THEOREM (Dirichlet). *If f is of bounded variation[†] in some neighborhood of x, the Fourier series of f converges to*

$$\frac{f(x+) + f(x-)}{2}.$$

[†]It is a classical result that a function of bounded variation is always expressible as a difference of two monotonic increasing functions and so the limits always exist.

THEOREM (Fejér). *Let f be integrable and let x be a point at which the limits $f(x\pm)$ exist. Then the Fourier series of f is summable $(C,1)$ to*

$$\frac{f(x+) + f(x-)}{2}.$$

If moreover f is continuous everywhere, then

$$\sigma_n(x) \to f(x) \qquad (n \to \infty)$$

uniformly for $x \in [-\pi, \pi]$.

We shall deduce Dirichlet's theorem from Fejér's theorem. The proof is a consequence of the following lemma:

LEMMA. *Let (k_n) be a sequence of bounded functions on (a, b) with the following properties:*

(i) *For all n,*

$$k_n \geq 0, \qquad \int_a^b k_n(v) dv = 1.$$

(ii) *For any $\delta > 0$,*

$$\sup_{v \in [a+\delta, b]} k_n(v) \to 0 \qquad (n \to \infty).$$

Then, for any integrable function g with $g(a+) = \ell$, we have

$$\int_a^b g(v) k_n(v) dv \to \ell \qquad (n \to \infty).$$

PROOF. We may assume that $\ell = 0$ by replacing g by $g - \ell$ in view of (i). Let $K = \int_a^b |g(v)|\, dv$. Given $\varepsilon > 0$, we choose $\delta > 0$ so that

$$|g(v)| < \varepsilon \qquad v \in (a, a + \delta).$$

Then, for any n,

$$\left| \int_a^b g(v) k_n(v) dv \right| \leq \int_a^{a+\delta} |g(v)| k_n(v)\, dv + \int_{a+\delta}^b |g(v)| k_n(v)\, dv$$

$$\leq \varepsilon + K \sup_{v \in [a+\delta, b]} k_n(v).$$

Letting $n \to \infty$ we get, as the second term goes to 0 by (ii),

$$\limsup_n \left| \int_a^b g(v) k_n(v) dv \right| \leq \varepsilon.$$

For Fejér's theorem the functions F_n satisfy the conditions of the lemma; indeed, they are continuous, hence bounded, and

$$\sup_{v \in [\delta, \pi/2]} F_n(v) \leq \frac{1}{n \sin^2 \delta} \to 0$$

as $n \to \infty$. Hence we have the theorem of Fejér.

PROOF. Only the second part needs to be proved. The proof is exactly the same. Since x is allowed to vary, we write $g(x, v)$ in place of $g(v)$ so that

$$g(x, v) = \frac{f(x + 2v) + f(x - 2v)}{2} - f(x).$$

As f is *uniformly continuous*, we can choose $\delta > 0$ such that $|g(x,v)| < \varepsilon$ for $0 \leq v \leq \delta$ and *all* x. Let $|f(x)| \leq K$ for a constant $K > 0$ for all x. Then $|g(x,v)| \leq 2K$ and

$$\left| \int_0^{\pi/2} g(x,v) F_n(v)\, dv \right| \leq \int_0^\delta |g(x,v)| F_n(v)\, dv + \int_\delta^{\pi/2} |g(x,v)| F_n(v)\, dv$$

$$\leq \varepsilon + \frac{L}{n}$$

where $L = K\pi(\sin^2 \delta)^{-1}$. The rest of the argument is the same as before; the uniform convergence is clear since the right side of the above estimate is *independent* of x.

REMARK. It should be noted that we have made f periodic, and so if f is given only on an interval of length 2π, the assumption of continuity means that we must have that f is continuous and *takes the same value at the end points of the interval*.

COROLLARY 1. *If the Fourier series of f converges at a point x where $f(x\pm)$ exist, then its sum has to be*

$$\frac{f(x+) + f(x-)}{2}.$$

COROLLARY 2. *If f is a continuous function on a finite closed interval I, there is a sequence of polynomials p_n converging uniformly to f over I.*

This is the famous *Weierstrass approximation theorem*. It is valid in higher dimensions also and plays an important role in classical analysis. It is deduced by replacing the trigonometric functions in the Fejér theorem by their power series expansions truncated suitably. But its full reach and power were understood only when M. H. Stone discovered a beautiful and far-reaching extension of it to any compact Hausdorff space. It is called the *Stone-Weierstrass theorem*. To state it, let X be any Hausdorff topological space. A set A of real-valued continuous functions on X is said to *separate the points of X* if, given any pair x, y of distinct points of X, there is an element f of A such that $f(x) \neq f(y)$.

STONE-WEIERSTRASS THEOREM. *Suppose X is a compact Hausdorff space and A is a set of real-valued continuous functions on X such that*

(i) $1 \in A$.

(ii) *A separates the points of X.*

Then, given any real continuous function f on X, there is a sequence of functions $f_n \in A$ uniformly converging to f where for each n, $f_n = p_n(a_1, \ldots a_{k_n})$ with p_n a polynomial.

The classical Weierstrass theorem for an interval I of the real line or a cube I^N of R^N is obtained if we take A to be the set consisting of the function 1 and the coordinate functions x_i. Then each f_n is a *polynomial*. The proof of the Stone-Weierstrass theorem is quite conceptual and contains the classical theorem as a special case (for a proof see [28]). The Stone-Weierstrass theorem is one of the central results of infinite-dimensional functional analysis. It has been generalized to the noncommutative context, and its usefulness appears to have no end.

Let us now return to Dirichlet's theorem.

Proof of Dirichlet's theorem. By the localization principle we can replace f by a function which coincides with f in a neighborhood of x and is 0 outside this neighborhood. Hence we may assume that f is of bounded variation over $[-\pi, \pi]$. In view of Fejér's theorem and Hardy's converse theorem for $(C, 1)$ summation, it is enough to prove that the Fourier coefficients of f are $O(1/n)$. For functions of bounded variation we have the theory of *Riemann-Stieltjes integration* that allows us to define integrals

$$\int_a^b g(t)df(t)$$

for continuous functions g. Among the usual properties of such integrals is the formula for integration by parts, when g has a continuous derivative, namely,

$$\int_a^b g(t)df(t) = g(b)f(b) - g(a)f(a) - \int_a^b f(t)g'(t)dt.$$

In particular we can take $g(t) = \cos nt,\ \sin nt$ to get

$$\int_{-\pi}^{\pi} f(t)\cos nt\ dt = -\frac{1}{n}\int_{-\pi}^{\pi} \sin nt\ df(t) = O\left(\frac{1}{n}\right)$$

$$\int_{-\pi}^{\pi} f(t)\sin nt\ dt = \frac{1}{n}\int_{-\pi}^{\pi} \cos nt\ df(t) = O\left(\frac{1}{n}\right).$$

REMARK. P. Chernoff discovered a very nice proof of convergence of Fourier series which presupposes very little; see [29].

The problem of pointwise convergence of Fourier series continued to attract attention. Early on, Kolmogorov constructed an example of a (Lebesgue) integrable function whose Fourier series diverged everywhere. The general question remained murky till Carleson proved in the 1960's that the Fourier series of a square integrable function f converges almost everywhere to f. The theorem of Carleson is thus a great climax of this classical question, settling it once and for all. In spite of many simplifications in its proof, Carleson's theorem still remains a monolith.

5.8. Fourier integral, Wiener Tauberian theorem, and Gel'fand transform on commutative Banach algebras

Even though it was Fourier who understood the true significance of the representation of functions as Fourier series in solving many problems of heat conduction, the formulae for the Fourier coefficients go back to Euler. So perhaps the priority of Fourier may be questioned so far as the creation of the theory of Fourier series is concerned. But he was without any doubt the sole creator of the technique of *Fourier integrals*. It was the problem of heat conduction in an *infinitely long linear rod* that led Fourier to the theory of Fourier integrals.

Fourier treated this problem by first starting with a rod of finite length 2ℓ and then allowing ℓ to become infinite. In the exponential form we have

$$f(x) \sim \sum c_n e^{in\pi x/\ell}, \qquad c_n = \frac{1}{2\ell}\int_{-\ell}^{\ell} f(x)e^{-in\pi x/\ell}dx.$$

We can rewrite these as

$$F\left(\frac{n\pi}{\ell}\right) = \frac{1}{\sqrt{2\pi}}\int_{-\ell}^{\ell} f(x)e^{-in\pi x/\ell}dx,$$

$$f(x) = \frac{1}{\sqrt{2\pi}}\sum F\left(\frac{n\pi}{\ell}\right) e^{in\pi x/\ell}\left(\frac{\pi}{\ell}\right).$$

When $\ell \to \infty$ and $n\pi/\ell \to \xi$, in the first formula *the range of integration becomes infinite*, while in the second, ℓ becomes an *infinitesimal and so one should replace the summations by integrals*. If one does this, one gets the remarkable formulae

$$F(\xi) = \frac{1}{\sqrt{2\pi}} \int_{-\infty}^{\infty} f(x)e^{-i\xi x}dx, \qquad f(x) = \frac{1}{\sqrt{2\pi}} \int_{-\infty}^{\infty} F(x)e^{i\xi x}dx.$$

These are of course the fundamental formulae of the theory of Fourier integrals. Actually, Fourier was working with even functions which have cosine transforms and so was led to the pair of equations

$$F(x) = \int_{0}^{\infty} dq\, Q \cos qx, \qquad Q(q) = \frac{2}{\pi} \int_{0}^{\infty} dx\, F(x) \cos qx.$$

Here is what he says about these equations (see [24], p. 336):

> ... *If we substituted for Q any function of q, and conducted the integration from $q = 0$ to $q = \infty$, we should find a function of x; it is required to solve the inverse problem, that is to say, to ascertain what function of q, after being substituted for Q, gives as the result the function $F(x)$, a remarkable problem whose solution demands attentive examination....*

There is no doubt that Fourier had a clear conception of the passage from the function $F(x)$ to the function $Q(q)$ as a transformation from one function space to another, the *Fourier Transform*. With the help of this tool he solved a remarkable number of problems which nowadays would be called boundary value and initial value problems for the heat equation

$$\frac{\partial u}{\partial t} = \frac{\partial^2 u}{\partial x_1{}^2} + \cdots + \frac{\partial^2 u}{\partial x_d{}^2}$$

for dimensions $d = 1, 2, 3$, not only when the space region was finite, but also otherwise (for instance, heat conduction on an infinite line).

In the modern theory there is no attempt to carry out any limiting process from finite intervals and Fourier series on them, and the starting point is a *direct definition* of the Fourier transform. The basic space is $L^1 = L^1(\mathbf{R})$, the space of integrable functions on $\mathbf{R}$. Even more than in the theory of Fourier series it is essential to work with the Lebesgue integral. L^1 is a linear space with norm

$$||f||_1 = \int |f(x)|dx.$$

Because we are working with the Lebesgue integral the space L^1 is a *Banach space*. If $f \in L^1$, all its translates and their averages are also in L^1; more precisely, if $f, g \in L^1$, then their *convolution* $f * g$ defined by

$$(f * g)(x) = \int_{\mathbf{R}} f(x - y)g(y)dy \qquad (x \in \mathbf{R})$$

is also in L^1 in the sense that the integral exists for almost all x and defines an element of L^1. We have

$$f * g = g * f, \qquad ||f * g||_1 \le ||f||_1 ||g||_1.$$

The operation $*$ converts L^1 into an *algebra*. In terminology to be introduced later, L^1 is a commutative Banach algebra. For any $f \in L^1$ its *Fourier transform* $\hat{f}$ is defined by

$$\widehat{f}(\xi) = \frac{1}{\sqrt{2\pi}} \int_{\mathbf{R}} f(x)e^{-i\xi x}\, dx \qquad (\xi \in \mathbf{R}).$$

The function $\widehat{f}$ is defined on all of $\mathbf{R}$, bounded, uniformly continuous, and vanishes at ∞; the vanishing at ∞, i.e.,

$$\lim_{\xi \to \pm\infty} \widehat{f}(\xi) = 0$$

is the Riemann-Lebesgue lemma.

Now we come to the L^2-theory. The main point is that if $f \in L^2$, we cannot define its Fourier transform as we did for functions in L^1 because the integral will not converge in general. This is analogous to the difficulty of ensuring pointwise convergence of Fourier series. But suppose that $f \in L^1 \cap L^2$. Then $\widehat{f}$ is well defined, and remarkably, one can show that it lies in L^2 in the variable ξ and satisfies the *Plancherel formula:*

$$\frac{1}{\sqrt{2\pi}} \int_{\mathbf{R}} |f(x)|^2\, dx = \frac{1}{\sqrt{2\pi}} \int_{\mathbf{R}} |\widehat{f}(\xi)|^2\, d\xi \qquad \textbf{(Pl)}.$$

It is obvious that $L^1 \cap L^2$ is dense in L^2, and it can be proved that the space of all $\widehat{f}$ for $f \in L^1 \cap L^2$ is also dense in $L^1 \cap L^2$. Hence the map

$$f \longmapsto \widehat{f} \qquad (f \in L^1 \cap L^2)$$

extends uniquely to a linear map of L^2 *onto* itself such that the Plancherel formula **(Pl)** given above is valid for all $f \in L^2$. We write $\widehat{f}$ for all elements f of L^2. In modern terminology, the L^2-Fourier transform is a *unitary map* of L^2 with itself.

The fundamental property of the Fourier transform is that *convolution goes over to multiplication*:

$$\widehat{f_1 * f_2} = \widehat{f_1}\widehat{f_2}.$$

In other words, the map $f \mapsto \widehat{f}$ represents the convolution algebra L^1 as a *multiplication algebra* of functions on $\mathbf{R}$, which are all uniformly continuous and vanish at ∞. Now, on the algebra of uniformly continuous functions vanishing at ∞ there is a natural norm, namely

$$\|g\|_\infty = \sup_{\xi} |g(\xi)|.$$

The Fourier transform is however not norm preserving but norm decreasing, i.e.,

$$\|\widehat{f}\|_\infty \leq \|f\|_1.$$

What makes the L^1-theory of Fourier transforms subtle is that *there is no precise and simple description of the functions of ξ which are Fourier transforms of elements of L^1*. However *if we assume that $\widehat{f} \in L^1$*, then we can show that

$$f(x) = \frac{1}{\sqrt{2\pi}} \int_{\mathbf{R}} \widehat{f}(\xi)e^{i\xi x}\, d\xi \qquad (x \in \mathbf{R})$$

and so the relation between f and $\widehat{f}$ becomes perfectly reciprocal for the space of functions f such that *both f and $\widehat{f}$ are in L^1*. Such f are dense in L^1, and the formula above is called the *inversion formula*. Notice that the problem of expressing f in terms of $\widehat{f}$ is the analogue of the problem of deciding when the Fourier series of a function converges to the function. It is therefore not surprising that in the theory of Fourier integrals there is an analogue of the theorem of Fejér which asserts that the inversion formula is valid *for all $f \in L^1$* provided the integral above is interpreted in the sense of $(C, 1)$ summability. Now, for any continuous function g, we have the easily proved identity

$$\int_{-u}^{u} \left(1 - \frac{|\xi|}{u}\right) g(\xi)\, d\xi = \frac{1}{u} \int_0^u du \int_{-\xi}^{\xi} g(\tau)\, d\tau.$$

This is in fact the integral analogue of the result for a series $\sum_n a_n$ with partial sums s_n that

$$\frac{s_0 + s_1 + \cdots + s_{n-1}}{n} = \sum_{r=0}^{n-1} \left(1 - \frac{r}{n}\right) a_r.$$

Hence, if we define, for any $u > 0$

$$f_u(x) = \frac{1}{\sqrt{2\pi}} \int_{-u}^{u} \left(1 - \frac{|\xi|}{u}\right) \widehat{f}(\xi) e^{i\xi x}\, d\xi,$$

then the assertion

$$f = \lim_{u \to \infty} f_u$$

can be interpreted as saying that

$$f(x) = \frac{1}{\sqrt{2\pi}} \int_{\mathbf{R}} \widehat{f}(\xi) e^{i\xi x}\, d\xi \qquad (C, 1)$$

and so is the analogue of Fejér's theorem. It is not difficult to show that

$$\|f - f_u\|_1 \to 0 \qquad (u \to \infty)$$

so that the limit relation is in the L^1-norm. It is actually true also that the limit is valid in the sense of almost everywhere convergence, i.e.,

$$f(x) = \lim_{u \to \infty} \frac{1}{\sqrt{2\pi}} \int_{-u}^{u} \left(1 - \frac{|\xi|}{u}\right) \widehat{f}(\xi) e^{i\xi x}\, d\xi$$

for almost all x. If $\widehat{f} \in L^1$ it is not difficult to deduce from this the inversion formula. It follows from the formula above that the map $f \mapsto \widehat{f}$ is one to one on L^1. Thus the *Fourier transform of a function $f \in L^1$ determines f uniquely.*

Let us write down some examples of pairs $(f, \widehat{f})$ which occur often in the theory and for which the inversion formula can be verified explicitly:

$$f(x) = e^{-x^2/2\sigma^2}, \qquad \widehat{f}(\xi) = e^{-\sigma^2 \xi^2/2},$$

$$f(x) = \sqrt{\frac{2}{\pi}} \frac{1 - \cos ax}{ax^2}, \qquad \widehat{f}(\xi) = \begin{cases} \left(1 - \frac{|\xi|}{a}\right) & \text{if } |\xi| \le a \\ 0 & \text{if } |\xi| > a. \end{cases}$$

In each of these cases $\widehat{f}$ is in L^1, and one can verify the inversion formula directly.

Let us now go back to the context of Wiener's theorem. Let us write $P(K)$ for the proposition

$$f(x) \to \ell\,(W(K))$$

where f is bounded. Let $\Sigma(K)$ be the set of all integrable K_1 such that $P(K) \Longrightarrow P(K_1)$. Clearly $\Sigma(K)$ is a linear subspace of L^1. If K_1 is a *translate* of K, i.e., if

$$K_1(x) = K(x + a)$$

for some $a \in \mathbf{R}$, then $K_1 \in \Sigma(K)$. Indeed,

$$\int_{\mathbf{R}} K_1(x - y) f(y)\, dy = \int_{\mathbf{R}} K(x_1 - y) f(y)\, dy \quad (x_1 = x + a)$$

and $x \to \infty$ is the same as $x_1 \to \infty$. We can then conclude that $\Sigma(K)$ is closed under convolution, i.e., for any $M \in L^1$, $K_1 \in \Sigma(K)$, $K_1 * M \in \Sigma(K)$. To see this, let $K_2 = K_1 * M$. Then (all integrals are over $\mathbf{R}$)

$$\int K_2(x-y)f(y)\,dy = \int f(y) \left(\int K_1(x-y-z)M(z)\,dz \right) dy$$

$$= \int M(z) \left(\int K_1(x-z-y)f(y)\,dy \right) dz$$

$$= \int M(z)F(x,z)\,dz.$$

We know

$$F(x,z) \to \ell \int K_1(t)dt \qquad (x \to \infty),$$

while

$$|M(z)F(x,z)| \le A|M(z)| \qquad A = \left(\sup_y |f(y)| \right) ||K_1||.$$

Hence, by Lebesgue's dominated convergence theorem we can pass to the limit for $x \to \infty$ under the integral sign and obtain

$$\int K_2(x-y)f(y)\,dy \to \ell \int \int M(z)K_1(u)\,du\,dz = \ell \int K_2(t)dt.$$

In the terminology of algebra, $\Sigma(K)$ is an *ideal* in L^1. Finally $\Sigma(K)$ is *closed* in L^1. If $(K_n)_{n\ge1} \in \Sigma(K)$ and K_0 is an integrable function such that

$$||K_n - K_0||_1 \to 0 \qquad (n \to \infty),$$

then $K_0 \in \Sigma(K)$. This is also simple. Write

$$f_n(x) = \int_{\mathbf{R}} K_n(x-y)f(y)\,dy, \qquad I_n = \int_r K_n(y)\,dy.$$

Then

$$|f_0(x) - \ell I_0| \le |f_n(x) - \ell I_n| + \ell|I_n - I_0| + \int_{\mathbf{R}} |K_n(x-y) - K_0(x-y)||f(y)|\,dy$$

and so, if $|f(y)| \le C$ for all y,

$$|f_0(x) - \ell I_0| \le |f_n(x) - \ell I_n| + \ell|I_n - I_0| + C||K_n - K_0||.$$

Letting $x \to \infty$ we get, for each n, as $|I_n - I_0| \le ||K_n - K_0||_1$,

$$\limsup_x |f_0(x) - \ell I_0| \le (C + \ell)||K_n - K_0||_1.$$

If we now let $n \to \infty$ we get the result.

Thus we have shown that $\Sigma(K)$ is a closed linear subspace of L^1 containing K which is closed under convolution, i.e., a closed ideal containing K. Wiener's great discovery was that *if the Fourier transform of K does not vanish anywhere on $\mathbf{R}$, then*

$$\Sigma(K) = L^1.$$

This is an immediate consequence of the following more general theorem which was also proved by him.

THEOREM (Wiener). *Let L^1 be the space of all integrable functions on* **R**. *Suppose that E is a subset of L^1 such that the Fourier transforms of elements of E do not have a common zero. Then the smallest ideal of L^1 containing E is dense in L^1. In particular, if $K \in L^1$ and $\widehat{K}$ does not vanish anywhere on* **R**, *then the ideal generated by K, namely the set of functions of the form $K * f$ ($f \in L^1$), is dense in L^1.*

REMARK. It can be shown that a closed linear subspace of L^1 is an ideal, i.e., is closed under convolutions if and only if it is closed under convolutions. Hence in the above theorem we can replace the property of being an ideal by the property of being closed under translations.

The idea behind the proof is very simple to illustrate when E consists of a single element K. The hypothesis then says that $\widehat{K}$ does not vanish anywhere on **R**. We would like to prove that if K_1 is in L^1, there is an $M \in L^1$ such that $K * M = K_1$ or at least, $K * M$ is arbitrarily close to K_1. Going over to Fourier transforms, we want to solve for M from the equation

$$\widehat{K_1} = \widehat{K}\widehat{M}.$$

There is only one possible solution for this equation, namely

$$\widehat{M} = \frac{\widehat{K_1}}{\widehat{K}}.$$

The right side is well defined because $\widehat{K}$ never vanishes, but it is not clear that it is the Fourier transform of an element of L^1. If however $\widehat{K_1}$ has *compact support*, then Wiener succeeded in proving this. So the set of functions of the form $K * M$ contains all functions *whose Fourier transforms have compact support*. It is not difficult to show that such functions are dense in L^1 and so the proof of Wiener's theorem is finished. So the key step of Wiener's proof is to show that when $\widehat{K_1}$ has compact support, $\widehat{K_1}/\widehat{K}$ is of the form $\widehat{H}$ for some $H \in L^1$.

The prime number theorem. The prime number theorem is the statement that if

$$\pi(n) = \{\text{the number of primes } \leq n\},$$

then

$$\pi(n) = \frac{n}{\log n} + o\left(\frac{n}{\log n}\right) \qquad (n \to \infty).$$

It was conjectured by Gauss and eventually proved by Hadamard and de la Vallée Poussin towards the end of the 19th century. Wiener deduced the prime number theorem as a consequence of his Tauberian theorem. His original derivation was somewhat involved, but later, simpler proofs were found by using the Ikehara Tauberian theorem. The Ikehara Tauberian theorem illustrates a basic principle: if the difference between two tempered positive measures has a Fourier transform which behaves locally nicely, then the two measures are asymptotic at infinity. To be more precise, let D be a lattice in $\mathbf{R}^d$ and let C be a compact set in $\mathbf{R}^d$ such that the translates of the interior of C by elements of the lattice D cover $\mathbf{R}^d$. Then we define **M** to be the *Wiener space* of all continuous functions g on $\mathbf{R}^d$ such that

$$\|g\| := \sum_{a \in D} \sup_{C+a} |g| < \infty.$$

Then $\mathbf{M}$ is a Banach space whose topology is independent of the set C used in its definition. It is in natural duality with the space $\mathbf{W}$ spanned by the positive Borel measures w such that

$$\sup_{x \in \mathbf{R}^d} w(E + x) < \infty$$

for some (hence every) compact set E. All measures in $\mathbf{W}$ are tempered. Then the principle mentioned above is described by the lemma [30a].

LEMMA. *Let p_1, p_2 be two tempered Borel measures on $\mathbf{R}^d$ such that $p_2 \in \mathbf{W}$ and $\widehat{p_1} - \widehat{p_2}$ is in $L^{1,\mathrm{loc}}(\mathbf{R}^d)$. Then $p_1 \in \mathbf{W}$ also and for any $h \in \mathbf{M}$,*

$$(h * p_1)(y) - (h * p_2)(y) \to 0 \qquad (|y| \to \infty).$$

COROLLARY (Wiener-Ikehara). *Suppose that $a_n \geq 0$ and*

$$f(s) = \sum_{n \geq 1} a_n n^{-s} \qquad no(s = \sigma + i\tau)$$

converges for $\Re(s) > 1$ and for some constant $A > 0$ we have

$$\lim_{\sigma \to 1+0} \left\{ f(\sigma + i\tau) - \frac{A}{\sigma + i\tau - 1} \right\} = g(\tau)$$

uniformly for bounded τ. Then

$$\sum_{r \leq x} a_r \sim Ax \qquad (x \to \infty).$$

PROOF (sketch). We take $d = 1$ and $D = \mathbf{Z}$ in the lemma. $p_1 = \beta$ is the measure with masses a_n/n at the points $\log n$, while $p_2 = \lambda$ is the measure which is Lebesgue measure on $(0, \infty)$ and 0 on $(-\infty, 0]$. The assumptions translate into the following:

(a) $\beta^\sim(s) = \int_{\infty}^{\infty} e^{-sx} d\beta(x) = \sum_n a_n n^{-(1+s)}$ is convergent for $\Re(s) > 0$
(b) $\lim_{\sigma \to 0+} \{\beta^\sim(\sigma + i\tau) - 1/(\sigma + i\tau)\} = g(\tau)$

uniformly for bounded τ. β and λ are tempered, and it is not difficult to show that $\widehat{\beta} - \widehat{\lambda} = g(-\tau)d\tau$. So the assumptions of the lemma are satisfied, and the result follows by applying the conclusion to

$$h(x) = \begin{cases} 0 & \text{if } x < 0 \\ e^{-x} & \text{if } x > 0. \end{cases}$$

h is not continuous, but it is easy to sandwich it uniformly between elements of the Wiener space $\mathbf{M}$ which converge to h in the limit so that the lemma is applicable (see [30a]).

Proof of the prime number theorem. The derivation of the PNT from the Wiener-Ikehara theorem uses standard tricks and the fact that

$$\zeta(1 + it) \neq 0 \qquad (t \in \mathbf{R})$$

which is the basis for the Hadamard-de la Vallée Poussin proofs. From the Euler product for $\zeta(s)$ (see Chapter 6) we get

$$\log \zeta(s) = -\sum_p p^{-s} + u(s) \qquad (\Re(s) > 1)$$

where u is holomorphic in $\Re(s) > 1/2$. Then

$$-\frac{\zeta'(s)}{\zeta(s)} = \frac{1}{s-1} + v(s)$$

where v is holomorphic on an open set U containing the half-plane $\Re(s) \geq 1$ and so

$$k(s) = \sum_p (\log p) p^{-s} - \frac{1}{s-1}$$

is holomorphic in U. This means that the Dirichlet series $\sum_p (\log p) p^{-s}$ satisfies the hypotheses of the Wiener-Ikehara Tauberian theorem, so that

$$\sum_{p \leq n} \log p \sim n \qquad (n \to \infty).$$

This implies PNT easily (see [30a] for more details).

This method applies without change to any number field K and yields the PNT on K: if

$$\pi_K(x) = \#\{\mathfrak{p} \mid N\mathfrak{p} \leq x\}$$

where $\mathfrak{p}$ are the prime ideals of K and $N\mathfrak{p}$ is the norm of $\mathfrak{p}$, then

$$\pi_K(x) = \frac{x}{\log x} + o\left(\frac{x}{\log x}\right) \qquad (x \to \infty).$$

Wiener's work marks him as one of the greatest figures in Fourier analysis. His conception of Fourier analysis was remarkably comprehensive, ranging from classical theory of summable functions to Brownian motion, spectral analysis, and cybernetics. A full treatment of his ideas is impossible here [30b].

In the 1930's, a miracle happened: Gel'fand developed his theory of commutative Banach algebras as the abstract version of the classical theory of Fourier series and integrals and thus brought the Wiener Tauberian theorem within the context of this theory [31]. The basic object of the Gel'fand theory is a *commutative* algebra A defined over the *complex* number field $\mathbf{C}$. We shall assume that it is a Banach space under a norm $\|\cdot\|$ that satisfies (these can be achieved by renorming)

$$\|ab\| \leq \|a\|\|b\| \qquad (a, b \in A).$$

We shall assume that A contains a unit element denoted by 1. This is a crucial assumption, but we must remember that L^1, with its norm, satisfies all the assumptions about A except that of having a unit. The theory of Banach algebras without a unit can be developed also in great depth, but we shall not do it here. Our main application is to L^1 where there are *local units*, as we shall see later, and so one can always come down to the case of Banach algebras with unit elements. The simplest examples of commutative complex Banach algebras are $\mathbf{C}$ and more generally the algebras $C(X)$ of continuous functions on a compact Hausdorff space X. We also note that if I is a proper ($\neq A$) closed ideal of A, then A/I is also a commutative Banach algebra where the norm of the coset $a + I$ is defined as $\inf_{x \in I} \|a + x\|$. It is a standard result that with this norm A/I is complete and so is a Banach algebra.

The basic discovery of Gel'fand was that one can define a *transform* on A, nowadays called the *Gel'fand transform*, that maps A into a multiplication algebra of continuous functions on a compact Hausdorff space S canonically associated to A, called the *spectrum of A*. We define S to be the set of all continuous homomorphisms of A into the complex numbers. S is then a subset of the dual of A

as a Banach space. It is easy to show that for any $\chi \in S$, we have $\chi(1) = 1$ and $|\chi(a)| \leq ||a||$ and so S is a closed subset of the unit ball of the dual of A. From the weak topology of the dual of A, S inherits a topology in which it becomes a compact Hausdorff space, metrizable if A is separable. In the case when A does not have a unit, the spectrum is defined as the set of all continuous nonzero homomorphisms of A into $\mathbf{C}$. In this case S is still locally compact.

For any $\chi \in S$, the kernel $M = M(\chi)$ is an *ideal* of A which is maximal and closed. The fundamental theorem in Gel'fand's theory is that the converse is true in a very strong form, and we shall formulate and sketch a proof of this.

THEOREM. *Any proper ideal of A is contained in a maximal ideal. Any maximal ideal is closed. If M is a maximal ideal, A/M is isomorphic to the field $\mathbf{C}$ of complex numbers and so M is the kernel of a unique element of S.*

REMARK. The key point in the proof that a complex Banach *field* is just $\mathbf{C}$ was also proved by **Mazur** and so it is called the *Gel'fand-Mazur theorem*. Its proof is a beautiful use of Liouville's theorem that a bounded entire function is a constant. If a is a nonzero element of a Banach field and $(1 - \lambda a)$ is never zero for $\lambda \in \mathbf{C}$, then $(1 - \lambda a)^{-1}$ is a bounded entire function.

The *Gel'fand transform* can now be defined as follows. Let S be the spectrum of A equipped as above with the compact Hausdorff topology it inherits from the unit ball of the Banach space dual of A equipped with its weak topology. From the theorem above we know that S is also (may be identified canonically with) the set of maximal ideals of A. For $u \in A$, its Gel'fand transform is the function $\widehat{u}$ defined on S by

$$\widehat{u}(\chi) = \chi(u) \qquad (\chi \in S).$$

Clearly $\widehat{u}$ is continuous on S and $\widehat{1} = 1$. The map

$$u \longmapsto \widehat{u} \qquad (u \in A)$$

is called the *Gel'fand transform*. It maps A into $C(S)$, the algebra of continuous functions on S. As $|\chi(u)| \leq ||u||$ we have

$$||\widehat{u}||_\infty \leq ||u|| \qquad (u \in A).$$

From this we get the following.

THEOREM. *Suppose I is an ideal of A such that there is no common zero for all the Gel'fand transforms of the elements of I. Then $I = A$. In particular, if $u \in A$ is such that $\widehat{u}$ does not vanish anywhere on S, then u is invertible.*

PROOF. If $I \neq A$, then there is a maximal ideal M of A containing I. If $\chi \in S$ is such that M is the kernel of χ, then χ vanishes on I and so $\widehat{u}(\chi) = 0$ for all $u \in I$, a contradiction.

This result is the crux of all proofs of Wiener's theorem as Gel'fand himself observed. This depends on the fact that for the group algebras the Gel'fand transform is exactly the Fourier transform.

The first example is when $A = L^1(\mathbf{Z})$ which we have introduced earlier. The Gel'fand transform maps A onto the multiplication algebra of all absolutely convergent Fourier series. Thus, in this example, the Gel'fand transform is the Fourier transform. The above theorem has as its consequence the following theorem of Wiener which played an essential role in his proof of the Tauberian theorem.

COROLLARY. *If an absolutely convergent Fourier series does not vanish anywhere, its reciprocal also has an absolutely convergent Fourier series expansion.*

We now take up L^1, which is strictly speaking not an example because even though it is a Banach algebra under convolution, it does not admit a unit element. Nevertheless the same method used above can be applied to it because of the following result:

LEMMA. *If $f * g$ is defined by*

$$f * g(x) = \frac{1}{\sqrt{2\pi}} \int f(x - y)g(y) \, dy,$$

then the continuous nonzero homomorphisms of $A = L^1$ are precisely the maps

$$\chi_\xi : f \longmapsto \frac{1}{\sqrt{2\pi}} \int_{\mathbf{R}} f(x)e^{-i\xi x} \, dx \qquad (\xi \in \mathbf{R}).$$

PROOF. It must be noted that we could use any other constant in place of 2π in the above theorem provided that *the same constant is used in the definition of the convolution.* We shall therefore omit the factor $\sqrt{2\pi}$ in what follows.

We now have everything needed to prove the Wiener theorem. Let I be an ideal in L^1 such that the Fourier transforms of elements of I do not have a common zero. Let $\mathbf{U}$ be the algebra of functions of ξ which are Fourier transforms of elements of L^1. The fact that I is dense in L^1 is a consequence of the following lemmas.

LEMMA 1. *If A, B are compact subsets of $\mathbf{R}$, then $1_A * 1_B$ is in $\mathbf{U}$, 1_E being the indicator function of the set E. If $C \subset G \subset \mathbf{R}$ with C compact, G open with compact closure, we can find $u \in \mathbf{U}$ such that $u = 1$ on C and $u = 0$ outside G. In particular, if C is compact and $\xi \notin C$, we can find $u \in \mathbf{U}$ such that $u(\xi) = 1$ and $u = 0$ on C.*

PROOF. Let $g^*(\xi) = \widehat{g}(-\xi)$ for any $g \in L^1$. Then $1_A, 1_B$ are both in $L^2 \cap L^1$ and so, by the L^2 theory, $1_A^*, 1_B^*$ are both in L^2; hence their product g is in L^1. But $\widehat{g} = 1_A * 1_B$, showing that the latter is in $\mathbf{U}$. This proves the first assertion. We now prove the second. For any $\delta > 0$ let $C(\delta)$ be the compact set of points which are at a distance $\leq \delta$ from some point of C. Then for some $\delta > 0$ we must have $C(\delta) \subset C(2\delta) \subset G$. Let 1_δ denote the indicator function of the interval $[-\delta/2, \delta/2]$. Let

$$u = 1_{C(\delta)} * (1/\delta)1_\delta.$$

We know from the first result that $u \in \mathbf{U}$. Moreover, from the definition of $*$, $1_{C(\delta)} * (1/\delta)1_\delta(x)$ is the average of the values of $1_C(\delta)$ in the interval of radius $\delta/2$ with center x. If $x \in C$, all the points of this interval are still inside $C(\delta/2)$ and so the average is 1. Similarly, u is 0 at all points of the complement of $C(2\delta)$, in particular at all points outside G.

Now fix a compact set C and let $\mathbf{U}_C$ be the restrictions to C of the functions in $\mathbf{U}$. The Fourier transform on L^1 followed by restriction to C maps L^1 onto $\mathbf{U}_C$ with kernel J_C which is a closed ideal of L^1. So we can view $\mathbf{U}_C$ as a commutative Banach algebra with *its norm given by this identification of it with L^1/J_C.* By Lemma 1 above, $\mathbf{U}_C$ has a unit element, and so we can apply the theorems proved above to $\mathbf{U}_C$.

LEMMA 2. *The spectrum of* $\mathbf{U}_C$ *is exactly* C, *and the corresponding homomorphisms are just evaluations at points of* C.

PROOF. Since the points of C define homomorphisms of $\mathbf{U}_C$ by evaluation, we must show that there are no other homomorphisms. Let χ be a continuous homomorphism of $\mathbf{U}_c$. This means that we can view χ as a nonzero homomorphism of L^1 which is 0 on J_C. So χ is of the form $f \longmapsto \widehat{f}(\xi)$ for some $\xi \in \mathbf{R}$. It is a question of showing that $\xi \in C$. Suppose $\xi \notin C$. By Lemma 1 we can find a $u \in L^1$ such that $\widehat{u}(\xi) = 1$ and $\widehat{u} = 0$ on C. But then $u \in J_C$ and so $\chi(u) = \widehat{u}(\xi) = 0$, a contradiction.

The images of the elements of I under the map

$$L^1 \longrightarrow L^1/J_C \simeq \mathbf{U}_C$$

then form an ideal I_C in $\mathbf{U}_C$.

LEMMA 3. $I_C = \mathbf{U}_C$; *i.e., there is an element* $v \in I$ *such that* $\widehat{v} = 1$ *on* C.

PROOF. We know that $\mathbf{U}_C$ is a commutative Banach algebra with unit and I_C is an ideal in it. The Gel'fand transform of any element of $\mathbf{U}_C$ is, by Lemma 2, itself, and so the transforms of the elements of the ideal I_C do not have a common zero on C. By our theorem above, this means that $I_C = \mathbf{U}_C$. So there is an element of I, say v, whose Fourier transform $\widehat{v}$ is 1 on C.

LEMMA 4. I *contains all elements of* L^1 *whose Fourier transforms have compact support.*

PROOF. Let $u \in L^1$ be such that $\widehat{u}$ has compact support and let C be the support of $\widehat{u}$. By Lemma 3 there is $v \in I$ such that $\widehat{v} = 1$ on C. As $\widehat{u}$ has C as its support, we have $\widehat{u} = \widehat{u}\widehat{v}$. If $w = u * v$, then $\widehat{w} = \widehat{u}$. Hence $w = u$. As $v \in I$ we must have $w = u * v \in I$. Hence $u = w \in I$.

LEMMA 5. *The elements of* L^1 *whose Fourier transforms have compact support are dense in* L^1.

PROOF. There are many ways of doing this. The following is a simple direct argument. Suppose $f \in L^1$. Then $f = gh$ where $g, h \in L^2$; indeed, if we denote for any complex number z, $sgn\ (z)$ as $z/|z|$ when $z \neq 0$ and 0 when $z = 0$, then we may take $g = sgn\ (f)|f|^{1/2}$ and $h = |f|^{1/2}$. Since the Fourier transform is an isomorphism of L^2 with itself and continuous functions with compact support are dense in L^2, we can find sequences $g_n, h_n \in L^2$ such that $\widehat{g_n}, \widehat{h_n}$ are continuous with compact support and

$$||g_n - g||_2 \to 0, \qquad ||h_n - h||_2 \to 0 \qquad (n \to \infty).$$

If $f_n = g_n h_n$, then $f_n \in L^1$, $\widehat{f_n} = \widehat{g_n} * \widehat{h_n}$ is continuous with compact support, and, by the Schwartz inequality,

$$||f_n - f||_1 \leq ||g_n h_n - gh_n||_1 + ||gh_n - gh||_1$$
$$\leq ||g_n - g||_2 ||h_n||_2 + ||g||_2 ||h_n - h||_2 \to 0$$

because $||h_n||_2$ is bounded. This finishes the proof.

In the 1930's, Weil developed the theory of Fourier integrals on an *arbitrary locally compact abelian group G*. In particular, using the structure theory of these groups which had been developed by Pontryagin and Van Kampen, Weil was able to develop the L^2 and L^1 aspects of Fourier analysis on G, including the Plancherel theorem and the L^1-inversion theorem. Later, after the appearance of Gel'fand's theory, Krein showed how the Fourier analysis on locally compact abelian groups could be obtained from Gel'fand's theory. As long as G was $\mathbf{R}^n$ or the torus $\mathbf{T}^n$, all this was just classical Fourier integral or Fourier series theory in many variables. But the needs of modern number theory forced one to bring in more general locally compact abelian groups such as the fields of p-adic numbers or the adele rings. Thus Fourier analysis on arbitrary locally compact abelian groups acquired a more fundamental place in mathematics. We shall see in Chapter 6 that precisely this theory in its full generality was needed to obtain the analytic continuation and functional equations for all the zeta and L-functions (and much more) defined by the classical number theorists.

If G is a locally compact abelian group, then one denotes by $\widehat{G}$ the *dual group* of G whose elements are the continuous homomorphisms of G into the unit circle. Then $\widehat{G}$ is also locally compact abelian, and we have a canonical isomorphism, the *Pontryagin duality isomorphism*,

$$G \simeq \widehat{\widehat{G}}.$$

Moreover, as in the case of $\mathbf{R}$, the elements of $\widehat{G}$ can also be thought of as the continuous nonzero homomorphisms of $L^1(G)$ into $\mathbf{C}$ by the correspondence that associates $\xi \in \widehat{G}$ to the map

$$f \longmapsto \int_G f(x)\xi(x)\,dx \qquad (f \in L^1(G)).$$

(The proof given for $\mathbf{R}$ goes over without change to this case.) If $f \in L^1(G)$, its Fourier transform is the function $\widehat{f}$ defined on $\widehat{G}$ by

$$\widehat{f}(\xi) = \int_G f(x)\xi(x)^{-1}\,dx \qquad (\xi \in \widehat{G}).$$

Then the proof given above for Wiener's Tauberian theorem goes through to the case of G with virtually no changes. We thus have:

THEOREM. *Let G be a locally compact abelian group and let I be an ideal in the algebra (under convolution) $L^1(G)$. Suppose that the Fourier transforms of elements of I do not have a common zero in $\widehat{G}$. Then I is dense in $L^1(G)$.*

In fact, one needs only minor modifications to formulate and prove a Wiener Tauberian theorem for arbitrary commutative Banach algebras. Here is one such version (see [28], p. 85).

THEOREM (Wiener Tauberian theorem for a commutative Banach algebra). *Let A be a commutative Banach algebra not necessarily having a unit element. Let S be its spectrum. Suppose it satisfies the following conditions:*

(a) *The Gel'fand transform is injective on A.*
(b) *For any point $\xi \in S$ and any compact set $C \subset S$ such that $\xi \notin C$, there is $u \in A$ such that $\widehat{u}(\xi) = 1$ and $\widehat{u} = 0$ on C.*

(c) *For any $\xi \in S$ there is a $u \in A$ such that $\widehat{u} = 1$ in a neighborhood of ξ.*[†]

(d) *The set of elements of A whose Gel'fand transforms have compact support is dense in A.*

If I is an ideal in A such that the Gel'fand transforms of elements of A do not have a common zero in S, then I is dense in A.

REMARK. It can happen that even for an arbitrary commutative Banach algebra with unit, one can have $\widehat{u} = 0$ without u being 0. In fact $\widehat{u} = 0$ if and only if $\lim_{n \to \infty} ||u^n||^{1/n} = 0$. One refers to u as a generalized nilpotent.

Even though we have emphasized the Banach algebra point of view in our discussion of the Wiener Tauberian theorem, it should be pointed out that Wiener's own proof was not that different from the proof of its Banach algebra version. Wiener indeed had a very conceptual view of his theorem, and it is this aspect of it that made it the springboard for Gel'fand's creation of the theory of Banach algebras.

The Gel'fand-Grothendieck point of view of commutative algebras as algebras of functions. However the most striking point in the Banach algebra point of view is the fact that an abstract commutative Banach algebra has canonically associated to it a topological space, namely its spectrum, and the algebra then becomes an algebra of functions on its spectrum in a very natural and simple manner. It turns out that in some respects this idea was already in the work of Hilbert on the foundations of complex algebraic geometry and the work of Stone on the structure of abstract Boolean algebras. Hilbert's famous nullstellensätz exhibited every affine algebraic variety as the set of maximal ideals of the algebra of regular functions on it and characterized these algebras as finitely generated over $\mathbf{C}$ and without nilpotents; Stone's theorem represented every Boolean algebra as the Boolean algebra of open-closed sets in its space of maximal ideals. In the 1950's Serre took over this point of view and developed a theory of algebraic varieties over an algebraically closed field of arbitrary characteristic very similar to the theory of differentiable manifolds, thus starting the unification of algebra and geometry. Thus, a Serre variety looks locally (in its Zariski topology) like the spectrum of a ring. However it was Grothendieck who had the genius and audacity to realize that Serre's approach did not go far enough and that for a full-fledged theory one should insist that *every commutative ring with unit* should be viewed as a ring of functions on its spectrum—except that the spectrum is now *not* the space of maximal ideals as in the Hilbert–Gel'fand–Serre theories, but is the space of all *prime* ideals of the ring, allowing it to vary *functorially* with the algebra. This meant that the map (the Grothendieck transform!) taking the ring into the ring of functions on the spectrum could have a kernel; this already could happen in the Hilbert or the Gel'fand theories. However Grothendieck realized that this is not a disadvantage but actually a crucial aspect of the theory. This kernel is the ideal of nilpotent elements and would serve as the source for the infinitesimal aspects of the geometry. The Grothendieck notion of a *scheme* is the result of this idea. It became the most general object that one could call a space.

[†]We can also state this in the form that for any compact set $C \subset S$ there is a $u \in \mathbf{U}$ such that $\widehat{u} = 1$ on C. But this follows from the assumption we have made. In fact, if $C = \cup_i C_i$ (finite union) where the C_i are compact, $\widehat{u_i}$ is 1 on C_i, and $u = \sum_i u_i - \sum_{i<j} u_i u_j + \sum_{i<j<k} u_i u_j u_k - \dots$, then $\widehat{u} = 1$ on C.

The Grothendieck point of view went even deeper. He realized that in actuality, the *category* of sheaves on a space was sufficient to get to wherever one wanted; one did not really need the points on the space to hang the sheaves on. His idea of a *topos* was born out of this, and it was this that led him to the cohomological theories that were capable of proving the Weil conjectures.

These ideas of Grothendieck enabled him to erect a monumental theory of schemes, toposes, and their cohomologies that completely revolutionized modern algebraic geometry. The solutions of all the major unsolved problems of algebraic geometry, such as the Weil conjectures (solved by Deligne) and the Mordell conjecture (solved by Faltings), are ultimately set in the Grothendieck framework. The Grothendieck view unified number theory and geometry and brought for the first time a systematic method of viewing arithmetic problems. Of course, the simplicity of the old ideas was lost in this revolution (as it happens in all revolutions), but a return to the older points of view is unthinkable.

In the 1970's the Gel'fand-Grothendieck point of view was subjected to another dramatic expansion when physicists introduced *supersymmetry*. To do quantum theory systematically with supersymmetry it became necessary to view spacetime as a space endowed with a sheaf of noncommutative but *supercommutative* algebras. The idea behind these developments has a long history. Its starting point is the remark that whenever it happens that a certain physical theory gets supplanted by a more general one, it is because the main objects of the old theory are some kinds of limits of the objects of the new theory. Thus, the transition to Einstein's special relativity from Galilei's special relativity could be explained as the consequence of the fact that the Galilean theory is obtained as a limiting case of the Einstein theory by letting the velocity of light (now treated as a parameter and not as a constant of nature) go to infinity. Similarly, the passage to quantum mechanics from classical mechanics could be seen as the consequence of the fact that classical mechanics is the result of letting Planck's constant become very small. In recent years many physicists also emphasized that at very small distances spacetime no longer has a point-like structure but becomes highly complex and subject to quantum fluctuations. If this is the case, then the coordinate functions on such objects cannot be commutative because commutative rings are ultimately rings of functions on spaces with points. Starting from such considerations as these, one can naturally hypothesize that for the new geometric objects, *the coordinate rings have to be noncommutative*. In the realm of supersymmetric quantum theory one thus has the Minkowski superspacetime acted on by the super Poincaré group. Then in the 1980's this view was extended at the basic level and spacetime itself was viewed as quantized, described not as a set of points but simply through its ring (now noncommutative) of coordinate functions. The symmetries of such quantum spacetimes are of course the *quantum groups*. There are many who hope that these kinds of ideas may be the way out of the difficulties that plague modern high-energy physics. However, in spite of the highly speculative nature of these remarks, one should not dismiss them as dreams of some deranged species of mathematician and physicist. These ideas have led to very concrete achievements in more conventional areas. For instance, recent work has shown that the viewpoint of quantum groups is a powerful new theme in the theory of algebraic groups over fields of *positive* characteristic.

5.9.   Generalized functions and smeared summation

We have discussed various methods of summation and applied them in a few cases to demonstrate their conceptual as well as practical importance. However these methods of summation are very much restricted to one-dimensional problems, and they do not generalize in any easy fashion to higher-dimensional situations. It turns out that there is a method, the so-called method of *smeared summation,* with origins in engineering and physics, which does not have this drawback. It can be formulated in all dimensions, its mathematical scope is vast, and it has revolutionized modern analysis. I am speaking of the theory of distributions or generalized functions.

One way to motivate the distribution point of view is to note that pointwise convergence and even L^2-convergence, of a Fourier series or a Fourier integral for example, is not very relevant to many situations arising in engineering and physics. This is due to the fact that physical observations often represent *not* sharp computations at a single space-time point but rather *averages* of fluctuations in small but finite spacetime regions. Such a view is quite decisive in signal theory, where there are limitations of determining pulses, and in quantum field theory, where the electromagnetic fields of elementary particles cannot be measured unless one uses a macroscopic *testing* body. From the mathematical point of view one can say that a measurement by a *test body* is just an average of the values of the physical quantity being measured in a very small region represented by a smooth function which is zero outside a small domain. One replaces the test bodies by these functions, which are naturally called *test functions*. Then the value measured is a *functional* on the space of test functions, and the interpretation of the measurement as an average makes it clear that this functional must be *linear*. Thus, if T is the space of test functions (unspecified at this point), physical quantities are elements of the *dual* of T. To do any sort of mathematics one has to exclude pathologies, and so we must assume that the space T has a reasonable topology and that dual means the topological dual. In keeping with our idea that measurements are averages, we recognize that sometimes things are not so bad and that actual point measurements are possible. Thus ordinary functions are also allowed to be elements of the dual of T, and if $f(x)$ is such an ordinary function, it represents the functional

$$\alpha \longmapsto \int \alpha(x) f(x)\, dx \qquad (\alpha \in T)$$

where dx is the integrating measure that represents our concept of averaging. However, since we admit measurements that are too singular to be represented by ordinary functions, we refer to the elements of the dual of T as *generalized functions.* If

$$T : \alpha \longmapsto T(\alpha)$$

is an element of the dual of T, we symbolically write

$$T(\alpha) = \int T(x) \alpha(x)\, dx$$

to emphasize the heuristic content of our considerations. The notation does not mean that T is a function; on the contrary, T makes sense only in this integrated sense, thus finally, only as a linear functional.

We have been extremely vague about what the space is in which we are operating, what the measure dx is in the above remark, and also what functions are chosen as test functions. This is actually a great strength of these ideas because it shows that these methods apply without any restriction on the nature or the dimensions of the spaces. Thus they make sense on $\mathbf{R}^n$, on the torus $\mathbf{T}^n$, on Lie groups and manifolds, on p-adic spaces, and even infinite-dimensional spaces. We shall of course presently give some examples, but it is good to keep the generality of this point of view in mind.

We now come to the theory of summation of generalized functions. Suppose that (T_n) is a sequence of generalized functions. Then we form the series

$$\sum_n T_n(\alpha) = T(\alpha).$$

Clearly T is a linear functional, and we should expect that it is also a generalized function (this is the case in all important situations). However, and this is the fact that is important to us, *even if the T_n are ordinary functions, the sum is often only a generalized function.* We shall say that the series

$$\sum_n T_n(x)$$

of ordinary functions *converges in the smeared sense* if for all test functions α the series

$$\sum_n \int T_n(x)\alpha(x)\,dx = T(\alpha)$$

converges. We write

$$\sum_n T_n = T.$$

In general T will not be an ordinary function. We shall see below by examples that this greatly enlarges the scope of Fourier series.

If T is a function of a real variable x with continuous derivative and the test function α has a continuous derivative and is of compact support, the formula for integration by parts shows that

$$\int T'(x)\alpha(x)\,dx = -\int T(x)\alpha'(x)\,dx,$$

and we have similar formulae when differentiability of higher order is assumed for the test functions. This suggests that one can *always* define derivatives of general functions: if T is a generalized function, its derivative T' is the generalized function defined by

$$\int T'(x)\alpha(x)\,dx = -\int T(x)\alpha'(x)\,dx$$

or more precisely,

$$T'(\alpha) = -T(\alpha').$$

This possibility is the reason for choosing $\mathcal{T}$ as the space $\mathcal{D}$ of all indefinitely differentiable functions with compact support. A generalized function is then a linear functional on $\mathcal{D}$ which is continuous in a certain topology for $\mathcal{D}$. It is not necessary to go into the definition of this topology, but it is sufficient to know that continuity for a linear functional T on $\mathcal{D}$ means that if (α_n) is a sequence of elements of $\mathcal{D}$, all of them vanishing outside a *fixed* compact set, and if for each $k = 0, 1, 2, \ldots$ the kth derivatives

$$\alpha_n^{(k)} \to 0 \qquad (n \to \infty)$$

uniformly, then

$$T(\alpha_n) \to 0.$$

This definition obviously extends to $\mathbf{R}^n$, and the generalized functions thus defined
are called *distributions*. Distributions can be differentiated arbitrarily many times
by the formula

$$\frac{\partial T}{\partial x_i}(\alpha) = -T\left(\frac{\partial \alpha}{\partial x_i}\right) \qquad (\alpha \in \mathcal{D}).$$

A distribution T can be multiplied by a smooth function f to give the distribution

$$fT : \alpha \longmapsto T(f\alpha).$$

It is thus possible to speak of distribution solutions to differential equations where
the differential operators have smooth coefficients. For instance a distribution T
satisfying the equation in $\mathbf{R}^2$ given by

$$\left(\frac{\partial^2}{\partial x^2} + \frac{\partial^2}{\partial x^2}\right) T = 0$$

is a *harmonic distribution*. Historically, such distribution solutions were called *weak
solutions*; in fact, if Δ is the Laplace operator above, a weak harmonic solution is
a distribution T satisfying

$$T(\Delta \alpha) = 0.$$

All the above considerations apply not only to $\mathbf{R}^n$ but also to the tori $\mathbf{T}^n$ and even
to general smooth manifolds. If we work on the torus, the space $\mathcal{T}$ is the space of
all smooth functions on the torus.

Thus, for a Fourier series

$$\sum_n c_n e^{inx}$$

convergence in the smeared sense or the weak sense means that the series

$$\sum_n \int_{-\pi}^{\pi} e^{inx} \alpha(x)\, dx$$

is convergent, α being a test function, namely a smooth function on the circle or
alternatively a smooth periodic function of period 2π. If this happened to be true,
the result would not in general be a point function but a generalized function. To
illustrate this let us see whether the series

$$\sum_{n \in \mathbf{Z}} e^{inx}$$

converges weakly and if so, what its sum is. Here and in what follows we are
working on the circle so that everything has to be viewed mod 2π. If

$$c_n = c_n(\alpha) = \frac{1}{2\pi} \int_{-\pi}^{\pi} \alpha(x) e^{inx}\, dx$$

are the Fourier coefficients of α, integration by parts shows that

$$c_n(\alpha) = \frac{1}{in} c_n(\alpha')$$

and so, by repetition,

$$c_n(\alpha) = \left(\frac{1}{in}\right)^k c_n(\alpha^{(k)}).$$

But the Fourier coefficients of any smooth function, in particular $\alpha^{(k)}$, are bounded trivially. Hence

$$c_n(\alpha) = O(n^{-k}) \text{ for every } k.$$

So the series in question converges, and in fact we have, writing d_0x for $(1/2\pi)dx$,

$$\sum_n \int e^{inx}\alpha(x)\, d_0x = \sum_n c_n(\alpha).$$

But, because α is smooth, its Fourier series converges to α absolutely and uniformly, and so

$$\sum_n c_n(\alpha) = \alpha(0).$$

Hence if we denote (finally!) by $\delta(x)$ the famous Dirac delta function which is the distribution

$$\delta : \alpha \longmapsto \alpha(0),$$

then we can say

$$\sum_{n \in \mathbf{Z}} e^{inx} = \delta(x).$$

Physicists and engineers used this long before generalized functions were invented. Indeed, this formula goes back to Euler(!), except that Euler found the sum to be 0. Actually Euler was very close to the correct result; the delta function is 0 if evaluated at any test function whose support does not contain 0, and so on the circle with the point 1 corresponding to $x = 0$ removed the sum is indeed 0! Euler of course used the Abel summation but after rewriting the sum as

$$1 + 2\sum_{n \geq 1} \cos nx.$$

If we note that for $r < 1$

$$1 + 2\sum_{n \geq 1} r^n \cos nx = \frac{1 - r^2}{1 - 2r\cos x + r^2} \to 0 \qquad (r \to 1 - 0, x \not\equiv 0(2\pi)),$$

we see that the Abel sum is 0. The weak summation result gives the full story, including what happens near 0.

The fact that the Fourier coefficients of a test function are rapidly decreasing, i.e., $O(n^{-k})$ for any $k \geq 0$, means that any trigonometric series

$$\sum_{n \in \mathbf{Z}} a_n e^{inx}$$

is weakly summable if $a_n = O(n^r)$ for some $r \geq 0$. It then converges to a distribution, and the series can be differentiated to all orders. Thus for example

$$\sum_{n \in \mathbf{Z}} n^k e^{inx} = (-i)^k \delta^{(k)}(x)$$

where $\delta^{(k)}(x)$ is the kth derivative of the delta function, namely, the functional

$$\alpha \longmapsto (-1)^k \alpha^{(k)}(0).$$

Unlike the classical summation theories, weak summation of Fourier series works in all dimensions. Thus we have, in $\mathbf{T}^d$,

$$\sum_{n \in \mathbf{Z}^d} e^{in \cdot x} = \delta(x)$$

where $x = (x_1, \ldots, x_d), n = (n_1, \ldots, n_d), n \cdot x = n_1 x_1 + \cdots + n_k x_k$.

To see how distribution theory affects Fourier integral theory, let us first consider the case of $\mathbf{R}$. The test functions are now no longer compactly supported and have rapid decay at infinity; i.e., they are smooth and each derivative is $O(|x|^{-k})$ for every $k \geq 0$. They form a space $\mathcal{S}$, the so-called Schwartz space of $\mathbf{R}$. It is wider than $\mathcal{D}$, its topology is stronger than that of $\mathcal{D}$, but $\mathcal{D}$ is dense in $\mathcal{S}$. Thus not every distribution can extend to a continuous linear functional on $\mathcal{S}$, but if it does, the extension is unique. The elements of the dual $\mathcal{S}'$ of $\mathcal{S}$ are called *tempered distributions*. Fourier transform

$$f \longmapsto \widehat{f}$$

is a *linear topological isomorphism* of $\mathcal{S}$ with itself, and this fact allows the concept of Fourier transform to be extended to all tempered distributions by

$$\widehat{T}(f) = T(\widehat{f}).$$

In some sense this definition gives the furthest extension of the notion of the Fourier transform. Thus, the Plancherel formula, expressing the unitarity of the L^2-Fourier transform, takes a very simple form in the language of distributions. Since

$$\int_{\mathbf{R}} |f|^2 \, dx = (f * \tilde{f})(0) \qquad (\tilde{f}(x) = f(-x)^{\mathrm{conj}})$$

the Plancherel formula is a consequence of the inversion formula

$$f(0) = \frac{1}{\sqrt{2\pi}} \int_{\mathbf{R}} \widehat{f}(\xi) \, d\xi \qquad (f \in \mathcal{S})$$

which can be written as

$$\delta = \frac{1}{\sqrt{2\pi}} \int_{\mathbf{R}} e^{i\xi x} d\xi$$

expressing δ, the Dirac delta function at the origin, as a linear combination of the basic exponentials. In this form this formula is commonly used by physicists. For Fourier series this becomes

$$\delta = \sum_{n \in \mathbf{Z}} e^{inx},$$

which we have seen above. As a final example we mention the Poisson Summation Formula, which is classically stated in the form

$$\sum_{n \in \mathbf{Z}} f(n) = \sum_{n \in \mathbf{Z}} \widehat{f}(2\pi n)$$

for functions f in the Schwartz space $\mathcal{S}$. Here the Fourier transform is defined by

$$\widehat{f}(\xi) = \int_{\mathbf{R}} f(x) e^{-i\xi x} \, dx.$$

Removing the test function f from the identity we get

$$\sum_{n \in \mathbf{Z}} e^{2\pi i n x} = \sum_{n \in \mathbf{Z}} \delta(x - n).$$

It is impossible here to give even an overview of the pervasive nature of the influence of the theory of distributions in modern analysis. I single out just two of the most striking to illustrate what I have in mind.

The first instance arises in the foundations of the theory of Riemann surfaces. Here the fundamental result is the theorem of Riemann asserting the existence of a nonconstant meromorphic function on a compact Riemann surface. This comes

down to constructing a harmonic function satisfying appropriate conditions, and
Riemann's argument was incomplete (as noted by Weierstrass) because the required
function, obtained by Riemann as a solution of a variational problem, could not
be guaranteed to be harmonic because it could not be ascertained that it had the
required smoothness. However it satisfied the Laplace equation in the *weak sense*
described above, and Hermann Weyl proved that weak solutions of the Laplace
equation are actually harmonic in the classical sense. He thus resurrected the
Riemann proof. Nowadays this result is imbedded in the general regularity theorem:
if D is an *elliptic* differential operator with smooth coefficients, any distribution
solution of the equation $DT = 0$ is necessarily smooth.

The second instance where distributions have played an indispensable role is in
the theory of irreducible unitary representations of Lie groups, especially semisimple
Lie groups. Among Lie groups the semisimple groups are particularly significant
because they are defined algebraically and are the only groups capable of acting
transitively on smooth projective varieties. This of course defines only the *complex*
semisimple groups; we may, for the present discussion, take the real semisimple
Lie groups to be the *real forms* of the complex groups. Their history is a long
one going back to Killing and E. Cartan, who classified all of them. They are
important in a variety of contexts, especially in physics and number theory. One of
the first applications of distribution theory to representation theory of semisimple
Lie groups was made by Bruhat to describe the intertwining operators between
two representations of a semisimple Lie group G or, more generally, the invariant
bilinear forms for a pair of representations. Now, if we take a space such as the
space $\mathcal{D}$ of test functions on $\mathbf{R}$, a bilinear form on $\mathcal{D} \times \mathcal{D}$ may be viewed as a
generalized kernel $k(x, y)$, i.e., a distribution on $\mathbf{R} \times \mathbf{R}$; this is in fact the famous
kernel theorem of Schwartz. Indeed, if we further assume that the bilinear form is
invariant under the (diagonal) action of $\mathbf{R}$ on $\mathbf{R} \times \mathbf{R}$, then the generalized function
k depends only on $x - y$ and so becomes a *convolution operator*. (This was in
fact the guiding principle of all of Wiener's work on harmonic analysis; see [31].
Using this idea, Bruhat reduced the problem of determining the invariant bilinear
forms for a pair of induced representations of G to the determination of certain
invariant distributions and proceeded to determine them. But the most profound
applications of the theory of distributions to representation theory of G were due
to Harish-Chandra. Harish-Chandra used distribution theory more systematically
in his study of the structure of the *characters* of irreducible unitary representations
of semisimple Lie groups.

If a representation L of a Lie group G is unitary and acts on a Hilbert space
$\mathcal{H}$, then as long as $\mathcal{H}$ is finite dimensional, one can calculate its *character* by the
formula

$$\Theta_L(x) = \operatorname{Tr}(L(x), \qquad (x \in G).$$

But if the representation is infinite dimensional, for instance if G is the Lorentz
group which has no finite-dimensional nontrivial unitary representations but plenty
of infinite-dimensional ones, any attempt to calculate the character of L by

$$\Theta_L(x) = \sum_n (L(x)e_n, e_n)$$

where (e_n) is an orthonormal basis of $\mathcal{H}$ is doomed to failure, because the sum
cannot converge even at the identity element and indeed seldom even anywhere
else because the terms oscillate too much. Nevertheless Harish-Chandra discovered

that when G is semisimple, the series above *always converges in the smeared sense* and so defines a distribution Θ_L:

$$\Theta_L(f) = \sum_n \int (L(x)e_n, e_n) f(x) \, dx.$$

This Θ_L is then called the *distribution character of L*, or, the *Harish-Chandra character of L*. These characters satisfy certain differential equations, and this fact allowed him to show that ultimately these characters are class functions just as in the compact case. However when the group is noncompact, the class functions are smooth only away from a subvariety of codimension 1, although they remain locally integrable everywhere; a simple example of such a situation is the function $|x|^{-1/2}$ on the real line, which is nice away from the origin but integrable around it. Using his character theory, Harish-Chandra erected his monumental theory of Fourier analysis on all semisimple Lie groups, including what is nowadays called the *Harish-Chandra character formula*. The character formula of Harish-Chandra is a far-reaching generalization of the Weyl character formula for compact Lie groups to the context of noncompact semisimple Lie groups and is perhaps the single most beautiful formula of infinite-dimensional representation theory.

This introduction to distributions and their role in Fourier analysis given at breakneck speed does not do much justice at all to the historical evolution of ideas. To trace this evolution in good detail is itself a monumental task. As a very weak compensation let me mention just a few of the many names of mathematicians and physicists who took some fundamental step: in the 19th century, the *operational calculus* of Heaviside, Mikusinski, and from quantum theory, Dirac, with his ubiquitous delta function; the idea of Hadamard of defining the *finite part* of a divergent integral, which he used effectively in his theory of the *wave equation*; the ideas of Bochner on Fourier integrals; the ideas of Sobolev, and earlier, of Weyl, on the concept of *weak solutions* of partial differential equations. Then in the 1950's, Laurent Schwartz carried out a great synthesis of these diverse themes and systematized the theory of distributions and their applications (especially to the Fourier transform theory) in his enormously influential book. After the appearance of Schwartz's book, the impact of the theory of distributions became tremendous, especially in the theory of differential equations and representation theory [32].

5.10. Gaussian integrals, Wiener measure and the path integral formulae of Feynman and Kac

Integrals like

$$\int e^{itx^2} \, dx$$

occur in optics. Over a finite range they exist, but over infinite ranges they do not exist even as conditionally convergent integrals and have to be viewed as distributions or understood in terms of some summability method. In higher dimensions the integrals analogous to the one above are of the form

$$\int_{\mathbf{R}^d} e^{itQ(x)} \, dx$$

where Q is a quadratic form. The classical summability methods do not work well in higher dimensions and so we have to use other methods. One effective way to deal with them is by the method of distributions. Here we shall discuss another way to treat them, namely by *analytic continuation*.

Gaussian integrals by analytic continuation. The idea behind the method of analytic continuation is very simple. If we make an analytic continuation in t and go over to the situation where t is *pure imaginary*, we get integrals of the form

$$\int_{-\infty}^{\infty} e^{-tx^2} \, dx, \qquad \int_{\mathbf{R}^d} e^{-tQ(x)} \, dx$$

which are convergent when $t > 0$ and Q is a positive definite quadratic form. They are in fact legitimate Gaussian integrals arising in the theory of probability. Let us now take the variable τ to be *complex* and consider

$$I(\tau) = \int_{\infty}^{\infty} e^{-\tau x^2} \, dx.$$

As

$$|e^{-\tau x^2}| \le e^{-\Re(\tau)x^2}$$

the integral is convergent when $\Re(\tau) > 0$; moreover, as the integral

$$\int_{\infty}^{\infty} e^{-ax^2} x^2 \, dx \qquad (a > 0)$$

is uniformly convergent when $a \ge a_0 > 0$, $I(\tau)$ can be differentiated with respect to τ under the integral sign. So the function I is analytic in τ. For τ real and > 0 we have

$$I(\tau) = \sqrt{\frac{\pi}{\tau}}$$

and so we have

$$I(\tau) = \frac{\sqrt{\pi}}{\sqrt{\tau}} \qquad (\Re(\tau) > 0)$$

where $\sqrt{\tau}$ is the analytic square root of τ in the half plane $\Re(\tau) > 0$ which is 1 when $\tau = 1$. Thus

$$I(\tau) = \sqrt{\frac{\pi}{|\tau|}} e^{-i\theta/2} \qquad (\tau = |\tau| e^{i\theta}, |\theta| < \pi/2).$$

Actually the analytic square root of τ we have determined lives on the sector where $|\theta| < \pi$ or on the cut plane obtained by removing the negative real axis and 0 from the complex plane. In particular, taking $\tau = -i = e^{-i\pi/2}$ we have

$$I(-i) = \int_{\infty}^{\infty} e^{ix^2} \, dx = \sqrt{\pi} e^{i\pi/4}.$$

The only delicate part of this calculation is the determination of the branch of the square root. This method generalizes easily to higher dimensions when the quadratic form Q is positive definite. Then in a suitable orthogonal basis it can be taken as

$$Q(x) = \lambda_1 x_1^2 + \cdots + \lambda_d x_d^2 \qquad (\lambda_j > 0)$$

so that the integral is a product of one-dimensional integrals. The result is

$$\int_{\mathbf{R}^d} e^{iQ(x)} \, dx = \pi^{d/2} e^{id\pi/4} (\det Q)^{-1/2}.$$

Wiener measure, Brownian motion, and Path integrals. Most branches of analysis deal almost exclusively with Lebesgue measure in Euclidean spaces or measures in the setting of locally compact spaces. The central problems of modern probability and physics however require the use of measure and integration in *function spaces*, which are usually infinite dimensional and so not locally compact. The mathematical theory of measure and integration in function space goes back to the pioneering work of Wiener in the early decades of the 20th century when he constructed rigorous mathematical models for *Brownian motion* and established a number of important properties for it which were till then understood only heuristically from the works of Einstein, Smouluchowsky, Perrin, and others. Although measure and integration in abstract spaces was developed by Frechet and Daniell before Wiener, it was Wiener's explicit construction of the measure on the space of continuous functions on the real line that was decisive. Wiener constructed a probability measure in the space of all continuous functions on $C[0, \infty)$, nowadays called the *Wiener measure*. It is a fundamental construct in infinite-dimensional analysis, exactly as the Lebesgue measure is the fundamental object in classical analysis. The method of integration and measure theory in spaces of functions is generally known as the method of *path integrals* among physicists [33].

As is well-known, Brownian motion refers to the continuous and irregular motion of a small particle suspended in a liquid, first observed by the English botanist Robert Brown in 1828. It was eventually realized that this motion arises from the collisions of the molecules of the liquid with the particles. Since approximately 10^{20} collisions occur per second, the movement of the particle could be treated only statistically. The work of Einstein and Smouluchowsky did exactly that. Let $x(t)$ be one of the coordinates of the position of the small particle at time t with $x(0) = 0$. Einstein, assuming essentially the independence of the *increments* of the process $x(t)$, showed that the function $p(t : u)$, which is the probability density for $x(t)$, satisfied the *diffusion equation*

$$\frac{\partial p}{\partial t} = \frac{A}{2} \frac{\partial^2 p}{\partial x^2}$$

and hence derived the formula

$$p(t : u) = \frac{1}{\sqrt{2\pi At}} \exp\left(-\frac{u^2}{2At}\right)$$

for p. Here A is a constant which can be explicitly written down in terms of the absolute temperature of the liquid, its coefficient of viscosity, the radius of the small particle (assumed to be spherical), and the *Avogadro number*. Since

$$A = \langle x(1)^2 \rangle := E(x(1)^2),$$

the mean value of $x(1)^2$, it follows that an experimental determination of A would (and did) yield the value of the Avogadro number. These ideas played a big role in the universal acceptance of the atomic hypothesis. It follows from these facts that the coordinates x_i at times t_i, $t_1 < t_2 < \cdots < t_k$, have the joint probability density

$$(\mathbf{D}) \qquad \frac{1}{(2\pi A)^{d/2} \prod_{1 \leq i \leq k} (t_i - t_{i-1})^{1/2}} \exp\left[-\frac{1}{2A}\left(\sum_{1 \leq i \leq k} \frac{(x_i - x_{i-1})^2}{(t_i - t_{i-1})}\right)\right]$$

with the convention that $t_0 = 0$ and $x_0 = 0$, that is, that the particle motion starts at the origin of coordinates. This formula is equivalent to the following properties (here and from now on we shall always suppose that $A = 1$):

(i) $x(0) = 0$ and the increment $x(t) - x(s)$ $(0 \leq s < t)$ is Gaussian with mean 0 and variance $(t - s)$.

(ii) The increments corresponding to disjoint intervals are independent.

The construction of Wiener measure can be carried out along the same lines as the construction of Lebesgue measure. In the Lebesgue theory, measures are assigned to certain *basic sets*, such as intervals in $\mathbf{R}$ or rectangles $I_1 \times \cdots \times I_d$ in higher-dimensional spaces. Then it is proved that these formulae are consistent and extend to a *countably additive* measure in the σ-algebra of all Borel sets in the space. In the case of Wiener measure the basic space is $\mathcal{C} := C[0, \infty)$. One can view $\mathcal{C}$ as a complete separable metric space under the collection of seminorms defined by

$$||x||_n = \sup_{0 \leq t \leq n} |x(t)| \qquad (n = 1, 2, \dots).$$

If we are working over $C[0, T]$ for some fixed time T, it is enough to use a single norm $||\cdot||_T$, and then $C[0, T]$ will be a Banach space, separable because of the Weierstrass theorem of polynomial approximation. For $\mathcal{C}$, the same argument shows that it is a separable Frechet space and hence a complete separable metric space. For the basic sets we take the sets which are *determined by conditions on a finite set of time points*, namely, sets of the form

$$\left\{ x \;\middle|\; (x(t_1), x(t_2), \dots, x(t_k)) \in E \right\}$$

where the $t_j \in [0, \infty)$ are arbitrary and E is a Borel set in $\mathbf{R}^k$. For these, the measure is calculated by using the density function $(\mathbf{D})$. The σ-algebra generated by the basic sets is defined as the class of Borel sets in the metric space $\mathcal{C}$; it coincides with the σ-algebra generated by the open sets. In the Lebesgue case the fact that the measure defined on the basic sets extends to a countably additive measure depends ultimately on the *compactness* of bounded closed subsets in Euclidean spaces. In $\mathcal{C}$ there is no local compactness, but Wiener overcame this huge obstacle and proved that there *is* a countably additive measure on the Borel sets with the values on basic sets determined by $(\mathbf{D})$. This is the *Wiener measure*, which we shall denote by W_0.

A full understanding of Wiener's epoch-making achievement was possible only after the foundations of modern probability theory were laid by **Kolmogorov** in the 1930's and the theory of *Stochastic Processes* was developed by Kolmogorov himself and many others. As a consequence of these developments it has now become accepted that Wiener's theory and the theory of *Markov Processes*, which is a deep and far-reaching generalization of it, form one of the central and active areas of mathematics today, with many applications beyond probability not only to mathematical physics but also surprisingly to geometry, such as a probabilistic proof of the *Atiyah-Singer Index Theorem*. I might remark that at the fundamental level, Wiener's construction is a special case of a whole class of probability measures not necessarily Gaussian even, where no hypothesis is made about the statistics of the process except a moment condition based on the *two-dimensional distributions,* namely, conditions on the moments of $(x(t_1), x(t_2))$. This is the famous criterion of Kolmogorov-Slutsky [34], namely, that if for every $T > 0$ there are some constants $K_T > 0, \alpha_T > 0, \delta_T > 0$ such that

$$E|x(s) - x(t)|^{\alpha_T} \leq K_T |s - t|^{1 + \delta_T} \qquad (0 \leq s < t \leq T)$$

(E is the expectation), then there is always a countably additive extension to the Borel sets of $\mathcal{C}$. Here the probability distributions at finite sets of points have been assumed to be prescribed in a consistent fashion. In the Wiener case it is enough to take $\alpha_T = 4$; then

$$E|x(s) - x(t)|^4 = 3|s - t|^2$$

so that the criterion is satisfied wih $\delta_T = 1$ and $K_T = 3$.

It turns out that with probability 1, a Brownian motion path satisfies a Lipschitz condition of the form

$$|x(t) - x(s)| \leq K_T|s - t|^{1/2} \qquad (0 \leq s < t \leq T).$$

It is easy to give a heuristic explanation of this fact. Indeed, the increment over an interval of length h is Gaussian with mean 0 and variance h, and so a typical increment will be of size $\sqrt{h}$. In his autobiography [35], Wiener recalls the great effect that the remarks of Perrin had on him. Perrin, who had made extensive numerical observations on Brownian motion and won the Nobel Prize for this work, made the remark that the highly irregular trajectories of a particle executing a Brownian motion reminded him of the nondifferentiable functions of the mathematicians. Inspired by this, Wiener proved not only that with probability 1 the Brownian path satisfies a local Lipschitz condition with $\alpha = 1/2$, but further that for any $\alpha > 1/2$, the probability is 1 that the Brownian path satisfies *at no point* a Lipschitz condition of order α in its neighborhood; in particular, with probability 1 the *Brownian path is nondifferentiable everywhere.*

One can make many conventional constructions once one has Wiener measure available on $\mathcal{C}$. For instance, instead of starting at 0 we can start the Brownian motion at any point a to get the measure W_a. Mathematically, it is the measure induced by the translation $x \longmapsto x + a$ in $\mathcal{C}$ (a is the constant function a). Equivalently, if f is a bounded continuous functional on $\mathcal{C}$,

$$\int_{\mathcal{C}} f[x - a] \, dW_a(x) = \int_{\mathcal{C}} f[x] \, dW(x).$$

We can then construct Wiener measure for Brownian motion on $\mathbf{R}^d$ by taking the products of d copies of W; i.e., we identify the space of continuous maps of $[0, \infty)$ into $\mathbf{R}^d$ with $\mathcal{C}^d$ and define $W_{\mathbf{a}}^d$ by

$$W_{\mathbf{a}}^d = W_{a_1} \times \cdots \times W_{a_d} \qquad (\mathbf{a} = (a_1, \ldots, a_d)).$$

Then $W_{\mathbf{a}}^d$ is Wiener measure for Brownian motion starting at $\mathbf{a}$. As a final example we discuss a more delicate construction. Suppose we start the Brownian motion at $a \in \mathbf{R}$. Then we can ask how the probabilities are to be calculated if we impose the condition that at some specified time point, say T, the Brownian path is constrained to assume the value b. This is just the construction of conditional distributions in probability theory. To build these measures, say, on $C[0, T]$, we use for the joint probability distribution of $(x(t_1), \ldots, x(t_k))$, $(0 < t_1 < \cdots < t_k < T)$, the Gaussian measure with density

$$w_{0,b}(x_1, \ldots, x_k) = \frac{w_0(x_1, \ldots, x_k, b)}{w_0(b)}$$

where $w_0(x_1, \ldots, x_k, x_{k+1})$ is the density of $(x(t_1), \ldots, x(t_k), x(T))$ defined by (**D**). This density in fact defines the conditional distribution of $(x(t_1), \ldots, x(t_k))$ given that $x(T) = b$. Now, if (X, Y) is Gaussian with $E(X^2) = \sigma_1^2$, $E(Y)^2 = \sigma_2^2$, $E(XY) = \rho\sigma_1\sigma_2$, then a simple calculation yields the formulae

$$E_b(X) = \rho\frac{\sigma_1}{\sigma_2}, \qquad E_b\left(\left(X - \rho\frac{\sigma_1}{\sigma_2}\right)^2\right) = \sigma_1^2(1 - \rho^2)$$

where E_b refers to the expectation under the condition $x(T) = b$. Applying this to $X = x(t) - x(s)$, $Y = x(T)$, we get

$$E_b\left((x(t) - x(s))^2\right) = (t - s)\left(1 - \frac{t - s}{T}\right) \leq (t - s) \qquad (0 \leq s < t \leq T).$$

Hence

$$E_b\left((x(t) - x(s))^4\right) = 3E_b\left((x(t) - x(s))^2\right) \leq 3(t - s)^2.$$

By the Kolmogorov-Slutsky criterion mentioned earlier we have a measure defined on the space $C_{0,b}[0, T]$ of continuous functions x with $x(0) = 0$, $x(T) = b$. This is the measure $W_{0,b,T}$. As before $W_{a,b,T}$ can be obtained from $W_{0,b-a,T}$ by translating by a. The measures $W_{a,b,T}$ satisfy

$$E_{W_a}(f[x]) = E\left(E_{W_{0,b,T}}f[x]\right)$$

where the outer expectation on the right side is with respect to the density $g_T(b)$ of $x(T)$. We can then call

$$W_{a,b,T}$$

the Wiener measure for the Brownian motion which starts from a at time 0 and ends at b at time T. Its extension to $\mathbf{R}^d$ is obvious and gives the measures

$$W_{\mathbf{a},\mathbf{b},T} = W_{a_1,b_1,T} \times \cdots \times W_{a_d,b_d,T}.$$

Once we have a measure we have the corresponding integration theory. This is the content of the great achievement of Lebesgue and others, namely that abstract measure and integration theory are independent of dimension. As an example that is extremely important for us let us take

$$f[x] = e^{-S[x]} \qquad (x \in C[0, T])$$

where

$$S[x] = \left(\int_0^T V(x(s))\, ds\right)$$

for some continuous function V on the real line which is nonnegative:

$$V : \mathbf{R} \longrightarrow \mathbf{R}, \qquad V \geq 0.$$

Clearly S is continuous and ≥ 0, and so f is bounded and continuous. Now, the Riemann integral of $V(x(s))$ can be approximated by Riemann sums, and so we can write

$$\int_0^T V(x(s))\, ds = \lim_N \sum_{1 \leq i \leq N} (t_{N,i} - t_{N,i-1})V(x(t_{N,i-1}))$$

where $(t_{N,0} = 0, t_{N,N} = T)$ and the limit is taken when

$$\sup_i |t_{N,i} - t_{N,i-1}| \to 0 \qquad (N \to \infty).$$

Then, by the bounded convergence theorem, for any probability measure μ on $C[0, T]$,

$$\int_{C[0,T]} e^{-S[x]} d\mu = \lim_N \int e^{-\left(\sum_{1 \le i \le N} (t_{N,i} - t_{N,i-1}) V(x(t_{N,i-1}))\right)} d\mu.$$

The integrands on the right are of functionals which *depend only on a finite number of time points* and so their integrals can be calculated from the definition of the densities at these points: for $\mu = W_0$ from (**D**) and for $\mu = W_{a,b,T}$ from the corresponding densities. So we have a very effective method of computing certain path integrals. Note however that the above formula *cannot* be made the basis of the definition of a path measure, just as the theory of Riemann sums and Riemann integration does not lead to a countably additive measure on the line. *One first has to have the measure, and then such formulae follow as a consequence.* With this we have completed our introduction to Wiener measure and taken care of the details that will be needed to connect with Feynman's work.

The Feynman approach to quantum mechanics. All the above developments, revolutionary so far as mathematics was concerned, had nothing to do with modern physics or the quantum theory that was revolutionizing physics at the same time. In 1942, Richard Feynman, a graduate student at Princeton University, wrote a dissertation in physics that contained a novel way to formulate quantum mechanics. Because of the war the publication of the results of this thesis was delayed and it was not until 1948 that they first appeared in print [36]. His idea was to start with a classical system and associate to it some sort of a *complex measure* on the space of classical trajectories in terms of which he gave prescriptions for calculating the probabilities of events in the associated quantum system. This measure was supposed to be *completely determined by the classical action functional.* Feynman's method is nowadays called the method of *path integrals;* it has since then spread into all aspects of quantum physics, and for many physicists it is the preferred mode of thinking and calculation. I shall now give a brief discussion of this method, explain the difficulties in formulating it as a rigorous mathematical instrument, and discuss some situations where this can be done.

The essence of Feynman's method is a formula that exhibits the *propagator* (= the operator that gives the time evolution) of a quantum dynamical system as an integral taken over the space of classical paths of the same system treated classically. The rigorous formulation of Feynman's ideas is rather subtle and is still possible only in certain limited circumstances. A big step was taken in the 1950's by Mark Kac, who obtained a variation of the Feynman formula for the propagator called the *Feynman-Kac formula* [37]. This formula is very similar in spirit to the evaluation of imaginary Gaussian integrals by analytic continuation that we discussed at the beginning of this section but is much more profound because the measure with respect to which the imaginary Gaussian integrals are taken does not even exist! Its starting point (roughly speaking) is the observation of Kac that the Schrödinger equation goes into the diffusion equation when time is made imaginary, and since the solutions to the diffusion equation generate the Wiener measure, the *propagator in imaginary time* must become a bona fide integral in function space with respect to the *Wiener measure.* A variety of physical and mathematical problems can be solved with this technique. Subsequently, this idea was generalized by Schwinger as a method of analytic continuation from the *Minkowski space time* to the *Euclidean space time* and played an important part in understanding some of the mathematical aspects of quantum field theory. I shall try to explain first the Feynman formula and then the Feynman-Kac formula. Although no complete proofs are possible or even attempted, extensive motivation can be gained by looking at *finite models*.

Let me first take up the Feynman formula. The starting point is that all point particles like the photon or the electron have both particle and wave aspects and only specific experimental arrangements can decide which particular aspect is being observed. The famous double slit experiment illustrates this point very well. In the idealized form of this experiment, electrons are made to fall on a screen and their paths are intercepted by a screen with two slits. The distribution of the electrons on the screen is *not* the sum of the distributions obtained when one or the other slit is kept closed; the explanation for this is that in some sense when both slits are open, the experimental arrangement does *not* single out the particle aspect and so wave aspects have to be allowed and sometimes the electron wave passes through both holes! If we try to make sure that the electron passes through one or the other hole by placing a scatterer at each hole which will send a signal each time an electron passes through that hole, we find that the distribution of the electrons on the screen *is indeed* the sum of the two distributions obtained when one of the holes is kept closed.

In the case of a general quantum situation the above heuristic explanation can be generalized to the following point of view. In classical mechanics, if we consider two points x, y in configuration space, then in general there is only one path the particle can follow in going from x to y in time t. This path is in fact the one that makes the action a minimum, or at least an extremum. But in quantum mechanics, because of Heisenberg's uncertainty principle, the position and momentum cannot both be exactly determined at any instant of time, so that the particle is *smeared out* and one can no longer say what its path is with certainty–the double slit experiments demonstrate this clearly. Feynman's idea was that *one should consider* all possible paths *that go from x to y in time t and* average over them *to get the quantum probability of going from x to y in time t.*

One can get a little better understanding of this definition of the quantum probability by taking a look at a *finite model*. We take the configuration space to be a *finite* set X and the Hilbert space of quantum theory to be the space $\mathcal{H}$ of complex functions on X. The time evolution of the particle will be assumed to be given by a one parameter unitary group

$$t \longmapsto U_t \qquad (t \in \mathbf{R})$$

with the interpretation that if the state of the system at a particular time is the unit vector $\psi \in \mathcal{H}$, then its state after t units of time is the unit vector $U_t\psi$. The key property is

$$U_{t_1+t_2} = U_{t_1}U_{t_2}.$$

For simplicity we consider time evolution to be also discrete, say, in steps of some unit τ. If we consider epochs of time $t_j = k_j\tau$ $(1 \leq j \leq N, k_1 < \cdots < k_N)$, then the unitary operator that takes the system from the state at time 0 to the state at time $t = k_N\tau$ is

$$U_t = U_{t_N-t_{N-1}}U_{t_{N-1}-t_{N-2}}\cdots U_{t_1}.$$

The function (often called *kernel*)

$$K_t(x, y) := (U_t \delta_y, \delta_x) \qquad (x, y \in X)$$

(the deltas denote delta functions) is called the *propagator* and satisfies the semi-group law

$$K_{t_1+t_2}(x, y) = \sum_{z \in X} K_{t_1}(x, z) K_{t_2}(z, y).$$

$K_t(x, y)$ is called the *complex probability amplitude* for going from x to y in time t. The quantity

$$P_t^q(x, y) = |K_t(x, y)|^2$$

(q for quantum) is the probability of finding the particle at y at time t given that it was at x at time 0. In general

$$P_{t_1+t_2}^q(x, y) \neq \sum_{z \in X} P_{t_1}^q(x, z) P_{t_2}^q(z, y).$$

If the evolution were classical the classical transition probabilities P^{cl} would satisfy the equation

$$P_{t_1+t_2}^{cl}(x, y) = \sum_{z \in X} P_{t_1}^{cl}(x, z) P_{t_2}^{cl}(z, y).$$

The $P^{cl}(x, y)$ will then determine a *Markov process* from which more complicated probabilities can be calculated. The difference between the classical and the quantum situations is striking: in the quantum case the probabilities do not satisfy the semigroup property, only the amplitudes do, and the probabilities are obtained as the absolute squared values of the complex amplitudes.

The semigroup property for the quantum amplitudes now gives us the formula

$$K_t(x, y) = \sum_{z_1, z_2, \dots, z_{N-1}} K_{t_1}(x, z_1) K_{t_2-t_1}(z_1, z_2) \dots K_{t_N - t_{N-1}}(z_{N-1}, y).$$

If we now regard the set of sequences

$$x, z_1, z_2, \dots, z_{N-1}, z_N = t$$

as a sort of path from x to y in time t, with the particle at z_i at time t_i, we see that the amplitude for going from x to y in time t is a sum over the paths of amplitudes attached to each path, the amplitude attached to the path

$$x, z_1, z_2, \dots, z_{N-1}, z_N = t$$

being

$$K_{t_1}(x, z_1) K_{t_2-t_1}(z_1, z_2) \dots K_{t_N - t_{N-1}}(z_{N-1}, y).$$

We can now take X to be a lattice of points on the real line with an interlattice distance that goes to zero as the number of points on the lattice go to infinity and the lattice approximates the whole line. We take the time evolution to be in steps of t/N and then pass to the limit as $N \to \infty$. Then we get the Feynman formulation at least heuristically:

$$K_t(x, y) = \int A[u] \mathcal{D}u$$

where $\mathcal{D}u$ is some kind of a measure on the space of paths u such that $u(0) = x, u(t) = y$ and $A[u]$ is the amplitude attached to the path. Feynman's stroke of genius was the insight that the amplitude associated to the path u is to be taken as

$$e^{(i/\hbar)S[u]}$$

where $S[u]$ is the *classical action* of the path u. Here $\hbar = h/2\pi$, h being Planck's constant. This was how Feynman obtained his famous formula

$$K_T(x, y) = \int_{\mathcal{P}(x,y,T)} e^{iS[u]/\hbar}\, \mathcal{D}u$$

where

$$\mathcal{D}u = \prod_{0 \leq t \leq T} du(t)$$

is the product of all Lebesgue measures corresponding to the various time points defined on the space $\mathcal{P}(x, y, T)$ of all classical paths, and the integration is over the space of all classical paths that start from a at time 0 and end up in b at time T. In his thesis Feynman used this as the *definition* of the propagator to derive the Schrödinger equation and to establish the equivalence of this approach to quantum mechanics to the usual one [36].

The reader does not have to be told that this is an absolutely astonishing recipe for calculating the propagator. The amazing thing about it was that Feynman was able to calculate with it, derive the Schrödinger equation, and prove that this method was *fully equivalent* to the Heisenberg–Schrödinger–Dirac–von Neumann way of doing quantum mechanics via operators, commutation rules, and spectral theory. Unfortunately, it is impossible to make any sense of it in a simple manner. For instance, it is impossible to find a candidate for the "Lebesgue measure" $\mathcal{D}u$ that takes the place of summation in the finite models. The attempts going back to Feynman to define the integral representation of the propagator by a limiting process similar to the ones used in the theory of Riemann integration are no good (as we have already observed) because one cannot define a theory of integration that way. The integration theory has to be defined first, and then one can use such approximation schemes to compute a well defined integral, but not the other way around. Moreover, even in the case of the finite models, there is no complex measure on the space of paths which extends the complex amplitudes associated to each "finite" path where we take the times at a finite number of points. Nevertheless this method has proved very popular among physicists. From the mathematician's point of view the method itself has thus to be treated as a formal instrument with great suggestive power, which can often lead to predictions that can be verified by other methods and which can be justified by a more sophisticated method than trying to define a complex measure. The method of Kac which we shall describe now is based on the observation that when we go to *imaginary time*, the ghostlike *Feynman measure* becomes the very concrete and nice *Wiener measure*, and the Feynman formulae can be obtained by analytic continuation. This is of course very similar in spirit to the method we used at the beginning of this section to evaluate imaginary Gaussian integrals by making the parameter t imaginary to get real Gaussian integrals. Actually this analogy is much deeper. In fact, in Feynman's original thesis, when he attempts to calculate the propagator for a single particle, he uses the Lagrangian

$$\mathcal{L}[u] = \dot{u}^2 - V(u)$$

so that the action is

$$S[u] = \int_0^T \left[\dot{u}(s)^2 - V(u(s)) \right] ds.$$

If we replace the path by an approximation at a finite set of time points with distance ε between them, then $S[u]$ can be approximated by

$$S[u] \approx S_N[u] = \sum_{k=1}^N \left[\left(\frac{u(t_k) - u(t_{k-1})}{t_k - t_{k-1}} \right)^2 - V(u(t_k)) \right] (t_k - t_{k-1})$$

and the corresponding finite-dimensional integrals

$$\int e^{iS_N[u]/\hbar}$$

are, apart from the factor involving V, imaginary Gaussian integrals. Thus the Kac method is *almost exactly the same* as the one we discussed. In other words, going to imaginary time may be viewed as *some sort of a summation method* that replaces an oscillatory undefinable complex measure into a genuine probability measure on path space. The transition from Feynman to Kac becomes even more remarkable if we noticed that whereas the Feynman integral is over a space of classical paths, the Wiener measure gives probability 0 for the classical trajectories because they are all differentiable!

Feynman-Kac formula. Mark Kac was at Cornell at the same time as Feynman when Feynman described some of his ideas in a physics colloquium in 1947. Kac realized that by going to imaginary time one can go over to Wiener integrals and do all calculations in a completely rigorous manner and that a good part of this had already been done by him. The formula that Kac obtained by pursuing this idea is known as the *Feynman-Kac formula* and is one of the landmarks of rigorous path integral theory [38]. I shall now briefly discuss this formula. In particular the derivation of this formula will explain at least partially Feynman's profound insight that the functional inside the integration for the propagator is proportional to the exponential of the action.

To motivate the formula let us again go to the finite model where the configuration space is a finite space X and the Hilbert space is $\mathcal{H}$, the space of complex functions on X; however we shall now assume that time evolves continuously. Let us assume that the Hamiltonian of the quantum system is of the form

$$H = H_0 + V, \qquad H_0 = -\Delta_X$$

where $H_0 = -\Delta_X$ is the "free" part and V is a "potential", V being a real function on X. The notation is clearly set up to highlight the analogy with the quantum mechanics of a particle in $\mathbf{R}^d$ moving in the field of a potential function V; there the Hamiltonian is $-\Delta_d + V$ where Δ_d is the Laplacian in $\mathbf{R}^d$. The dynamical group is (with $\hbar = 1$)

$$U_t = e^{-itH},$$

and we wish to get a path integral representation for the kernel

$$K_t(x, y) = (U_t \delta_y, \delta_x) \qquad (x, y \in X).$$

The interchange of x and y from our previous discussion is made for minor reasons. Because of the finite dimensionality of $\mathcal{H}$ the *complex dynamical group*

$$U_z = e^{-izH} \qquad (z \in \mathbf{C}).$$

In particular we shall be interested in

$$e^{-tH}.$$

The basic assumption now is that the free part H_0 has the following property: the semigroup

$$e^{-tH_0} = e^{t\Delta_X}$$

generates a Markov semigroup; i.e., the function

$$p_t(x, y) = (e^{t\Delta_X}\delta_y, \delta_x) \qquad (t > 0)$$

is the transition probability of a Markov particle being at y at time t if at time 0 it was at x. This is the same as requiring that

$$p_t(x, y) \geq 0 \qquad \sum_y p_t(x, y) = 1.$$

It can be shown that this is possible only if Δ_X satisfies the following conditions:

(1) $\Delta_X 1 = 0$.
(2) If f is a real function on X and x_0 is a point at which f is minimum, then $(\Delta_X f)(x_0) \geq 0$.

The second condition is known as the *minimum value principle* for Δ. This continues the analogy with the Laplacian one step further. Let

$$a(x, y) = (\Delta_X \delta_y, \delta_x), \qquad A = (a(x, y))_{x,y \in X}$$

so that A is the matrix of Δ_X in the basis (δ_x). Then the above conditions are also equivalent to the following:

(1) $\sum_y a(x, y) = 0$
(2) $a(x, y) \geq 0$ $(x \neq y)$ $(\Longrightarrow a(x, x) \leq 0)$

We shall also add a third condition:

(3) $0 < a(x, x) < 1$ for all $x \in X$.

One can then build a family

$$P_x (x \in X)$$

of genuine probability measures on the space $\mathcal{F}$ of all step functions ω on the positive time axis $[0, \infty)$ which are right continuous and take values in X such that the probability

$$P_x \left(\omega \mid \omega(t_1) = x_1, \omega(t_2) = x_2, \ldots, \omega(t_N) = x_N \right)$$

is equal to

$$p_{t_1}(x, x_1) p_{t_2 - t_1}(x_1, x_2) \ldots p_{t_N - t_{N-1}}(x_{N-1}, x_N).$$

Notice the resemblance to the amplitude formula described in the beginning of this section. We also note that the elements of the space $\mathcal{F}$ are precisely all functions

$$\omega : [0, \infty) \longrightarrow X$$

such that for some s_i $(i = 1, 2, \ldots,)$ and $y_i \in X$,

$$s_0 = 0 < s_1 < s_2 < \cdots < s_m < \ldots, \qquad s_m \to \infty$$

and we have

$$\omega(t) = y_j (s_j \leq t < s_{j+1})(j \geq 0).$$

The probability measure on $\mathcal{F}$ corresponds to a Markov evolution of the particle which jumps from element to element according to the following rules: if it is at any given "site" $x \in X$, it stays there a random time t with a density

$$a(x, x)e^{-ta(x,x)}dt$$

so that the mean waiting time at the site x is

$$\frac{1}{a(x, x)}.$$

When it jumps from the site it moves to the site $y \neq x$ with probability

$$\frac{a(x, y)}{\sum_{z \neq x} a(x, z)}.$$

The probability measure is thus completely determined by Δ_X, and moreover Δ_X itself has a probabilistic interpretation as above. The meaning of the condition $0 < a(x, x) < 1$ is now clear; it means that the mean waiting time at a site is finite, so that the paths are step functions.

As an interesting special case let us now assume that the sites are *linearly ordered*, say,

$$x_1 < x_2 < \cdots < x_N,$$

and let $a > 0$ be a positive constant. Let us take Δ_X to be the matrix

$$a(x_i, x_j) = \begin{cases} \frac{1}{a^2} & \text{if } j = i \pm 1 \\ -\frac{1}{a^2} & \text{if } j = i \end{cases}$$

for $1 < i < N$; $a(x_1, x_1) = a(x_N, x_N) = -1/a^2$; $a(x_1, x_j)(j > 1)$ and $a(X_N, x_j)(j < N)$ are defined to be ≥ 0 and satisfy $\sum_{j>1} a(x_1, x_j) = \sum_{j<N} a(x_n, x_j) = 1/a^2$. Then Δ_X satisfies the required conditions and its action on a function f on X is

$$(\Delta_X f)(x_i) = \frac{f(x_{i-1}) - 2f(x_i) + f(x_{i+1})}{2a^2} \qquad (1 < i < N)$$

unless $i = 1, N$, in which case $(\Delta_X f)(x_i) = \sum_j a(x_i, x_j)f(x_j)$. The corresponding Markov process is a *random walk* where the particle jumps after a random waiting time to either of the two adjacent sites with probability $1/2$ unless it is at the boundary, in which case it goes into one of the remaining points with specified probabilities. We may also define the boundary values by periodicity, i.e., by identifying x_{N+1} with x_1. If we take the x_i to be equispaced with $x_i - x_{i-1} \sim a \sim N^{-1/2}$, it is clear that for large N the operator Δ_X is very close to d^2/dx^2.

Given the measures P_x we can define the conditional probability measures $P_{x,y,T}$ by

$$P_{x,y,T}(E) = P_x\left(\omega \in E \mid \omega(T) = y\right).$$

We then have the Feynman-Kac formula in this situation:

$$\left(e^{-TH}\delta_y, \delta_x\right) = \int_{\mathcal{F}} e^{-\int_0^T V(u(s))ds} \, dP_{x,y,T} \cdot p_t(x, y).$$

The original propagator e^{-iTH} is obtained by analytic continuation of the imaginary time propagator constructed above.

We shall give a sketch of the proof of this formula. We start with the easily proved result that if A, B are finite square matrices of complex entries,

$$e^{A+B} = \lim_{N \to \infty} \left(e^{A/N}e^{B/N}\right)^N.$$

To see this we write

$$e^{A/N}e^{B/N} = 1 + \frac{A+B}{N} + O\left(\frac{1}{N^2}\right) = 1 + \frac{C_N}{N} \qquad (C_N \to A + B).$$

An easy argument shows that

$$\left(1 + \frac{C_N}{N}\right)^N \to e^{A+B}.$$

Applying this result with $A = T\Delta_X$ and $B = -TV$ we have

$$e^{-TH} = e^{T(\Delta_X - V)} = \lim_{N \to \infty} \left(e^{T\Delta_X/N} e^{-TV/N}\right)^N.$$

In the infinite-dimensional case this formula is much more difficult to establish; it is the so-called *Trotter product formula*. Now a simple calculation shows that the matrix of

$$e^{T\Delta_X/N} e^{-TV/N}$$

is

$$p_{T/N}(x,y) e^{-TV(y)/N}$$

and so the matrix of

$$\left(e^{T\Delta_X/N} e^{-TV/N}\right)^N$$

is

$$\sum p_{T/N}(x, z_1) p_{T/N}(z_1, z_2) \ldots p_{T/N}(z_{N-1}, y) e^{-(T/N)[V(z_1) + \cdots + V(z_N)]}$$

where the sum is over all possible $z_1, z_2, \ldots, z_{N-1}$. Now the expression

$$p_{T/N}(x, z_1) p_{T/N}(z_1, z_2) \ldots p_{T/N}(z_{N-1}, z_N)$$

can be written as

$$P_{x,y,T}\left(\omega(T/N) = z_1, \ldots, \omega((N-1)T/N) = z_{N-1}\right) \cdot p_T(x, y)$$

and so the matrix of

$$\left(e^{T\Delta_X/N} e^{-TV/N}\right)^N$$

becomes

$$\int_{\mathcal{F}} e^{-(T/N)\sum_{1 \le r \le N} V(\omega(rT/N))} dP_{x,y,T}(\omega) \cdot p_T(x, y).$$

The exponent is a Riemann sum which, when $N \to \infty$, goes to

$$\int_0^T V(\omega(s)) ds$$

and so the whole expression has the limit

$$\int_{\mathcal{F}} e^{-\int_0^T V(\omega(s)) ds} \, dP_{x,y,T}(\omega) \cdot p_T(x, y)$$

which is the formula of Feynman-Kac in this very special case.

One can also prove the formula by starting from the integral and showing that it is a semigroup with infinitesimal generator $\Delta_X - V$ so that the matrix satisfies

$$\frac{dM}{dt} = (\Delta_X - V)M.$$

This is how Feynman proved his own formula, by proving that it satisfies the semigroup property and that its infinitesimal form gives the Schrödinger equation.

So far we have treated only the "toy" situation when the configuration space is finite. Let us now go over to the general case but in dimension 1 (for simplicity); that is, the configuration space is **R**. The Hamiltonian is

$$H = -\frac{1}{2}\frac{d^2}{dx^2} + V(x)$$

where we assume that the potential is ≥ 0. It is known that if V is continuous, this is an essentially self-adjoint operator on the space of smooth functions with compact support and is ≥ 0. The dynamical group is

$$e^{-itH}.$$

Since $H \geq 0$, the *complex semigroup*

$$e^{-izH}$$

is defined for $\Re(z) > 0$. In particular, we can speak of the semigroup

$$e^{-tH} \qquad (t \geq 0).$$

By analogy with what we did in the finite case we must associate a probability measure with the semigroup

$$e^{(t/2)d^2/dx^2} \qquad (t \geq 0).$$

We can use Fourier transform theory to give sense to this operator. At the Fourier transform level d^2/dx^2 is multiplication by $-\xi^2$ and so $e^{(t/2)d^2/dx^2}$ is multiplication by $e^{-(t/2)\xi^2}$; in the original space it is then *convolution* by the inverse Fourier transform of $e^{-(t/2)\xi^2}$ which is the Gaussian density

$$\frac{1}{\sqrt{2\pi t}}e^{-x^2/2t}.$$

So

$$e^{(t/2)d^2/dx^2} \qquad (t \geq 0)$$

is an integral operator with the Gaussian kernel

$$g_t(x,y) = \frac{1}{\sqrt{2\pi t}}e^{-(x-y)^2/2t}.$$

We have already seen that this is the transition probability density of the *Wiener measure* which models Brownian motion. The corresponding probability measures W_x are defined on the space $C[0,\infty)$ of all continuous functions on $[0,\infty)$ and model Brownian motion starting from x at time 0. It is not surprising that the free part of the Hamiltonian corresponds to Brownian motion, because in the approximating finite case it was a random walk and typically the random walks have the Brownian motion as their limits. The final formula of Feynman-Kac is that e^{-TH} is an integral operator on $L^2(\mathbf{R})$ with kernel

(FK) $$K_T(x,y) = \int_{C[0,T]} e^{-\int_0^T V(u(s))ds} \, dW_{x,y,T} \cdot g_T(x,y).$$

Here $W_{x,y,T}$ are the conditional Wiener measures that model Brownian motion which start from x at time 0 and reach y at time T.

The modifications introduced by Kac in Feynman's heuristic formula are profound and unexpected. They led to a formula in which the potential is *separated from the free part*, and the integration is determined by a measure that is independent of the potential and involves only the free part. None of these is obvious from the Feynman heuristics and so the Feynman-Kac formula is indeed a major advance.

The Feynman path integral as I have described it is anchored in nonrelativistic quantum mechanics. However, as it can be written down as soon as one knows the classical Lagrangian, it became possible to adapt it to relativistic situations. Physicists have used it in all contexts where it is known as *the sum over histories*. The mathematical underpinning for such generalizations is still very much under construction.

Feynman himself contributed enormously and fundamentally to the way path integrals were used in quantum electrodynamics. In these applications the relativistic propagator was formally expressed as an infinite power series in the coupling constant whose terms are certain integrals in products of Minkowski spaces. These integrals are all divergent and have to be normalized by some regularization procedure, but even then the series is only asymptotic and not convergent. Feynman described these procedures by certain combinatorial schemes called *Feynman diagrams* which encode particular physical processes. These diagrams are extraordinarily effective and have permeated all of modern high-energy physics. This whole structure is called *renormalization*, and only recently have its mathematical features been emerging.

In some sense Feynman resembles Euler in the scope and daring of his imagination and the insight into situations that have never been studied before. For a marvelous comparison of the two, see [8].

Notes and references

[1] See the article by Raymond Ayoub in Amer. Math. Monthly, 81 (1974), 1067-1087.

[2] See *Divergent Series*, by G. H. Hardy, Oxford, 1973, 5-6. This is the classic book on divergent series, summability methods, and Tauberian theorems. The first couple of chapters of the book are historical and expository and are wonderful to read.

[3] Euler's paper *De seriebus divergentibus* is in *Opera Omnia*, I, 14, 585-617.

[4] See the article *Euler's 1760 paper on divergent series*, by E. J. Barbeau and P. J. Leah, Historia Mathematica, **3** (1976), 141-160.

[5] Letter to Goldbach, p. 323 in P.-H. Fuss (see [1], Ch. 2.).

[6] Letter to Nicolaus Bernoulli I, in *Opera Omnia* IV, Vol. 2, p. 628.

[7] *Theory and Application of Infinite Series*, by K. Knopp, 1928.

[8] P. Cartier, *Mathemagics (A Tribute to L. Euler and R. Feynman)*, Preprint, July 2000.

[9] See Riemann's *Collected Papers*, p. 177, Springer, 1990.

[10] See *Two lectures on number theory, past and present*, in Weil's Collected Papers III, Springer, 1979, p. 279.

[11] *Remarques sur un beau rapport entre les series des puissances tant directes que reciproques*, in Opera Omnia I-15, Leipzig and Berlin, 1927, p. 70.

[12] E. Landau, *Euler und die Funktionalgleichung der Riemannschen Zetafunktion*, Bibliotheca Math., 7 (1906), p. 69. Landau gives references to other works related to this.

[13] Euler's treatment of infinite continued fractions is scattered throughout his work. See for instance Opera Omnia I-14, p. 187 and p. 291 as well as the Chapter entitled *De Fractionibus Continuis*, in Opera Omnia I-8 (*Introduction in Analysin Infinitorium*, Chapter 18.

[14] Poincaré and Stieltjes were the first to work with asymptotic series. For Poincaré's work see Acta Math., 8 (1886), 295. For Stieltjes, see *Oeuvres*, Vol. 2, p. 402.

[15] See G. N. Watson, Phil. Trans. Roy. Soc. (A), 211 (1912), p. 279.

[16] See J. Tucciarone, *The Development of the Theory of Summable Divergent Series from 1880 to 1925*, Ann. Hist. Exact Sci., 10 (1973), p. 1, where the reference to the letter is given as N. H. Abel, *Oeuvres d'Abel*, Vol. 2, p. 256.

[17] See *Über die Leibnitzsche Reihe*, in the Frobenius *Gesammelte Abhandlungen*, II, Springer, 1968, p. 8.

[18] Hölder, Math. Ann., 20 (1882), p. 535; Cesàro, Bull. des Sci. Math., 14 (1890), p. 114.

[19a] E. Borel, *Lecons sur les Series Divergentes*, Éditions Jacques Gambay, 1988 (reprinting of the original 1928 work).

[19b] J. P. Ramis, *Séries divergentes et théories asymptotiques*, Publication l'Institut de Recherche Mathématique Avanceé, Strasbourg, 1991 (preprint).

[20] Tauber's paper is in Monats. für Math., 8 (1897), p. 273.

[21] J. E. Littlewood, Proc. Lond. Math. Soc., 9 (1910), p. 434.

[22] N. Wiener, *Tauberian Theorems*, Ann. Math., 33 (1932), p. 1. See also *Norbert Wiener: Collected Works*, Vol. II, The MIT Press, 1979, p. 519.

[23] See the long survey of G. W. Mackey, *Harmonic analysis as the exploitation of symmetry—a historical survey*, Bull. Amer. Math. Soc., 3 (1980), p. 543.

[24] Fourier's treatise, *Théorie analytique de la chaléur*, Paris, 1822. English translation, *The Analytical Theory of Heat*, by A. Freeman, Dover 1955.

[25] See Dirichlet's *Werke*, Chelsea, 1969, p. 117. Dirichlet's paper was a watershed because it heralded a new era of clarity and rigor in analysis.

[26] See [9] above, p. 227.

[27] Fejér's paper appeared in Math. Ann., 58 (1903), p. 51. Fejér was only 19 when he proved this result! For a very interesting and thoroughly enjoyable account of Fourier series, see *Fourier Series*, by T. W. Körner, Cambridge University Press, 1988. The book ranges over a large number of topics related to and originating from Fourier series, with many applications and historical vignettes.

[28] I have in mind a proof such as the one in the book of Loomis, *Abstract Harmonic Analysis*, 1953, p. 9.

[29] P. Chernoff, Amer. Math. Monthly, 87 (1980), p. 399.

[30a] V. S. Varadarajan, *Some remarks on the analytical proof of the prime number theorem*, Nieuw Archief Voor Wiskunde, 16(3) (1998), 153-160.

[30b] For a complete perspective of Wiener's work and its impact, the reader should consult his *Collected Works* ([22] above) as well as the biography of Wiener by P. Masani: *Norbert Wiener: 1894-1964*, Vita Mathematica, 5, Birkhäuser, 1990.

[31] For Gel'fand's original papers on Banach algebras see his *Collected Papers*, I, pp. 169-203.

[32] L. Schwartz, *Théorie des distributions*, 1973. This is a revised version of the two volumes which were originally published in 1950-1951. Also highly recommended is the series of 6 volumes entitled *Generalized Functions* written by Gel'fand with various collaborators. Originally written in Russian, they have been translated into English. These volumes contain an enormous number of formulae as well as a panoramic sweep of applications of distribution theory from classical analysis to the theory of group representations. For a very beautiful and prophetic paper on the potential of distribution theory for solving problems of analysis as well as a general survey of problems of functional analysis, see Gel'fand's address to the International Congress of Mathematicians in 1954 (*Collected Papers*, I, p. 3). See also in the same volume some of his other addresses, pp. 27-76.

[33] The papers of Einstein and Smouluchowski were written around 1905. Perrin wrote a book on the atomic theory which was widely influential. But the first rigorous description of Brownian motion through a measure on path space was done by Wiener in his great paper *Differential space* (cf. Wiener's *Collected Works*, I, p. 455). In a whole series of subsequent papers he carried out a deep study of its properties, among them the nondifferentiability of its paths.

[34] For a proof, see K. R. Parthasarathy, *Probability Measures on Metric Spaces,* 1967. This book is a very good reference for the basic theory of construction of measures on path spaces.

[35] *I am a mathematician. The later life of a prodigy,* 1956. It is impossible to overstate the towering stature of Wiener as a mathematician, scientist, and humanist. The reader who wants to get a deep perspective on all aspects of Wiener's life, science and humanism should refer to his collected works [22] above and Masani's biography [30] above.

[36] R. P. Feynman, *Space-time approach to nonrelativistic quantum mechanics,* Rev. Mod. Phys., 20 (1948), p. 267. This paper is still inspiring to read. Feynman was led to his ideas by a paper of Dirac, *The Lagrangian in quantum mechanics,* Phys. Zeit. Sowjetunion. It is reprinted along with Feynman's paper, as well as all the fundamental papers on quantum electrodynamics in *Selected Papers on Quantum Electrodynamics,* edited by Julian Schwinger, Dover, 1958. In the book *QED and the Men Who Made It: Dyson, Feynman, Schwinger, and Tomonaga,* 1994, S. S. Schweber gives an account of the circumstances that led to Feynman's discovery; see pp. 389-397.

[37] The paper of Kac where the formula is established is in *Mark Kac: Probability, Number Theory, and Statistical Physics. Selected Papers,* 1979, p. 268. The basic text for the connections between Wiener integrals and mathematical physics is the book of B. Simon, *Functional Integration and Quantum Physics,* 1979. For a very nice treatment of Feynman and Wiener integrals, see the paper of E. Nelson in Jour. Math. Phys., 5 (1964), 332.

[38] In his autobiography *Enigmas of Chance,* 1985, Kac describes the circumstances of his discovery of the Feynman-Kac formula and has this to say (see pp. 115-116):

> ... In its various guises the F–K formula is ubiquitous throughout much of quantum physics on the one hand and probability theory on the other. It is probably safe to say that I am better and more widely known for being the K in the F–K formula than for anything else I have done during my scientific career.
>
> But first a little history. At the end of the war Hans Bethe returned to Cornell from Los Alamos and brought with him a group of brilliant young experimental and theoretical physicists. Among them was Richard Feynman who, to no one's surprise, went on to become a leading physicist of our day. It must have been in the spring of 1947 that Feynman gave a talk at the Cornell Physics Colloquium based on some material from his 1942 Ph. D. dissertation, which had not yet been published.
>
> A fundamental concept of quantum mechanics is a quantity called the propagator, and the standard way of finding it (in the nonrelativistic case) is by solving the Schrödinger equation. Feynman found another way based on what became known as the Feynman path integral or "the sum over histories" which Dyson alludes to in his book *Disturbing the universe.*
>
> During his lecture Feynman sketched the derivation of his formula and I was struck by the similarity of his steps to those I had encountered in my work. In a few days I had my version of the formula, although it took some time to complete a rigorous proof.
>
> My formula connected solutions of certain differential equations closely related to the Schrödinger equation with Wiener integrals.

It is only fair to say that I had Wiener's shoulders to stand on. Feynman, as in everything else he has done, stood on his own, a trick of intellectual contortion that he alone is capable of.

I find Feynman's formula to be very beautiful. It connects the quantum mechanical propagator, which is a twentieth century concept, with the classical mechanics of Newton and Lagrange, in a uniquely compelling way. . . .

CHAPTER 6

Euler Products

6.1. Euler's product formula for the zeta function and others

In a paper [1] presented in 1737 and published in 1744, Euler reported on his stunning discovery that connected the zeta function

$$\zeta(s) = 1 + \frac{1}{2^s} + \frac{1}{3^s} + \ldots \qquad (s > 1)$$

with the sequence of primes. More precisely, he discovered the remarkable product formula

$$\zeta(s) = \prod_p \frac{1}{1 - \frac{1}{p^s}} \qquad (s > 1)$$

where $\prod_p$ means that the product is taken over all the primes. In Euler's notation this would appear as

$$1 + \frac{1}{2^s} + \frac{1}{3^s} + \cdots = \frac{2^s.3^s.5^s.7^s.\ldots}{(2^s - 1)(3^s - 1)(5^s - 1)(7^s - 1)\ldots} \qquad (s > 1).$$

This identity is just the analytical way of expressing the unique prime factorization property of the natural numbers. This was the first time that such a product taken over the set of primes occurred in mathematics, and it was an event to be marked. Going to $\zeta(s)^{-1}$ and taking $s = 1$, Euler concluded that

$$0 = \prod_p \left(1 - \frac{1}{p}\right)$$

where the product is over all the primes $2, 3, 5, 7, 11, \ldots$. This of course implies immediately that

$$\sum_p \frac{1}{p} = \infty$$

which sharpens Euclid's result that there are infinitely many primes. In fact Euler had discovered an analytical proof of Euclid's theorem.

Arithmetic Euler products. In the same paper mentioned above Euler considered a number of products of the form

$$\prod_p \frac{1}{\left(1 - \frac{c(p)}{p^s}\right)}.$$

Here c is a function defined on the set of positive integers satisfying

$$c(mn) = c(m)c(n), \qquad c(1) = 1.$$

We shall call these *fully multiplicative*. The fully multiplicative functions are completely determined once their values at the primes are known. By the same argument as in the case of the zeta function, Euler obtained the identity

$$\sum_{n=1}^{\infty} \frac{c(n)}{n^s} = \prod_p \left(1 - \frac{c(p)}{p^s}\right)^{-1}.$$

However, with an instinct that is remarkable, he began looking into cases where $c(n)$, while being fully multiplicative, *depends only on the residue class of n modulo a fixed integer.* For example, he took

$$c(n) = \begin{cases} \pm 1 & \text{if } n \equiv \pm 1 (4) \\ 0 & \text{if } n \text{ is even} \end{cases}$$

and obtained the result

$$1 - \frac{1}{3^s} + \frac{1}{5^s} - \cdots = \prod_p \left(1 - \frac{\pm 1}{p^s}\right)^{-1}$$

where the ± 1 is according to whether p is of the form $4m \pm 1$. At $s = 1$ this gives

$$\frac{\pi}{4} = 1 - \frac{1}{3} + \frac{1}{5} - \frac{1}{7} + \cdots = \prod_p \left(1 - \frac{\pm 1}{p}\right)^{-1}$$

from which he concluded that

$$\sum_p \frac{\pm 1}{p}$$

is finite, the primes being arranged in increasing order. Thus, if we write

$$S_{\pm} = \sum_{p \equiv \pm 1(4)} \frac{1}{p},$$

then

$$S_+ + S_- = \infty, \qquad S_+ - S_- \text{ is finite}$$

showing that

$$S_{\pm} = \infty.$$

Thus primes of each of the form $4m \pm 1$ are infinite. He discussed these matters in his letters to Goldbach on August 5 and October 28, 1752 [2], where he gave the approximate numerical value

$$S_+ - S_- = 0.334980.$$

In his *Introductio in Analysin Infinitorum* (*Omnia Opera*, I-8), Euler devotes a whole chapter to similar questions, treats some other series and products:

$$\sum_{n \geq 1} \frac{c(n)}{n^s} = \prod_p \left(1 - \frac{c(p)}{p^s}\right)^{-1}$$

with c always fully multiplicative, and mentions that he can treat "innumerable" other products of this nature. Thus he treats the cases $c = c_3, c_8$ where

$$c_3(n) = \begin{cases} \pm 1 & \text{if } n = 3m \pm 1 \\ 0 & \text{if } n \text{ is divisible by } 3 \end{cases}$$

and

$$c_8(n) = \begin{cases} +1 & \text{if } n \text{ is of the form } 8m+1 \text{ or } 8m+3 \\ -1 & \text{if } n \text{ is of the form } 8m+5 \text{ or } 8m+7 \\ 0 & \text{if } n \text{ is even.} \end{cases}$$

In each of these cases c is fully multiplicative. He verifies that the products remain finite at $s = 1$, evaluating them explicitly using his theory of partial fraction expansions of the circular functions. From his formula

$$\frac{\pi}{n} \cot \frac{m\pi}{n} = \frac{1}{m} - \sum_{k=1}^{\infty} \left(\frac{1}{nk - m} - \frac{1}{nk + m}\right)$$

he obtains, taking $m = 1, n = 3$,

$$\sum_{n=1}^{\infty} \frac{c_3(n)}{n} = 1 - \frac{1}{2} + \frac{1}{4} - \frac{1}{5} + \frac{1}{7} - \frac{1}{8} + \frac{1}{10} - \cdots = \frac{\pi}{3\sqrt{3}}$$

where the parentheses have been removed to exhibit the formula the way he wrote it. Thus

$$\prod_p \left(1 - \frac{c_3(p)}{p}\right)^{-1} = \frac{\pi}{3\sqrt{3}}.$$

From the identity

$$\frac{\pi}{n} \operatorname{cosec} \frac{m\pi}{n} = \frac{1}{m} + \sum_{k=1}^{\infty} (-1)^{k-1} \left(\frac{1}{kn - m} - \frac{1}{kn + m}\right),$$

which he writes as

$$\frac{\pi}{n} \operatorname{cosec} \frac{m\pi}{n} = \frac{1}{m} + \frac{1}{n - m} - \frac{1}{n + m} - \frac{1}{2n - m} + \frac{1}{2n + m} + \frac{1}{3n - m} - \cdots,$$

he gets, taking $m = 1, n = 4$, the result

$$\prod_p \left(1 - \frac{c_8(p)}{p}\right)^{-1} = \frac{\pi}{2\sqrt{2}}.$$

For c_3 one can deduce as before that the number of primes in the various residue classes is infinite. Let us write p for a generic prime, q (resp. r) for a generic prime of the form $3m + 1$ (resp. $3m - 1$). Euler worked mod 6, but for a prime $p > 3$,

$p \equiv \pm 1$ (6) is the same as $p \equiv \pm 1$ (3). Then, taking the values at $s = 1$, we can write

$$P = \prod_p \left(1 - \frac{1}{p}\right)^{-1} = \infty$$

$$R = \prod_q \left(1 - \frac{1}{q}\right)^{-1} \prod_r \left(1 + \frac{1}{r}\right)^{-1} = \frac{\pi}{3\sqrt{3}} < \infty.$$

We also write

$$M = \prod_q \left(1 - \frac{1}{q}\right)^{-1} \qquad N = \prod_r \left(1 - \frac{1}{r}\right)^{-1}.$$

Then

$$PR = M^2 \prod_r \left(1 - \frac{1}{r^2}\right)^{-1} = \infty \Longrightarrow M = \infty$$

$$\frac{P}{R} = \prod_r \left(1 - \frac{1}{r^2}\right) N^2 = \infty \Longrightarrow N = \infty$$

since in both cases the extra factor is finite because

$$\prod_p \left(1 - \frac{1}{p^2}\right) = \frac{6}{\pi^2} < \infty.$$

So both M and N are ∞; hence there is an infinity of primes of each of the forms $3m \pm 1$.

The case of modulus 8 is more involved. Here there are 4 classes of primes, of the form $8m + k$ ($k = 1, 3, 5, 7$). Since their squares are all of the form $8m + 1$ and $3 \times 5 \equiv 7$ (8), we can distribute the signs of all fully multiplicative functions defined mod 8 as follows:

$$\begin{pmatrix} 1 & 3 & 5 & 7 \\ + & + & + & + \\ + & + & - & - \\ + & - & + & - \\ + & - & - & + \end{pmatrix}.$$

Let us write c_{jk} ($j = 1, 2, 3, 4, k = 1, 3, 5, 7$) for the entries of this matrix and define, following Euler's method,

$$P_j = \prod_k \prod_{p \equiv k(8)} \left(1 - c_{jk} p^{-1}\right)^{-1}.$$

Let

$$A_k = \prod_{p \equiv k(8)} \left(1 - p^{-1}\right)^{-1} \qquad (k = 1, 3, 5, 7)$$

and let us write, taking our cue from the previous example, $\alpha \sim \beta$ for two infinite products if they differ by a *convergent* product. Then

$$P_1 P_2 P_3 P_4 \sim A_1^4, \qquad \frac{P_1 P_2}{P_3 P_4} \sim A_3^4$$

$$\frac{P_1 P_3}{P_2 P_4} \sim A_5^4, \qquad \frac{P_1 P_4}{P_2 P_3} \sim A_7^4$$

where we repeatedly use the fact that the product of $(1 - p^{-1})^{-2}$ over any set of primes is finite. Now $P_1 = \infty$; and so if we knew that P_2, P_3, P_4 are all finite, then it would follow that each of the A_k would be ∞, showing that there is an infinity of primes, each of the form $8m + k$ ($k = 1, 3, 5, 7$). As was mentioned earlier Euler knew the formula

$$P_2 = \frac{\pi}{2\sqrt{2}};$$

on the other hand,

$$P_3 = \prod_p (1 - c_3(p)p^{-1})^{-1} = \sum_{n=1}^{\infty} \frac{(-1)^{2n-1}}{2n - 1} = \frac{\pi}{4}.$$

It will follow from Dirichlet's explicit evaluations of these sums that

$$P_4 = \frac{1}{\sqrt{8}} \log(3 + 2\sqrt{2}).$$

So there are infinitely many primes in all these classes. Although the evaluation of P_4 is well within the scope of what Euler knew at that time, I have not found it among his papers.

These themes of Euler—namely, fully multiplicative functions $c(n)$ and the associated series (called Dirichlet series) with the product representations (Euler products)

$$\sum_{n=1}^{\infty} \frac{c(n)}{n^s} = \prod_p (1 - c(p)p^{-s})^{-1},$$

and their behavior at $s = 1$—were taken up by his successors and continually generalized and extended in scope. With Ramanujan, Artin, Hecke, and Langlands, Euler products of the form

$$\prod_p \frac{1}{P_p(p^{-s})}$$

where P_p is a polynomial of degree $d > 1$ with constant term 1,

$$P_p(X) = 1 + c_{p1}X + \cdots + c_{pd}X^d,$$

were introduced. We shall call them Euler products of *degree d*. The (Eulerian) factors P_p are of course not arbitrarily chosen; already, even in Euler's work, the $c(p)$ are determined by the residue class of p mod N, where N is a fixed integer. This is the big clue in the choice of P_p: they must somehow be derived from some *global object*—be it geometric, analytic, or number theoretic—for the Euler product to have arithmetic significance. They have been called *L*-functions because the first extended family of Euler products considered were the functions $L(s : \chi)$ of Dirichlet. The analytic *L*-functions are capable of being studied in depth, have meromorphic continuation to the whole *s*-plane and satisfy functional equations, of which the functional equation of the Riemann zeta function is prototypical. The *L*-functions coming from geometry and algebraic number fields are more mysterious and very difficult to study directly. However, and this is the central point of the theme of Euler products, these various *L*-functions are not all distinct; in particular it happens, generally for very deep reasons, that a geometric or arithmetic *L*-function coincides with an analytic one. In such a case one can use this identification to obtain the properties of meromorphic continuation and functional equation from the corresponding properties of the analytic *L*-function. Quite often no other way of obtaining these properties for the geometric or arithmetic L-function is known. Classical reciprocity laws (quadratic, cubic, power, Artin, Hasse, etc.) are equivalent to such equalities between arithmetico-geometric and analytic L-functions. From the modern perspective, the classical L-functions, being of degree 1, pertain to *abelian* objects, while the Euler products of degree > 1 arise from *nonabelian* objects and so are technically much harder to study. For these, the equalities of the arithmetico-geometric L-functions with the analytic ones represent the ultimate forms of *nonabelian reciprocity laws*.

In the following sections I shall briefly review the story of Euler products. The story of this development is an astonishing one and is the result of contributions of a large number of mathematicians. It is not linear, and there are many subplots and twists. To use Weil's analogy, it is perhaps like a symphony[†] with many themes that come in and go out; moreover it is unfinished. However, one thing remains constant: the L-functions, i.e., the Euler products, have revealed fundamental aspects of number theory. More than this, some of the most central conjectures of number theory can be formulated in terms of L-functions. It is the complex fabric of these results, ideas, and conjectures that is known today as the *Langlands program*.

I begin in §6.2 with Dirichlet's work. After him, Riemann brought in complex variables and related the zeta function to the problem of distribution of primes, leading eventually to the proof of the prime number theorem by Hadamard and de la Vallée Poussin, and the analytic continuations and functional equations of the Dirichlet L-functions by Hurwitz and de la Vallée Poussin. After that, with the discovery of ideals and unique factorization in number fields, it was natural to extend the Euler products to the context of number fields. I have therefore included a brief introduction to this part of algebraic number theory. This direction would culminate in the discovery by Hecke of Euler products associated to a grössencharakter. I have used the adelic language in describing the Hecke theory, not only because it is the most elegant, but also since all modern work is in this language, so I have also included a brief account of this point of view.

After this, propelled by the discoveries of Ramanujan, Mordell, and above all Hecke (again), Euler products of degree 2 entered the picture, associated to modular forms. §6.3 is concerned with this development. By a device discovered by Gel'fand and Fomin, modular forms could be viewed as the highest weight vectors of representations of the discrete series of the group $\mathrm{SL}(2, \mathbf{R})$, so that representation theory was brought into the picture. At the same time, the Hecke operators could be viewed as describing the action of the groups $\mathrm{SL}(2, \mathbf{Q}_p)$ on the space of modular forms, so that modular forms could be replaced by irreducible representations of the *adelic group* $\mathrm{GL}(2, \mathbf{A})$. It is to discuss this that I included earlier a discussion of the adelic point of view. This point of view led to the association of Euler products to such representations (automorphic) by Godement, Jacquet and Langlands. It is here that the essential nature of the adelic point of view emerges, because it leads naturally to Euler products of degree n associated to automorphic representations of $\mathrm{GL}(n)$, and these have no classical analogues for $n > 2$, although such Euler products would arise in the Siegel theory of symplectic groups. The big question, formulated in the preface to the monumental work by Jacquet and Langlands [21a], was whether the automorphic Euler products of $\mathrm{GL}(2)$ and even $\mathrm{GL}(n)$ for $n > 2$ are identical with those that were defined decades earlier by Artin, the so-called Artin

---

[†]More like Mahler's than Mozart's!

nonabelian L-functions. This question, which could be paraphrased as the existence of a correspondence between Galois representations of degree n and automorphic (irreducible) representations of the adelic group associated to $\mathrm{GL}(n)$, has been the central driving force in the Langlands program. But for $n = 1$ this is just class field theory, and for this purpose I have devoted §6.4 to a brief review of abelian extensions and class field theory, and §6.5 to the Artin L-functions. The stage is then set for describing the main outlines of the Langlands program. I do this very briefly in the last section.

6.2. Euler products from Dirichlet to Hecke

The first extended family of Euler products were those of Dirichlet, introduced in his great Berlin memoir of 1837 [3], namely

$$L(s:\chi) = \prod_p \frac{1}{1 - \chi(p)p^{-s}}$$

for χ a character of the group G_N of residue classes mod N prime to N. Dirichlet considered these as functions of a *real* variable s. By defining $\chi(n)$ to be zero if n is not prime to N, we get a fully multiplicative function. However, χ is in general complex-valued, unlike the cases treated by Euler; for instance, for $N = 5$, G_N is cyclic of order 4 and is generated by the residue class of 2, so that we get 4 characters corresponding to the choices $\pm i, \pm 1$ for the value of $\chi(2)$. The $L(s:\chi)$ converge for $s > 1$, and for Dirichlet, the key to understanding the distribution of primes in residue classes mod N was the behavior of the $L(s:\chi)$ as $s \to 1 + 0$. Dirichlet was able to show that for any integer a prime to N,

$$\sum_{p \equiv a(N)} \frac{1}{p^s} = \frac{1}{\varphi(N)} \log \frac{1}{s-1} + O(1) \qquad (s \to 1+0)$$

where $\varphi(N) = |G_N|$ is the Euler number. In particular

$$\lim_{s \to 1+0} \sum_{p \equiv a} \frac{1}{p^s} = \infty,$$

and so each residue class contains infinitely many primes. On the other hand, Euler's result can be reformulated in a more rigorous way as the statement that

$$\sum_p \frac{1}{p^s} = \log \frac{1}{s-1} + O(1),$$

and so it appears natural to say that the primes are *equidistributed* among the residue classes. To relate the $L(s:\chi)$, which involves all the primes, to primes in residue classes, Dirichlet used the fact that the characters form an orthonormal basis for the space of functions on G; i.e., for any complex valued function f on a finite abelian group G,

$$f = \sum_{\chi \in \widehat{G}} (f, \chi)\chi, \qquad (f, \chi) = \frac{1}{|G|} \sum_{x \in G} f(x)\overline{\chi(x)}.$$

From this one gets

$$\sum_{p \equiv a\ (N)} \frac{1}{p^s} = \frac{1}{\varphi(N)} \sum_\chi \sum_p \overline{\chi(a)} \frac{\chi(p)}{p^s}.$$

Hence the theorem of infinitude of primes would follow if we prove that the right side $\to \infty$ when $s \to 1 + 0$ through real values. We shall now sketch briefly Dirichlet's argument, still beautiful from across the gulf of over 150 years separating us from his work, for proving this.

We begin by noting that the series

$$\sum_{n \geq 1} \frac{\chi(n)}{n^s}$$

defining $L(s : \chi)$ is uniformly convergent for $s \geq 1 + \delta > 1$ so that $L(s : \chi)$ is continuous on $s > 1$. But if χ is not the trivial character χ_0, the series converges uniformly as before even in the region $s \geq \delta > 0$ so that $L(s : \chi)$ is defined and continuous in the region $s > 0$. In particular $L(1 : \chi)$ $(\chi \neq \chi_0)$ is finite. To emphasize that we are interested in quantities becoming infinite when $s \to 1 + 0$, let us introduce, for two functions $A(s), B(s)$ defined for $s > 1$, the notation $A(s) \approx B(s)$ to mean that $A(s) - B(s)$ is bounded as $s \to 1 + 0$. As an example consider the zeta function. From Euler's formula

$$(1 - 2^{1-s})\zeta(s) = 1 - \frac{1}{2^s} + \frac{1}{3^s} - \dots \qquad (s > 0)$$

we see that

$$\zeta(s) = \frac{1}{s - 1} + I(s)$$

where I is continuous in $s > 0$, and so, taking logarithms (observe that the product formula implies that $L(s : \chi)$ does not vanish anywhere in $s > 1$),

$$\log \zeta(s) \approx \log \frac{1}{s - 1}.$$

This implies also that

$$\log L(s : \chi_0) \approx \log \frac{1}{s - 1}$$

since $L(s : \chi_o)$ differs from $\zeta(s)$ only by the product of a finite number of factors, namely, the $(1 - p^{-s})^{-1}$ for the primes p that divide N. On the other hand, for $\chi \neq \chi_0$,

$$\log L(s : \chi) = -\sum_p \log(1 - \chi(p)p^{-s}).$$

Since χ is in general complex valued we use the principal branch of the log for the terms on the right side. Expanding the logs, we get

$$\log L(s : \chi) = \sum_p \frac{\chi(p)}{p^s} + \sum_{p,m} \frac{\chi(p)^m}{mp^{ms}} \approx \sum_p \frac{\chi(p)}{p^s}.$$

The last step follows from the fact that

$$\sum_{m \geq 2, \, p} \frac{1}{mp^{ms}}$$

is uniformly convergent for $s \geq 1/2 + \delta$ for every $\delta > 0$. So

$$\sum_{p \equiv a \, (N)} \frac{1}{p^s} \approx \frac{1}{\varphi(N)} \log \frac{1}{s - 1} + \frac{1}{\varphi(N)} \sum_{\chi \neq \chi_0} \overline{\chi(a)} \log L(s : \chi).$$

If $L(1 : \chi) \neq 0$ for all $\chi \neq \chi_0$, then *the second term on the right above stays finite as $s \to 1 + 0$*, and so we get

$$\sum_{p \equiv a \ (N)} \frac{1}{p^s} \approx \frac{1}{\varphi(N)} \log \frac{1}{s-1}.$$

This proves that there are infinitely many primes in the residue class of a! This is the Dirichlet proof. The reader should compare this argument with the one given for $N = 8$ in the previous section.

The proof that

$$L(1 : \chi) \neq 0 \qquad (\chi \neq \chi_0)$$

is therefore the climax of Dirichlet's work. The evaluation of $L(1 : \chi) \neq 0$ for the few special χ that Euler had treated already depended on his deep theory of partial fraction expansions of the circular functions, and so it was to be expected that the evaluation of $L(1 : \chi)$ in the general case would prove difficult. Dirichlet did this by finding an explicit expression for

$$L(1 : \chi) = \prod_p \frac{1}{1 - \chi(p)p^{-1}} = \sum_{n=1}^{\infty} \frac{\chi(n)}{n}.$$

The essence of the difficulty resides in the case when χ is *quadratic*, i.e., $\chi(m) = \pm 1$ for all $m \in G_N$. The product formula for $L(1 : \chi)$ shows that $L(1 : \chi) \geq 0$ in this case. However to see that $L(1 : \chi)$ is positive for quadratic χ required quite a bit of further work. Dirichlet used Gauss's work on cyclotomy to settle the case when N is an odd prime and completed a little later the proof in the general case by interpreting $L(1 : \chi)$ for quadratic χ as a *class number*. Here we must remember that although he lacked the theory of quadratic fields and their class numbers, he could work with the equivalent theory of classes of quadratic form, familiar to him from Gauss's work. For an account that preserves Dirichlet's original ideas, see [4]. Subsequently de la Vallée Poussin and Landau found simple proofs of the positivity of $L(1 : \chi)$ for quadratic χ based entirely on the theory of Dirichlet series, although this argument is essentially equivalent to the one based on class number [4]; indeed both proofs use the product $\zeta(s)L(s : \chi)$ in an essential manner. We have stuck to real s to keep the flavor of Dirichlet's argument, but one could treat s as a complex variable. Then the $L(s; \chi)$ are holomorphic and nowhere vanishing on $\Re(s) > 1$, and for $\chi \neq \chi_0$, $L(s : \chi)$ is analytic on $\Re(s) > 0$.

It may be of interest for later purposes to sketch the analytical argument for showing that $L(1 : \chi) \neq 0$. Let us consider the function

$$Z(s) = \prod_{\chi} L(s : \chi) = L(s : \chi_0) \prod_{\chi \neq \chi_0} L(s : \chi)$$

where the product is over all the characters of $G(N)$. We know that $Z(s)$ is a holomorphic function in $\Re(s) > 1$ and further that the product over the nontrivial characters χ is a holomorphic function in $\Re(s) > 0$. The point now is that the expression for Z can be simplified considerably. For any fixed prime p it is easy to show that

$$\prod_{\chi} \left(1 - \frac{\chi(p)}{p^s}\right)^{-1} = \left(1 - p^{-sf(p)}\right)^{-g(p)}.$$

Here $g(p) = |G_N|/f(p)$, $f(p)$ being the order of p in G_N. Hence

$$\prod_\chi L(s : \chi) = \prod_p \left(1 - p^{-sf(p)}\right)^{-g(p)}.$$

In particular

$$\left|\prod_\chi L(s : \chi)\right| \geq 1 \qquad (s > 1).$$

Since $L(s : \chi_0)$ has a simple pole at $s = 1$ it follows that there can be no more than one character χ such that $L(s : \chi) = 0$, and if there is one such, that χ has to be real, and further, that $L(s : \chi)$ must have only a simple zero at $s = 1$. A character which is real is ± 1–valued and is called a *quadratic character*.

The proof that $L(1 : \chi) \neq 0$ for a quadratic χ is however highly nontrivial. In order to complete the proof at this stage we shall give a purely analytical proof depending on the basic elements of the theory of Dirichlet series.

LEMMA. *Let $a_n \geq 0$ for all n and let the Dirichlet series $F(s) = \sum_n a_n n^{-s}$ be convergent for a real $s = \sigma_1$. Suppose F has an analytic continuation to a half plane $\Re(s) > \sigma_0$ where $\sigma_0 < \sigma_1$. Then the Dirichlet series F is absolutely convergent for $\Re(s) > \sigma_0$.*

PROOF. Since the a_n are ≥ 0, the Dirichlet series F is absolutely convergent for $\Re(s) > \sigma_1$ and is a holomorphic function in this half plane. Suppose now that $\sigma_0 < \tau < \sigma_1$. We must prove that the series $\sum_n a_n n^{-\tau}$ is convergent. Take a $\sigma_2 > \sigma_1$. Then

$$F(s) = \sum_n \frac{a_n}{n^s} \qquad (s > \sigma_1)$$

and we have, for all integers $k \geq 1$,

$$F^{(k)}(\sigma_2) = (-1)^k \sum_n \frac{a_n (\log n)^k}{n^{\sigma_2}}.$$

Since F is holomorphic on $\Re(s) > \sigma_0$, it must have a convergent power series expansion in the disk $|s - \sigma_2| < \sigma_2 - \sigma_0$ so that

$$F(\tau) = \sum_k \frac{F^{(k)}(\sigma_2)}{k!}(\tau - \sigma_2)^k = \sum_k \frac{(\sigma_2 - \tau)^k}{k!} \sum_n \frac{a_n(\log n)^k}{n^{\sigma_2}}.$$

Since everything in sight on the right side is positive we can reverse the order of summation and obtain

$$F(\tau) = \sum_n \frac{a_n}{n^{\sigma_2}} \sum_k \frac{(\log n)^k (\sigma_2 - \tau)^k}{k!} = \sum_n \frac{a_n}{n^{\sigma_2}} n^{\sigma_2 - \tau} = \sum_n \frac{a_n}{n^\tau}$$

as we wanted to show.

We now return to $L(s : \chi)$ for a nontrivial quadratic χ. Let us write q (resp. r) for a generic prime p with $\chi(p) = 1$ (resp. $\chi(p) = -1$). If we define

$$F(s) := \frac{L(s : \chi_0)L(s : \chi)}{L(2s : \chi_0)} = \prod_q \frac{(1 + q^{-s})}{(1 - q^{-s})},$$

then F is holomorphic on $\Re(s) > 1$, since $L(2s : \chi_0)$ is holomorphic and nowhere 0 in $\Re(s) > 1/2$. If $L(s : \chi)$ now has a simple zero at $s = 1$, that zero will balance the simple pole of $L(s : \chi_0)$ at $s = 1$ and so the numerator will be holomorphic at $s = 1$.

Hence the numerator is analytic in $\Re(s) > 0$ and so F is analytic in $\Re(s) > 1/2$. Moreover, as the denominator has a pole at $s = 1/2$, $F(\tau) \to 0$ if $\tau \to 1/2 + 0$. If we write

$$F(s) = \sum_n \frac{a_n}{n^s} \qquad (s > 1)$$

it is clear that the a_n are all ≥ 0 and cannot all be 0, since then F will be 0. By the lemma above, the above series will converge for any $\tau > 1/2$ so that

$$F(\tau) = \sum_n \frac{a_n}{n^\tau} > 0 \qquad (\tau > 1/2).$$

As τ decreases to $1/2$, $F(\tau)$ will *increase* and so cannot go to 0, a contradiction. This completes the proof of the nonvanishing at $s = 1$ of the L-series of a quadratic character, and hence of the proof of Dirichlet's theorem also.

Only after the development of the theory of algebraic number fields did it become possible to obtain a more fundamental view of Dirichlet's discovery that $L(1 : \chi) \neq 0$ for $\chi \neq \chi_0$ and that $L(1 : \chi) > 0$ for quadratic χ. In fact, the product

$$\prod_\chi L(s : \chi)$$

is just the *Dedekind zeta function* of the cyclotomic field

$$K_N = \mathbf{Q}(e^{2\pi i/N})$$

except for a finite number of Eulerian factors; from the theory of number fields one knows that the Dedekind zeta function has a simple pole at $s = 1$ with a positive residue, and the same is true of $L(s : \chi_0)$ since it is essentially the same as the Riemann zeta function. So

$$\prod_{\chi \neq \chi_0} L(1 : \chi) > 0.$$

Riemann. In 1859 Riemann introduced a new theme, namely that $\zeta(s)$ should be studied for *complex* s. In his famous memoir [5], the only one he wrote on number theory, he proved that $\zeta(s)$ continues analytically to the whole s-plane, with a simple pole at $s = 1$ where the residue is 1, and with no other singularity, and further that the extended function satisfies a functional equation that relates $\zeta(s)$ and $\zeta(1 - s)$. Riemann was interested in the zeta only for the problem of the distribution of primes, and [5] contains his celebrated conjectures on the location and distribution of the zeroes of $\zeta(s)$ and their relationship to the problem of the distribution of primes. If we define

$$Z(s) = \pi^{-\frac{s}{2}} \Gamma\left(\frac{s}{2}\right) \zeta(s),$$

then $Z(s)$ extends meromorphically to the whole s-plane with simple poles at $s = 0, 1$ and no other singularities, and Riemann's functional equation takes the form

$$Z(s) = Z(1 - s).$$

The continuation to $\Re(s) > 0$ follows from Euler's formula for $(1 - 2^{1-s})\zeta(s)$; if one proves the functional equation on $0 < \Re(s) < 1$, then one can use the functional equation itself to continue ζ analytically to the whole s-plane, and the full function will automatically satisfy the functional equation. To do this Riemann introduced another fertile idea, already implicit in Dirichlet, that the zeta function is essentially the *Mellin transform* of a *modular form*, indeed, a *theta function*. (The Mellin transform is just the Fourier transform in multiplicative disguise.) From

$$\pi^{-\frac{s}{2}}\Gamma\left(\frac{s}{2}\right)n^{-s} = \int_0^\infty e^{-\pi n^2 x}x^{\frac{s}{2}-1}dx \qquad (\Re s > 0)$$

we get

$$Z(s) = \int_0^\infty x^{\frac{s}{2}-1}\sum_{n\geq 1}e^{-\pi n^2 x}dx \qquad (\Re s > 1).$$

We now write

$$\vartheta(\tau) = \frac{1}{2}\sum_{-\infty}^{\infty}e^{i\pi n^2\tau} = \frac{1}{2} + \sum_1^{\infty}e^{i\pi n^2\tau} \qquad (\Im\tau > 0).$$

Then ϑ is holomorphic on the Poincaré upper half plane

$$\mathcal{H} = \{\tau \in \mathbf{C} \mid \Im(\tau) > 0\}.$$

Moreover

$$Z(s) = \int_0^\infty \left(\vartheta(iy) - \frac{1}{2}\right)y^{\frac{s}{2}}\frac{dy}{y} \qquad (\Re(s) > 1).$$

We have already encountered ϑ while discussing Jacobi's work on representations of numbers as sums of squares. This exhibits ζ as the Mellin transform of ϑ. (We note that dy/y is the Haar measure on the multiplicative group $(0,\infty)$ and $y^{\frac{s}{2}}$ are the characters of this group.) The key properties of ϑ are:

(a) $\vartheta(\tau + 2) = \vartheta(\tau)$.
(b) $\vartheta(-1/\tau) = (-i\tau)^{1/2}\vartheta(\tau)$ where $z^{1/2}$ is the holomorphic square root on $\{-\pi < \arg z < \pi\}$ which is 1 at $z = 1$.
(c) For every $\sigma > 1$ there is a constant $A(\sigma) > 0$ such that

$$\left|\vartheta(\tau) - \frac{1}{2}\right| \leq A(\sigma)y^{-\sigma/2}e^{-\pi y/2} \qquad (\tau = x + iy,\ y > 0).$$

Properties (a) and (b) express the fact that ϑ is a *modular form* with respect to a suitable subgroup of the modular group. Property (a) is obvious while (b) is classical and follows from the Poisson summation formula which states that if g is a Schwartz function on $\mathbf{R}$ with Fourier transform $\hat{g}$ defined by

$$\hat{g}(\xi) = \int_{\mathbf{R}} g(x)e^{2\pi i\xi x}dx \qquad (\xi \in \mathbf{R}),$$

then

$$\sum_{n\in\mathbf{Z}}g(n) = \sum_{n\in\mathbf{Z}}\hat{g}(n).$$

Taking

$$g(x) = e^{-\pi u x^2}, \qquad \hat{g}(\xi) = u^{-1/2}e^{-\pi u^{-1}\xi^2} \qquad (u > 0)$$

will give

$$(-i(iu))^{1/2}\vartheta(iu) = \vartheta(iu^{-1}) \qquad (u > 0).$$

We get (b) by analytic continuation.

For proving (c) note that for any $\alpha > 0, k > 0, y > 0$, we have the estimate

$$e^{-ky} \leq \left(\frac{\alpha}{ek}\right)^\alpha y^{-\alpha}$$

obtained by noting that for $y \geq 0$ the maximum of $y^\alpha e^{-ky}$ is attained at $y = \alpha/k$. Then

$$\sum_{n \geq 1} e^{-\pi n^2 y} \leq e^{-\pi y/2} \sum_{n \geq 1} e^{-\pi((n^2 - 1/2)y}$$

from which (c) follows easily. The estimate (c) makes the convergence of the integral for $Z(s)$ manifest at both end points. If we write, still for $\Re(s) > 1$,

$$Z(s) = \int_0^1 \left(\vartheta(iy) - \frac{1}{2} \right) y^{\frac{s}{2}} \frac{dy}{y} + \int_1^\infty \left(\vartheta(iy) - \frac{1}{2} \right) y^{\frac{s}{2}} \frac{dy}{y} = I_1(s) + I_2(s),$$

then, $I_2(s)$ is *entire* in s. Let us now apply to I_1 the transformation

$$y \longmapsto \frac{1}{y}$$

under which dy/y is invariant. Then, using the functional equation for ϑ, one is led to

$$I_1(s) = \int_1^\infty \left(\vartheta(iy^{-1}) - \frac{1}{2} \right) y^{-\frac{s}{2}} \frac{dy}{y}$$

$$= \int_1^\infty \left(y^{1/2}\vartheta(iy) - \frac{1}{2} \right) y^{-\frac{s}{2}} \frac{dy}{y}$$

$$= \int_1^\infty \left(\vartheta(iy) - \frac{1}{2} \right) y^{-\frac{s}{2}+\frac{1}{2}} \frac{dy}{y} + \int_1^\infty \left(\frac{1}{2}y^{\frac{1}{2}} - \frac{1}{2} \right) y^{-\frac{s}{2}} \frac{dy}{y}$$

$$= I_2(1-s) + \frac{1}{s-1} - \frac{1}{s}.$$

Thus

$$Z(s) = I_2(s) + I_2(1-s) - \frac{1}{s} - \frac{1}{1-s}$$

from which the analytic continuation and functional equation follow immediately.

It is of course well known that Riemann's theory of ζ was the point of departure for Hadamard and de la Vallée Poussin for their proof of the prime number theorem. For a proof of the prime number theorem via Tauberian arguments, see Chapter 5. The method applies without change to lead to the prime number theorem in algebraic number fields, as remarked there.

Hurwitz in 1882 and de la Vallée Poussin in 1896 took up the question of the functional equation for the Euler products constructed by Dirichlet. We begin by noting that for $\chi \neq \chi_0$, $L(s : \chi)$ continues analytically to an entire function. This can be seen by a formula for $L(s : \chi)$ that goes back to Dirichlet's work. We have, with Γ as Euler's gamma function,

$$\frac{\Gamma(s)}{n^s} = \int_0^\infty t^{s-1} e^{-nt} dt$$

and so

$$\Gamma(s)L(s : \chi) = \int_0^\infty t^{s-1} \left(\sum_{n \geq 1} \chi(n) e^{-nt} \right) dt \qquad (\Re(s) > 1).$$

Now χ has period N while $\chi(1) + \cdots + \chi(N-1) = \sum_{n \in G_N} \chi(n) = 0$ because $\chi \neq \chi_0$ (orthogonality relation for a nontrivial character). So, writing

$$g(y) = \chi(1)y + \cdots + \chi(N-1)y^{N-1},$$

we have,

$$\sum_{n=1}^{\infty} \chi(n)y^n = \frac{g(y)}{1-y^N} \qquad (0 < y < 1),$$

where $g(y)$ has both y and $1-y$ as factors. Write $g(y) = y(1-y)h(y)$ where h is a polynomial. Then

$$\sum_{n\geq 1} \chi(n)y^n = y\left(\frac{h(y)}{1+y+\cdots+y^{N-1}}\right) = yk(y)$$

where k is C^{∞} on $[0,1]$ and is bounded, together with all of its derivatives, over the same interval. Thus

$$L(s:\chi) = \frac{1}{\Gamma(s)} \int_0^{\infty} t^{s-1} k(e^{-t}) e^{-t} dt \qquad (\Re(s) > 1).$$

This formula makes it clear that $L(s:\chi)$ has a continuation to the whole s-plane as an entire function. Indeed, the function $\ell(t) = k(e^{-t})$ is smooth and is bounded, together with all of its derivatives, over $[0, \infty)$. The proof that the integral

$$\int_0^{\infty} t^{s-1} \ell(t) e^{-t} dt$$

continues meromorphically to the whole s-plane with simple poles at $s = 0, -1, -2, \ldots$ and no other singularities now follows the same lines as the proof for the gamma function, namely, by repeated integration by parts. Since $\Gamma(s)^{-1}$ is entire and *vanishes* at the points $s = 0, -1, -2, \ldots$, it is clear that $L(s:\chi)$ is entire.

The main new feature now is that in order to get the simplest functional equations we must assume that χ is *primitive*; this means that χ is not the lift to G_N of any character of G_M through the map $G_N \longrightarrow G_M$ for any divisor $M < N$ of N. More generally, given a character of G_N there is a unique divisor M of N, called the *conductor* of χ, with the property that χ is the lift of a character χ_M of G_M and that for any divisor M' of N such that χ is the lift of a character of of $G_{M'}$, we have $M|M'$. Thus χ is primitive if and only if it has conductor N and for any χ, χ_M is a primitive character of G_M. There is a simple relation between $L(s:\chi)$ and $L(s:\chi_M)$ so that it is enough to restrict oneself to primitive χ. One now has to make a distinction between the cases $\chi(-1) = \pm 1$, which has already been encountered in the Dirichlet theory and expresses the difference between real and imaginary quadratic fields. Let

$$\Lambda(s:\chi) = \begin{cases} \pi^{-\frac{s}{2}} \Gamma\left(\frac{s}{2}\right) L(s:\chi) & \text{if } \chi(-1) = 1 \\ \pi^{-\frac{s+1}{2}} \Gamma\left(\frac{s+1}{2}\right) L(s:\chi) & \text{if } \chi(-1) = -1. \end{cases}$$

Then, with $\overline{\chi}$ denoting the character conjugate to χ and $G(\chi)$ denoting the Gaussian sum

$$G(\chi) = \sum_{x \bmod N} \chi(x) e^{2\pi i \frac{x}{N}},$$

we have the functional equation

$$\Lambda(s:\chi) = \varepsilon(\chi) N^{\frac{1}{2}-s} \Lambda(1-s:\overline{\chi})$$

where

$$\varepsilon(\chi) = \begin{cases} \frac{G(\chi)}{\sqrt{N}} & \text{if } \chi(-1) = 1 \\ -i\frac{G(\chi)}{\sqrt{N}} & \text{if } \chi(-1) = -1. \end{cases}$$

Now

$$|G(\chi)| = N^{1/2}$$

so that $\varepsilon(\chi)$, the so-called *root number*, is a complex number of absolute value 1 and is essentially a Gauss sum, thus has considerable arithmetic significance. When χ is quadratic (the case first treated by Hurwitz; de la Vallée Poussin then treated the general case), $\chi = \overline{\chi}$, and we have a functional equation for $\Lambda(s:\chi)$ itself, with $\varepsilon(\chi) = \pm 1$.

One can extend more or less everything that Riemann did for the zeta function to the $L(s:\chi)$ and build a parallel analytic theory for the Dirichlet L-series. We do not go into this here; see [6], [7].

Algebraic number fields. After Riemann and Hurwitz the development of Euler products was based on the general theory of (algebraic) number fields created by Kummer for cyclotomic fields, and extended later on to *all* number fields by Dedekind and Kronecker. Once it was understood how to replace the unique factorization in the ring of ordinary integers by the unique factorization of *ideals* in the ring of integers of a number field, the Riemann zeta function and the Dirichlet L-functions could be generalized to the more general context of an arbitrary number field, and it was natural to try to extend the theory of Dirichlet series and Euler products to this more general framework. To explain this let me take a brief detour and take a look at the discovery of unique factorization in number fields.

The central fact of arithmetic in the field $\mathbf{Q}$ of rational numbers is the property of unique factorization of nonzero numbers as products of primes numbers. Mathematicians like Euler, Lagrange, Dirichlet, and Gauss had already encountered fields which are *finite extensions* of $\mathbf{Q}$. Thus Euler (essentially) worked in $\mathbf{Q}(\sqrt{-3})$ for his proof, in the special case $n = 3$, of Fermat's last theorem, namely, the impossibility of finding integer solutions of the equation $x^n + y^n + z^n = 0$ where none of x, y, z is 0. One might even say that this work of Euler marked the time when algebraic number theory entered in questions of rational number theory. Lagrange, Dirichlet and Gauss had worked on the theory of classes of quadratic forms with integer coefficients, which is essentially equivalent to the theory of quadratic fields from the modern point of view. But none of them ventured into a fundamental study of more general algebraic number fields. Here, by an *algebraic number field*, or simply a *number field*, we mean an extension of finite degree over $\mathbf{Q}$. Gauss had studied the field $\mathbf{Q}(\sqrt{-1})$ and the ring $\mathbf{Z}[\sqrt{-1}]$ in which the unique factorization into prime factors is valid, a fact that allowed him to prove an extension of his quadratic reciprocity to this field. It was however generally known that unique factorization was not always valid in more general quadratic fields if one adopted the simple-minded definition of a prime element as one which is divisible only by itself and the unit element, up to units. For instance $\mathbf{Z}[\sqrt{-5}]$ does not have the unique factorization property, as can be verified from

$$6 = 2 \times 3 = (1 + \sqrt{-5}) \times (1 - \sqrt{-5})$$

where we write $\sqrt{-5}$ for a specific choice of the square root.

The person who overcame this great barrier was Kummer. He had been working on Fermat's last theorem, and as part of his study he was looking at the structure of $\mathbf{Z}[\omega]$ where $\omega_N = e^{2\pi i/N}$ is a primitive N^{th} root of unity. This of course is the ring of integers in the cyclotomic field $K_N = \mathbf{Q}(\omega_N)$, although this fact is only implicit in Kummer's work. His stroke of genius was to realize that one must give up the notion of *prime elements* in the formulation of the law of unique factorization and that one must create *ideal numbers* to play the role of prime divisors if unique factorization is to be preserved; his great achievement was to show that this is in fact possible for K_N. His ideal numbers would eventually become the ideals we know today as a consequence of the work of Dedekind and Kronecker. Kummer considered at first only the prime cyclotomic fields, namely the fields K_λ where λ is an odd prime integer; only much later did he take care of the general cyclotomic case. For accounts of the remarkable circumstances surrounding Kummer's discovery, the reader should look at Weil's introduction to Kummer's *Collected Papers* [8] and also [9], [10]. There is another point: $\lambda = 23$ is the first prime for which unique factorization fails in K_λ. This means that examples illustrating the new ideas require much computational work. Kummer was a prodigious calculator, and his work, with its flights of imagination, was nevertheless grounded on the most solid computations. After Kummer's work Kronecker and Dedekind extended Kummer's theory to *all* number fields.

Let K be any field and R a subring whose elements may be thought of as the integers in K. We shall suppose that K is the quotient field of R and that R is integrally closed in K (this is true if R has unique factorization). Assume now for a moment that R has unique factorization. If θ is a prime element of R, then one can define for any nonzero element $a \in K$ the integer r such that θ^r is the precise power of θ that divides a; i.e., $a = \theta^r b$ where b is prime to θ in the sense that $b = c/d$ where $c, d \in R$ and both are prime to θ. If we denote this integer by $v_\theta(a)$, then v_θ is a map of $K^\times = K \setminus \{0\}$ into the set of integers; we extend it to all of K by the convention that $v_\theta(0) = \infty$. Then v_θ has the following properties:

(a) $v_\theta(1) = 0$, $v_\theta(0) = \infty$, $v_\theta(K^\times) = \mathbf{Z}$.
(b) $v_\theta(ab) = v_\theta(a) + v_\theta(b)$ for all $a, b \in R$.
(c) $v_\theta(a + b) \geq \min(v_\theta(a), v_\theta(b))$.

If $K = \mathbf{Q}$, then one can show that the *only functions on $\mathbf{Q}$ with the above properties are the functions v_p, p a rational prime*, defined as above. So, for $\mathbf{Q}$, the properties given above capture completely the concept of a prime divisor. For general (K, R) let us call a *valuation* any function from K to $\mathbf{Z}$ with the properties listed above. Kummer's idea was to think of the valuations as *ideal numbers* that play the role of divisors in K, a role that was hitherto reserved only for the elements of K. The essence of Kummer's work was the explicit construction of *all the valuations of the cyclotomic field* and the restoration of unique factorization to the fields K_N by using ideal numbers. The last observation needs a little amplification. The valuations generate a free abelian group, the group $\mathfrak{D}$ of divisors, with a distinguished subsemigroup $\mathfrak{D}^+$ of positive elements. What one must show (roughly speaking) is that there is an imbedding of $K^\times$ into $\mathfrak{D}$ such that divisibility in $K^\times$ (relative to R) is compatible with divisibility in $\mathfrak{D}$ (relative to $\mathfrak{D}^+$). The imbedding is unique but may not always exist; when it does, we call the ring R a *ring with divisors*. Thus Kummer's work is essentially the proof that the ring

$$\mathbf{Z}[\omega_N], \qquad \omega_N = e^{2\pi i/N}$$

of integers in $K_N = \mathbf{Q}(\omega_N)$ is a ring with divisors. The work of Kronecker and Dedekind showed that it is possible to extend these ideas and results of Kummer to a very extensive class of rings which includes the rings of integers of any algebraic number field. The main results are as follows [11], [12], [13].

THEOREM. *If R is a Noetherian domain which is integrally closed, then R is a ring with divisors. In particular, the ring of integers in a number field is a ring with divisors.*

THEOREM. *If r is a ring with divisors with quotient field k and if K is an extension of finite degree over k, then the integral closure R of r in K is a ring with divisors. In particular, the ring of integers in a number field is a ring with divisors.*

Although the first theorem is more general, for number theory it does not add anything. Also when the ring is of dimension > 1 (in a certain sense) the theory of divisors is not nearly as decisive as in the case of rings of dimension 1.

For algebraic number fields everything can be done directly in the language of ideals. Let $\mathfrak{r}$ be the ring of integers of an algebraic number field K. For ideals $\mathfrak{a}, \mathfrak{b} \subset \mathfrak{r}$ we define their *product* $\mathfrak{a}\mathfrak{b}$ as the set (ideal) of numbers of the form $\sum_i a_i b_i$ where $a_i \in \mathfrak{a}, b_i \in \mathfrak{b}$. Let $\mathbf{P}$ be the set of *prime* ideals of $\mathfrak{r}$. Then every ideal $\mathfrak{a} \subset \mathfrak{r}$ can be *factorized* in a unique fashion as

$$\mathfrak{a} = \prod_{\mathfrak{p} \in \mathbf{P}} \mathfrak{p}^{a(\mathfrak{p})}$$

where the $a(\mathfrak{p})$ are integers ≥ 0 and almost all of them are 0. If $a \in \mathfrak{r}$ is nonzero, then the above formula applied to the principal ideal $(a) = a\mathfrak{r}$ gives

$$(a) = \prod_{\mathfrak{p} \in \mathbf{P}} \mathfrak{p}^{\nu_{\mathfrak{p}}(a)}$$

where

$$\nu_{\mathfrak{p}}(a) \text{ is an integer } \geq 0 \text{ and is 0 for almost all } \mathfrak{p}.$$

The $\nu_{\mathfrak{p}} (\mathfrak{p} \in \mathbf{P})$ are precisely all the valuations of K, and $\mathfrak{p}$ is just the set of elements divisible by $\nu_{\mathfrak{p}}$. Thus there is no distinction between the valuations or ideal numbers of Kummer and the prime ideals of Dedekind. From this point of view it is also clear what happens when a ring does not have unique factorization in the old sense: prime elements do not always generate a prime ideal, and prime ideals are not always generated by a single element. When both of these properties are possessed by the ring of integers, we shall refer to it as a unique factorization domain. For a number field K the factorization of the ideal (p) where p is a rational prime then yields all the prime divisors of K. The central problem in number theory is then to write down the factorization

$$(p) = \prod_{\mathfrak{p} \in \mathbf{P}} \mathfrak{p}^{\nu_{\mathfrak{p}}(p)}$$

of rational primes in K. The primes p for which some $\nu_{\mathfrak{p}}(p) > 1$ are *ramified*; there are only finitely many of them, namely those that divide the *discriminant* ideal of K, which is the principal ideal (d) for a unique positive integer d. For any $K \neq \mathbf{Q}$ there are always such primes. The prime p *splits completely* if

$$(p) = \mathfrak{p}_1 \mathfrak{p}_2 \ldots \mathfrak{p}_n$$

where

$$\mathfrak{p}_j \text{ are distinct and } n = [k : \mathbf{Q}].$$

These considerations also apply when we consider two number fields k, K with $k \subset K$. The primes of k that ramify in K are those that divide the *discriminant ideal* $\mathfrak{d}$ of k.

Just as in the rational number field we can extend unique factorization to *fractional ideals*, namely finitely generated $\mathfrak{r}$-modules contained in K, or equivalently, those of the form xA where $A \subset \mathfrak{r}$ is an ideal and $x \in K^{\times}$. The fractional ideals form a *group* under multiplication; the quotient of this group by the group of principal fractional ideals $x\mathfrak{r}(x \in K^{\times})$ is a *finite* group. It is the *class group* of K. It is a deep-lying invariant of K. Its cardinality is the *class number* of K. The field has class number 1 if and only if its ideals are all principal; this is equivalent to saying that it is a unique factorization domain.

Given K, the determination of the prime ideals in K is then nothing other than the factorization of the rational primes in K. From hindsight we can see that Kummer was also lucky because the factorization of rational primes in the cyclotomic fields K_N is especially simple and essentially depends only on the residue class of the prime mod N; this is the reason why he was able to construct the cyclotomic valuations explicitly. Eventually the role of the cyclotomic fields would become even clearer when it was realized that they are *class fields* and contain all abelian extensions of $\mathbf{Q}$. This can be seen in its simplest form when we take N to be an odd prime and identify its unique quadratic subfield, leading to the law of quadratic reciprocity; this is perhaps the most natural way to prove this law (see [13]).

Once the foundations of the theory of ideals was laid down, it was natural for people to try to extend to algebraic number fields everything that had been done for number theory over the rational numbers. In particular it became possible to introduce the *ideal class groups* as a direct generalization to a number field of the residue class groups G_N of Dirichlet and to formulate and prove the appropriate generalizations of the Dirichlet-Riemann-Hurwitz theory and in fact to erect what is known as *class field theory*, the theory of abelian extensions of a number field.

Algebraic number fields are not the only ones on which arithmetic can be done. In characteristic $p > 0$ we have *function fields* which are finite extensions of the fields $F(T)$ where F is a *finite* of characteristic p and T is an indeterminate. The arithmetic of such fields got going in a serious way with Artin's thesis [14a]. By a *global field* we mean either a number field or a function field. The function fields are the fields of rational functions on smooth algebraic curves defined over F, and so it will come as no surprise that the deeper aspects of such fields involve the use of algebraic geometry. If we change F to $\mathbf{C}$, we obtain the function fields of Riemann surfaces. In this way we obtain a *geometric* counterpart to classical arithmetic. The origins of the algebraic theory of function fields over general fields goes back to a famous paper of Dedekind-Weber, extended and generalized by Artin and Chevalley [14b], [14c], [14d].

Ideles and the adelic language. In 1941, Chevalley [15] found a very elegant formulation of class field theory by introducing the concepts of *ideles* and the *idele class group*. These are generalizations of the classical notions of ideals and ideal class groups, and in terms of these the central theorems of class field theory could be formulated most succinctly. More importantly, Chevalley's approach made class field theory compatible with the developments in topological groups and functional analysis. Soon afterwards Artin and Whaples introduced the additive analogue of the idele, the so-called valuation vector, eventually named an *adele* [14a]. The method of adeles turned out to be very general and powerful and led to a much richer view of number theory in general and class field theory in particular. Moreover it allows one to view classical number theory as a special case of the arithmetic of algebraic groups. Since this language is the one in which almost all current work is formulated, we shall take a brief look into it.

The basic idea is to introduce *absolute values* on a number field K. These are maps $|\cdot|$ of K into nonnegative reals such that $|0| = 0, |1| = 1, |ab| = |a||b|, |a + b| \leq |a| + |b|$. Absolute values which are powers of one another are equivalent and such equivalence classes of absolute values are called *places*. If v is a valuation on K and $c > 1$, the function $x \mapsto |x|_v = c^{-v(x)}$ is an absolute value of K and the corresponding place is independent of the choice of c. This is *ultrametric* in the sense that $|a+b|_v \leq \max(|a|_v, |b|_v)$. The places thus defined are called *nonarchimedean* or *finite*. Any ultrametric absolute value is equivalent to one coming from a valuation, and so finite places are essentially the same as valuations. But K has additional absolute values. For instance, $\mathbf{Q}$ has the usual absolute value. To obtain these additional ones we densely imbed the number field in $\mathbf{R}$ or $\mathbf{C}$, and we can use the ambient absolute value to define absolute values on K; the corresponding places are called the *archimedean* or *infinite*. The place is *real* or *imaginary* according to whether the imbedding is into $\mathbf{R}$ or $\mathbf{C}$. The number field can be completed with respect to any of these absolute values, and these completions are *local fields*, namely locally compact nondiscrete fields. Consideration of real or complex imbeddings occur prominently in classical number theory: in Minkowski's geometry of numbers, in Dirichlet's theorem of units, and so on. In those days these arguments were considered foreign to number theory. In the more inclusive view that has developed since then, the real or complex imbeddings are seen as being analogous to the imbeddings into nonarchimedean local fields, and the basic results of number theory come out of a method that treats all these imbeddings on the same footing. To work *globally* over K one then has a technique that relates global problems over K to problems over the product $\prod_v' K_v$ where the product is over *all* places of K, finite as well as infinite; K_v is the completion of K at the place v; and the product is a so-called *restricted* one. This will be explained presently.

We start with the notion of a *restricted product* of a family of locally compact groups. Let $(G_v)_{v \in F}$ be a family of locally compact groups, F being a countably infinite set. Then the product $\prod_v G_v$ is locally compact if all but a finite number of the G_v are compact, but in general, not otherwise. For applications to number theory it is necessary to consider the case where all the G_v are only locally compact, not compact. The basic concept is that of a *restricted product*, and it requires us to assume that for almost all $v \in F$, i.e., $\forall v \in F \setminus F_0$ where $F_0 \subset F$ is a finite set, one is given a distinguished *compact open subgroup* $U_v \subset G_v$. The *restricted product* of the G_v denoted by $\prod_v' G_v$ is the subgroup of the full product $\prod_v G_v$ of all elements (x_v) such that $x_v \in U_v$ for almost all v. Before describing the topology that makes this locally compact, let us mention the two special cases of this construction: (1) $U_v = G_v$ for almost all v, so that the restricted and full products are the same:

$$\prod_v' G_v = \prod_v G_v.$$

(2) All the G_v are discrete and $U_v = (1)$ for almost all v. In this case the restricted product is the algebraic direct sum:

$$\prod_v' G_v = \bigoplus_v G_v.$$

Returning to the general case, let F_0 be the exceptional finite set of v's such that for all $v \in F \setminus F_0$ we have the compact open subgroup $U_v \subset G_v$, and for any finite set E with $F_0 \subset E \subset F$ write

$$G(E) = \prod_{v \in E} G_v \times \prod_{v \in F \setminus E} U_v.$$

The second factor in the definition of $G(E)$ is a product of compact groups and so $G(E)$ is *locally compact in the product topology*. We have the set inclusion

$$G(E) \subset \prod_v G_v.$$

If $E \subset E' \subset F$ where E' is also finite, the fact that the U_v are open in G_v implies easily that in the natural inclusion

$$G(E) \subset G(E')$$

$G(E)$ is an *open subgroup* of $G(E')$. So if we consider

$$\bigcup_E G(E)$$

there is a unique way to topologize this union so that it becomes locally compact and contains each $G(E)$ as an open subgroup. Since

$$\prod_v{}' G_v = \bigcup_E G(E)$$

set theoretically, we shall equip $\prod_v' G_v$ with the topology defined on the right side. The natural inclusion

$$\prod_v{}' G_v \hookrightarrow \prod_v G_v$$

is an injective morphism of topological groups but is not a homeomorphism. If each G_v is a topological ring and each U_v (whenever defined) is a subring of G_v, it is immediate that $\prod_v' G_v$ is also a topological ring.

The emergence of locally compact fields and rings allows one to bring in Haar measure and the apparatus of harmonic analysis, including the theory of Fourier transforms. For a G_v with $v \in F \setminus F_0$ we can normalize the left Haar measure μ_v by requiring that $\mu_v(U_v) = 1$. Then, for given choices of left Haar measures μ_v on $G_v (v \in F_0)$ we have the product measure $\otimes_{v \in F} \mu_v$ on $\prod_v' G_v$, which is a *standard* Haar measure. Often there are other ways of standardizing the Haar measure. In any case as long as the μ_v are chosen such that $\mu_v(U_v) = 1$ for almost all v, the product measure above is defined. We can also use convergence factors, which is equivalent to asking that

$$\prod_{v \in F \setminus F_0} \mu_v(U_v)$$

converge.

Example: The adele ring and idele group of a number field K. Let us fix a number field K, and let v be a valuation of it with prime ideal $\mathfrak{p}$. Then $\mathfrak{r}/\mathfrak{p}$ is a finite field with $q_v = N\mathfrak{p}$ elements. The (normalized) restriction of v to $\mathbf{Q}$ is a valuation of $\mathbf{Q}$ and so defines a prime p; we say that $\mathfrak{p}$ *lies above* p. Then $q_v = p^{d_v}$. We write

$$|x|_v = q_v^{-v(x)} \quad (x \in K^\times), \qquad |0|_v = 0.$$

Then $|\cdot|_v$ is a *standard* nonarchimedean absolute value for K, and we can pass to its metric completion K_v, which will be a totally disconnected field—in fact a finite extension of the field $\mathbf{Q}_p$ of p-adic numbers. If $K = \mathbf{Q}$ and v is defined by p, then $K_v = \mathbf{Q}_p$. For $\mathbf{R}$ the standard absolute value is the usual one, while for $\mathbf{C}$ it is the *square* of the usual one. The standard absolute values satisfy the relation

$$d(cx) = |c|_v dx$$

where dx is a Haar measure of the additive group of K_v (this relation explains the squaring in the complex case). For a finite place v, the closure $\mathfrak{r}_v$ of $\mathfrak{r}$ is an open compact subring (even maximal) of K_v and is in fact the set of elements x with $|x|_v \le 1$.

Write $P = P(K)$ for the set of its places of K, and P_f the subset of finite places. With respect to the family $(K_v, \mathfrak{r}_v)$ we form the restricted product

$$\mathbf{A}(K) = {\prod_v}' K_v$$

which is called the *ring of adeles of K*. The restricted product

$$\mathbf{A}(K)_f = {\prod_{v \text{ finite}}}' K_v$$

is called the ring of *finite adeles*.

If we take the family $(K_v^\times, \mathfrak{r}_v^\times)$ we get the restricted product

$$\mathbf{J}(K) = {\prod_v}' K_v^\times$$

which is called the *group of ideles of K*. For any idele $x = (x_v)$ we associate the fractional ideal

$$\mathrm{id}(x) = \prod_v \mathfrak{p}_v^{\mathrm{ord}_v(x_v)};$$

then the map $x \longmapsto \mathrm{id}(x)$ maps $\mathbf{J}(K)$ *onto the group of fractional ideals*. Thus an idele generalizes the notion of the ideal (hence the name!).

The basic map in this setting is the *diagonal map*, which takes any element $x \in K$ (resp. $x \in K^\times$) to the adele (resp. idele) (x_v) where $x_v = x$ for all v, giving us imbeddings

$$K \hookrightarrow \mathbf{A}(K), \qquad K^\times \hookrightarrow \mathbf{J}(K).$$

It is a fundamental result that the images of these maps are *discrete* subgroups in their respective ambient groups. We can therefore identify K (resp. $K^\times$) as a discrete subgroup of $\mathbf{A}(K)$ (resp. $\mathbf{J}(K)$). The group

$$\mathbf{C}(K) = \mathbf{J}(K)/K^\times$$

is the *idele class group*. The map $x \mapsto \mathrm{id}(x)$ defined above gives rise to maps $\mathbf{C}(K)$ onto the ideal class groups mentioned earlier.

There are two key facts: the first is that $\mathbf{A}(K)/K$ *is compact*. Thus K is a *lattice* in $\mathbf{A}(K)$. On $\mathbf{A}(K)$ one can define an absolute value $|\cdot|$ by

$$|x| = \prod_v |x_v|_v, \qquad x = (x_v)_v.$$

Then $|\cdot|$ is a continuous multiplicative map into the positive reals and we have the *product formula*

$$|x| = 1 \qquad (x \in K^\times).$$

Let
$$\mathbf{J}_1(K) = \left\{ x \in \mathbf{J}(K) \mid |x| = 1 \right\}.$$
The second key fact is that $\mathbf{J}_1(K)/K^\times$ *is compact, and* $\mathbf{J}(K)/\mathbf{J}_1(K) \simeq \mathbf{R}_+^\times$. One can show that $\mathbf{J}(K) \simeq \mathbf{J}_1(K) \times \mathbf{R}_+^\times$ (noncanonically).

All the basic results of algebraic number theory can be proved in this setting; in fact, in [16a] the idea of a locally compact nondiscrete field is taken as the basis for a completely independent development of algebraic number fields, including the ideal theory. Moreover, as mentioned already, class field theory has its most elegant formulation in this setting due to Chevalley.

The standard Haar measures on K_v are the ones with $\mu_v(\mathfrak{r}_v) = 1$, and on $K_v^\times$ are the ones with $\mu_v^\times(U_v) = 1$. On $\mathbf{R}$ it is dx, the usual Lebesgue measure, and on $\mathbf{C}$ it is $2dxdy = idz \wedge d\bar{z}$. The standard Haar measure on $\mathbf{A}(K)$ then induces a Haar measure on the compact group $\mathbf{A}(K)/K$, and it is an interesting calculation to compute vol $(\mathbf{A}(K)/K)$, its *volume*. Similarly vol $(\mathbf{J}_1(K)/K^\times)$ also makes sense. These volumes can be evaluated explicitly in terms of the invariants of K.

Example: Adelic groups attached to $\mathrm{GL}(n)$ and algebraic groups. The idea of imbedding a number field inside the restricted product of its completions can be carried over to objects that are defined *functorially* over the number field. For instance if we have an associative algebra M over K of finite dimension, we can use the above process of completion and diagonal imbedding to define the adele ring $M(\mathbf{A}_K)$; the same process applied to the groups of units of the completions gives the adele group $M(\mathbf{A}_K)^\times$ of units of $M(\mathbf{A}_K)$. Taking M to be the full matrix algebra in n dimensions, we obtain the restricted direct product of the completions of $\mathrm{GL}(n, K)$ and obtain the adelic group

$$\mathrm{GL}(n, \mathbf{A}_K) = \prod_v{}' \mathrm{GL}(n, K_v).$$

Here we must note that the $\mathrm{GL}(n, K_v)$ are locally compact groups and that if v is finite with $\mathfrak{r}_v$ as the compact open subring of integers of K_v, then $\mathrm{GL}(n, \mathfrak{r}_v)$ is a compact open subgroup, even maximal, of $\mathrm{GL}(n, K_v)$. We again obtain the imbedding

$$\mathrm{GL}(n, K) \hookrightarrow \mathrm{GL}(n, \mathbf{A}_K).$$

$\mathrm{GL}(n, K)$ *is discrete in* $\mathrm{GL}(n, \mathbf{A}(K))$. This construction can be generalized to the case where $\mathrm{GL}(n)$ is replaced by an algebraic group defined over K.

These constructions are important for many reasons. Thus, for $n = 1$,

$$\mathrm{GL}(1, \mathbf{A}(K)) = \mathbf{J}(K),$$

and this fact allows us to view class field theory from this unified perspective. Second, for $n = 2$, one can formulate the Hecke theory of modular forms and their Euler products in the context of the homogeneous space

$$\mathrm{GL}(2, \mathbf{A}(K))/\mathrm{GL}(2, K)$$

and the action of $\mathrm{GL}(2, \mathbf{A}(K))$ on the space of functions on it. Finally, this framework permits the introduction of entirely new ideas in the arithmetic theory of algebraic groups. For instance, if G is *semisimple*, in particular for $G = \mathrm{SL}(n)$, the homogeneous space

$$G(\mathbf{A}(K))/G(K),$$

although not compact for $n \geq 2$, has finite volume. This is a famous result of Borel and Harish-Chandra. For general semisimple G defined over K, the volume

$$\text{vol} \, (G(\mathbf{A}(K))/G(K),$$

computed with respect to a specific Haar measure coming from a rational volume form over G, has a deep arithmetic interpretation. It is called the *Tamagawa number* of G. For $G = \text{SO}(n)$ the Tamagawa number is 2, and this fact is equivalent to some of the deeper results of Minkowski and Siegel in the arithmetic theory of quadratic forms; see [16b].

Grössencharakters of Hecke and the associated Euler products. Hecke added a new theme to the story of Euler products when he introduced a new type of character, the *Grössencharakter*, and associated an *L*-function to it. In the adelic terminology, the Grössencharakters are simply characters of the idele class group $\mathbf{C}(K) = \mathbf{J}(K)/K^{\times}$. The subgroup of Grössencharakters of finite order can then be identified with the group of characters of the ideal class groups, which are generalizations of the Dirichlet characters and coincide with them when $K = \mathbf{Q}$. Hecke proved both analytic continuation and functional equation for the *L*-functions associated with the Grössencharakters [17]. These Hecke *L*-functions are thus far-reaching generalizations of the Dirichlet ones.

The groups $\mathbf{C}(K), K_v^{\times}$ are examples of locally compact abelian groups G which are (noncanonically) of the form $G_1 \times H$ where G_1 is compact and H is either $\mathbf{R}$ or $\mathbf{Z}$. We call them *quasicompact* and of type $\mathbf{R}$ or $\mathbf{Z}$ according to whether H is isomorphic to $\mathbf{R}$ or $\mathbf{Z}$. Then G_1 is canonical and is the unique maximal compact subgroup of G, and we have a homomorphism $|\cdot|$ of G into $G/G_1 \simeq H$ whose kernel is exactly G_1. It is therefore possible to look at the *quasicharacters* of G, namely continuous homomorphisms of G into $\mathbf{C}^{\times}$. They form a group $G^{\sim}$ containing the annihilator $G_1^{\perp}$ as an open subgroup so that $G^{\sim}/G_1^{\perp} \simeq \widehat{G}_1$. For each $\xi \in \widehat{G}_1$, the preimage $G^{\sim}(\xi)$ of ξ in $G^{\sim}$ is connected, and the $G^{\sim}(\xi)$ are the connected components of $G^{\sim}$. The connected components are isomorphic to the group of quasicharacters of H and so isomorphic to $\mathbf{C}$ or $\mathbf{C}^{\times}$ in the two cases, where s in $\mathbf{C}$ or $\mathbf{C}^{\times}$ defines the quasicharacter

$$t \longmapsto e^{st} \quad (t \in \mathbf{R}) \qquad \text{or} \qquad n \longmapsto s^n \quad (n \in \mathbf{Z}).$$

In our applications we consider $G = K_v$ or $G = \mathbf{C}(K) = \mathbf{J}(K)/K^{\times}$. Then $G_1 = U_v$ or $G_1 = \mathbf{J}_1(K)/K^{\times}$. In this case there is a canonical choice for $|\cdot|$, namely, the local or adelic absolute value, so that the characters in $G^{\sim}(\xi)$ are of the form $\xi|\cdot|^s$ for $s \in \mathbf{C}$. The case when G/G_1 is $\simeq \mathbf{Z}$ arises when v is a finite place and $G = K_v$ or when K is a function field and $G = \mathbf{C}(K)$. Thus, in all cases, the $G^{\sim}(\xi)$ are *complex manifolds*.

The Hecke *L*-functions are built as Euler products. Let χ be a quasicharacter of $\mathbf{C}(K)$; then $\chi = \otimes_v \chi_v$ where the χ_v are quasicharacters of the $K_v^{\times}$ with the property that for almost all v, $\chi_v|U_v = 1$ where U_v is the group of units of K_v, i.e., the subgroup of elements y with $|y|_v = 1$, thus the maximal compact subgroup of $K_v^{\times}$; we write $\chi = (\chi_v)$ and call the χ_v the *local factors* of χ. A quasicharacter of $K_v^{\times}$ is said to be *unramified* if its restriction to U_v is trivial; otherwise it is called *ramified*. In the unramified case if ϖ_v is a *uniformisant* of K_v, namely a generator for the maximal ideal of the ring of integers of K_v, then $\chi(\varpi_v)$ is uniquely defined independent of the choice of ϖ_v. Then

$$L(s:\chi) \stackrel{\text{def}}{=} L(\chi|\cdot|^s) = \prod_{v \text{ finite}} L_v(\chi_v|\cdot|_v^s)$$

where the local Eulerian factor L_v is defined for the finite places v as follows. For v finite,

$$L_v(\omega) = \begin{cases} \frac{1}{1-\omega(\varpi_v)} & \text{if } \omega \text{ is unramified,} \\ 1 & \text{if } \omega \text{ is ramified.} \end{cases}$$

Clearly

$$L_v(\omega|\cdot|^s) = \frac{1}{1 - \omega(\varpi_v)q_v^{-s}}$$

if ω is unramified and 1 otherwise. To complete the definition we should also define the Eulerian factors coming from the infinite places. For the infinite places v the Eulerian factor is a gamma function. We write

$$\Gamma_{\mathbf{R}}(s) = \pi^{-s/2}\Gamma\left(\frac{s}{2}\right), \quad \Gamma_{\mathbf{C}}(s) = 2(2\pi)^{-s}\Gamma(s) = \Gamma_{\mathbf{R}}(s)\Gamma_{\mathbf{R}}(s+1).$$

If v is a real place and $K_v = \mathbf{R}$, we define

$$L_v(x^{-N}|\cdot|_v^s) = \Gamma_{\mathbf{R}}(s) \qquad (N = 0, 1, s \in \mathbf{C}),$$

and for v complex with $K_v = \mathbf{C}$, we put

$$L_v(z^{-N}|\cdot|^s) = L_v(\bar{z}^{-N}|\cdot|^s) = \Gamma_{\mathbf{C}}(s) \qquad (N \geq 0 \text{ an integer, } s \in \mathbf{C}).$$

Then one defines

$$\Lambda(s:\chi) = \left(\prod_{v \text{ infinite}} \Gamma_v(s)\right) \cdot L(s:\chi).$$

The Euler product converges on some right half plane and defines an analytic function nowhere vanishing there; if χ is *unitary*, the convergence is for $\Re(s) > 1$. If $\chi = \chi_0$, the trivial character, $L(s:\chi)$ is the Dedekind zeta function $\zeta_K(s)$ and

$$\Lambda(s:\chi_0) \stackrel{\text{def}}{=} Z_K(s) = \left(\prod_{v \text{ infinite}} \Gamma_v(s)\right) \zeta_K(s).$$

Clearly the exact definition of the $\Lambda(s:\chi)$ is quite delicate, since the object is to do this in such a way that the functional equation becomes as simple as possible. Hecke's theorem is as follows.

THEOREM (Hecke). (i) *For any quasicharacter χ of the idele class group $C(K)$, $L(s:\chi)$ extends meromorphically to the whole s-plane; the extension is entire for $\chi \neq \chi_0$, and for $\chi = \chi_0$ it has a single pole at $s = 1$ which is simple and the residue at which is a positive number explicitly computable in terms of the invariants of the number field.*

(ii) **(Hadamard-de la Vallée Poussin)** *If χ is unitary,*

$$L(1+it:\chi) \neq 0 \qquad (t \in \mathbf{R}).$$

(iii) **(Hecke)** *We have*

$$\Lambda(s:\chi) = \varepsilon(\chi|\cdot|^s)\Lambda(1-s:\chi^{-1})$$

where ε is holomorphic on $C(K)^{\sim}$ and is of the form

$$\varepsilon(\omega|\cdot|^s) = A(\omega)B(\omega)^s \qquad (A(\omega) \neq 0, B(\omega) > 0 \text{ are constants}).$$

REMARK. The Hadamard-de la Vallée Poussin nonvanishing in (ii) for the Riemann zeta function, namely

$$\zeta(1 + it) \neq 0 \qquad (t \in \mathbf{R} \setminus \{0\}),$$

is the property needed to prove the prime number theorem (PNT). The most transparent proof of PNT is via Fourier transforms based on the above and the Wiener-Ikehara Tauberian theorem (see Chapter 5). Actually, the above property can be generalized to the construction of zero-free regions for the zeta function and other L-functions, the regions containing the line $\Re(s) = 1$ in their interior but tapering to it asymptotically when $\Im(s) \to \infty$ (see [4], [7]).

Tate's theory. In the late 1940's a new theme arising out of the adelic point of view of number theory entered the story. This theme would prove to be extraordinarily influential when noncommutative generalizations of arithmetic were made. Following the suggestion of Artin that systematic use should be made of *harmonic analysis* on the adele ring, Tate, a student of Artin, generalized vastly the notion of zeta functions and obtained for them both the analytic continuations and functional equations by methods of harmonic analysis on the adele ring $\mathbf{A}(K)$ and the Mellin transform on $\mathbf{J}(K)$. His results were part of his Princeton thesis in 1950 [18a]. Written with exceptional clarity and a minimum of fuss, this thesis was extremely influential. For recent expositions of the material in it, the reader may look up [7], [16a], [18b], [18c].

The main idea of Tate was to use the theory of Fourier transforms in an essential manner to define an extended class of zeta functions both at the local and global levels. More precisely, he started with a nice (from the point of view of harmonic analysis) function on the *additive groups* K_v and $\mathbf{A}(K)$ and defined a corresponding zeta function by taking the Fourier transform of the restriction of this function to the *multiplicative groups* $K_v^\times, \mathbf{J}(K)$, i.e., the *Mellin transform*. These Tate zetas have meromorphic continuations and functional equations; the local theory is very simple, and the global theory uses *Poisson summation* on $\mathbf{A}(K)$ with respect to the lattice K imbedded in it. By choosing the original function on $K_v, \mathbf{A}(K)$ suitably the Tate zetas become the Eulerian factors locally and the Hecke L-functions globally. In particular, this method gives the entire Hecke theory at one stroke.

Local zeta functions. We choose a nontrivial additive character ψ_v on K_v. One knows that ψ_v induces an isomorphism of the additive group of K_v with its dual group if we identify $a \in K_v$ with the character $x \mapsto \psi_v(ax)$. This allows the Fourier transform of a function on K_v to be viewed as a function on K_v itself. Let $\mathcal{S}(K_v)$ be the Schwartz space of K_v if $K_v = \mathbf{R}$ or $\mathbf{C}$, and the space of Schwartz-Bruhat functions on K_v if v is a finite place, namely the space of locally constant complex functions on K_v with compact supports. For $f \in \mathbf{S}(K_v)$ we define its Fourier transform $\widehat{f}$ on K_v by

$$\widehat{f}(\xi) = \int_{K_v} f(x)\psi(-\xi x)dx \qquad (\xi \in K_v).$$

The theory of Fourier transforms tells us that the map $f \mapsto \widehat{f}$ is a linear isomorphism of $\mathcal{S}(K_v)$ with itself. One choice for $d_v x$ could be the *self dual* measure, namely one such that the Fourier inversion formula is valid with respect to it, i.e.,

$$f(x) = \int_{K_v} \widehat{f}(\xi)\psi_v(\xi x)d_v x \qquad (x \in K_v, f \in \mathcal{S}(K_v)).$$

We now choose a multiplicative Haar measure $d_v^\times x$ on $K_v^\times$; for instance we can choose $d_v^\times x = |x|_v^{-1}|d_v x$. The local Tate zeta functions are defined as the *Mellin transforms* of the restriction of elements of $\mathcal{S}(K_v)$ to $K_v^\times$:

$$Z_v(f,\omega) = \int_{K_v^\times} f(x)\omega(x)d_v^\times x \qquad (f \in \mathcal{S}(K_v), \omega \in K_v^{\times\sim}).$$

The integral converges if $\sigma(\omega) > 0$ where $|\omega| = |\cdot|_v^{\sigma(\omega)}$. The basic result of Tate is that the Z_v have analytic continuations to the manifold of all quasicharacters and satisfy the functional equation

$$Z_v(f,\omega) = \varepsilon_v(\omega, \psi_v, d_v x)Z_v(\widehat{f}, \omega^{-1}|\cdot|).$$

Note that (here and later in the global theory) the transition $\omega \to \omega^{-1}|\cdot|$ reduces to the classical transition $s \to 1 - s$ when $\omega = |\cdot|^s$. The factor $\varepsilon_v(\omega, \psi_v, d_v x)$ is independent of f and meromorphic in ω. It is independent of the choice of $d_v^\times x$. Its dependence on the choices of ψ_v and $d_v x$ are very simple so that ε_v is essentially a function of ω. There are standard choices for ψ_v, dx_v, and $d_v^\times x$ in terms of which ε can be computed explicitly by choosing special f.

Global zeta functions. The theory of the global Tate zetas is quite similar. We start with a character ψ of the additive group of $\mathbf{A}(K)$, the adele ring of K, which is nontrivial but trivial on K; there is a standard way to construct such a ψ. Then $\psi = (\psi_v)$ and ψ_v is nontrivial for *all* places v. We choose the additive Haar measure $d_v x$ to be self dual with respect to ψ_v for *all* v. It can then be shown that

$$\int_{\mathfrak{r}_v} d_v x = 1$$

for almost all v, $\mathfrak{r}_v$ is the compact open ring of integers of K_v, that the product measure $dx = \prod_v d_v x$ is well defined on $\mathbf{A}(K)$ and a Haar measure for $\mathbf{A}(K)$, and finally that

$$\int_{\mathbf{A}(K)/K} dx = 1.$$

Moreover dx is the self dual measure on $\mathbf{A}(K)$ with respect to the choice of ψ. Notice that the above property defines a Haar measure on $\mathbf{A}(K)$ *canonically* and is independent of the choices made; it is sometimes called the *Tamagawa measure*. The choice of the Haar measure on $\mathbf{J}(K)$ also needs more care than in the local case because $\mathbf{J}(K)$ has measure 0 in $\mathbf{A}(K)$. We define $d_1^\times x$ as the Haar measure on $\mathbf{J}_1(K)/K^\times$ so that $\mathbf{J}_1(K)/K^\times$ has volume 1 and choose the Haar measure on $\mathbf{J}(K)$ so that under the isomorphism of $\mathbf{J}(K)/\mathbf{J}_1(K) \simeq \mathbf{R}_+^\times$, the measure on $\mathbf{J}(K)/\mathbf{J}_1(K)$ goes over to dt/t. On the other hand, there is a Haar measure γ on $\mathbf{J}(K)$ of the form

$$\gamma = \otimes_v \gamma_v$$

where γ_v is the Haar measure on $K_v^\times$ for finite v defined by $\gamma_v(U_v) = 1$, $\gamma_v = |x|^{-1}dx$ for v real, and $\gamma_v = (x\bar{x})^{-1}(idx \wedge d\bar{x})$ for imaginary v. Then

$$\gamma = c(K)d^\times x$$

where $d^\times x$ is the Haar measure on $\mathbf{J}(K)$ defined previously and $c(K)$ is the invariant of K given by

$$c(K) = \gamma(\mathbf{J}_1(K)/K^\times) = \frac{2^{r_1}(2\pi)^{r_2}hR}{w}$$

where r_1 (resp. r_2) is the number of real (resp. imaginary) places of K, h is the class number of K, R is the regulator of K and w is the number of roots of unity in K.

The functions considered on the adele ring are linear combinations of functions of the form $\otimes_v f_v$ where the product is over all places; f_v for infinite v is a Schwartz function; and for finite v a Schwartz-Bruhat function, with $f_v = 1_{\mathfrak{r}_v}$ for almost all v, $1_{\mathfrak{r}_v}$ is the characteristic function of the ring $\mathfrak{r}_v$ of integers of K_v. Then for such functions we define the global Tate zeta function as

$$Z(f, \omega) = \int_{\mathbf{J}(K)} f(x) d^\times x$$

where ω is a quasicharacter of the idele class group $\mathbf{C}(K)$. If $\sigma(\omega)$ is defined by $|\omega| = |\cdot|^{\sigma(\omega)}$, the integral converges for $\sigma(\omega) > 1$. For suitable choices of f this will reduce to the Hecke Euler product associated to the quasicharacter ω, completed by the gamma factors. These general zeta functions have meromorphic continuations to the whole manifold of quasicharacters and satisfy the functional equation

$$Z(f, \omega) = Z(\widehat{f}, \omega^{-1}|\cdot|)$$

where the Fourier transform is taken with respect to the Tamagawa measure. They are holomorphic everywhere except at $\omega_0 = \chi_0$ and $\omega_1 = |\cdot|$ where they have simple poles and residues $-f(0)$ and $\widehat{f}(0)$ respectively. The Hecke L-functions are obtained by choosing f in special ways. The simplicity of the functional equation is however slightly misleading: when we make specific calculations we will need to specify the Haar measures more explicitly, and then constants will appear in the functional equations.

As an illustration of this method let us take the simplest example, when $K = \mathbf{Q}$. The first task is to choose the additive character ψ of $\mathbf{A}$ (we omit the reference to $\mathbf{Q}$). We note that

$$\mathbf{A} = A_\infty + \mathbf{Q}, \quad A_\infty \cap \mathbf{Q} = \mathbf{Z} \quad \text{where } A_\infty = \mathbf{R} \times \prod_p \mathbf{Z}_p.$$

The set A_∞ is an *open* subgroup of $\mathbf{A}$ and so any character of A_∞ trivial on $\mathbf{Z}$ extends uniquely to a character of $\mathbf{A}$ trivial on $\mathbf{Q}$. Thus there is a unique character of ψ of $\mathbf{A}$ trivial on $\mathbf{Q}$ such that

$$\psi((x_\infty, x_2, x_3, \dots)) = e^{-2\pi i x_\infty} \qquad (x_\infty \in \mathbf{R}, x_p \in \mathbf{Z}_p).$$

We write $\psi = (\psi_v)$ and compute the local characters ψ_v. It is obvious that

$$\psi_\infty(x) = e^{-2\pi i x} \qquad (x \in \mathbf{R}).$$

If p is a prime, let us define for any $x \in \mathbf{Q}_p$ the element $x' \in \mathbf{Q}$ as follows. Let us write x in its usual p-adic expansion as

$$x = \sum_{m=-r}^\infty a_m p^m, \qquad a_m \in \{0, 1, 2, \dots, p-1\}.$$

Then

$$x' = a_{-r} p^{-r} + a_{-(r-1)} p^{-(r-1)} + \cdots + a_{-1} p^{-1}.$$

Clearly $x - x' \in \mathbf{Z}_p$ and x' is in $\mathbf{Z}_{p'}$ for all primes $p' \neq p$. Now if $x \in \mathbf{Q}_p$, we have

$$\psi_p(x) = e^{2\pi i x'}.$$

Indeed, if $a(x)$ is the adele all of whose components are 0 except the one corresponding to p which is x and if $b(x')$ is the adele all of whose components are equal to the rational number x' defined as above, then the component of $a(x) - b(x')$ corresponding to any prime q is $-x'$ and so is in $\mathbf{Z}_q$, while the component in $\mathbf{R}$ is $-x'$. Hence $a(x) - b(x') \in A_\infty$ and so

$$\psi(a(x) - b(x')) = 1.$$

But then

$$\psi_p(x) = \psi(a(x)) = \psi(a(x) - b(x')) = \psi_\infty(-x') = e^{2\pi i x'}.$$

In particular note that ψ_v is nontrivial for all v and unramified for all p. This done, it is now an easy calculation that the self dual measures are dx (usual Lebesgue measure) for $\mathbf{R}$ and the measure on $\mathbf{Q}_p$ that gives $\mathbf{Z}_p$ the measure 1, for each p. For f we take

$$f = \otimes_v f_v, \qquad f_\infty = e^{-\pi x^2}, \qquad f_p = 1_{\mathbf{Z}_p}.$$

All the functions f_v are equal to their own Fourier transforms so that at the adelic level we have $f = \hat{f}$. Thus the Tate functional equation becomes

$$Z(f : |\cdot|^s) = Z(f : |\cdot|^{1-s}).$$

But as

$$\int_{\mathbf{R}} e^{-\pi x^2} |x|^s \frac{dx}{|x||} = \pi^{-s/2} \Gamma\left(\frac{s}{2}\right) = \Gamma_{\mathbf{R}}(s)$$

and

$$\int_{\mathbf{Z}_p} |x|_p^s d_p^\times x = \frac{1}{1 - p^{-s}} \qquad \int_{\mathbf{Z}_p^\times} d_p^\times x = 1,$$

we have

$$Z(f : |\cdot|^s) = \pi^{-s/2} \Gamma\left(\frac{s}{2}\right) \prod_p \frac{1}{1 - p^{-s}} = \Gamma_{\mathbf{R}}(s) \zeta(s)$$

and so we have the Riemann functional equation! For K other than $\mathbf{Q}$ the same method leads to the functional equation of the Dedekind zeta function; the calculations are a little more delicate and involve doing some arithmetic on K, but there is no difficulty in principle. The method extends with minor modifications to the case of the Hecke L-series, but the calculations are somewhat more tedious; see [16a], [18a].

It was this application to the deeper parts of number theory that justified the development of harmonic analysis on general locally compact abelian groups, beyond the euclidean spaces and tori of analysis. In addition to the book of Weil the reader may also look at the paper of Cartan and Godement [18d], where a very elegant account of abelian harmonic analysis is given.

6.3. Euler products from Ramanujan and Hecke to Langlands

The history of Euler products took a major turn with a remarkable paper of the great Indian mathematician Ramanujan [19a], published in 1916, interestingly enough before Hecke had discovered his Grössencharakters and worked out the theory of the L-functions associated to them.

For functions on the natural numbers it is perhaps more natural to weaken the concept of full multiplicativity. We shall say that a function f defined on the positive integers is *multiplicative* if

$$f(mn) = f(m)f(n) \text{ whenever } m \text{ and } n \text{ are mutually prime.}$$

The Möbius function $\mu(n)$; $d(n)$, the number of divisors of n; $\sigma_k(n)$, the sum of kth powers of the divisors of n; and so on, are multiplicative in this sense. As a more subtle example we have the *Ramanujan sums* $c_q(n)$ defined by

$$c_q(n) = \sum_{r \bmod q} e\left(\frac{rn}{q}\right) \qquad e(x) = e^{2\pi i x}$$

where the summation is over the residue classes $[r]$ mod q. It even makes sense to speak of multiplicative functions with values in an arbitrary commutative algebra. To know a multiplicative function f it is necessary to know not only $f(p)$ for all primes but $f(p^k)$ for all primes p and all integers $k \geq 1$. The classical multiplicative functions mentioned above of course generate Dirichlet series with Euler products of degree > 1, but these factorize into products of Euler factors of degree 1. Here is a list of some of the formulae that arise in this way:

$$\sum_{n=1}^{\infty} \frac{\mu(n)}{n^s} = \frac{1}{\zeta(s)}, \qquad \sum_{n=1}^{\infty} \frac{d(n)}{n^s} = \zeta^2(s),$$

$$\sum_{n=1}^{\infty} \frac{\sigma_k(n)}{n^s} = \zeta(s)\zeta(s-k), \qquad \sum_{n=1}^{\infty} \frac{c_n(m)}{n^s} = \frac{\sigma_{s-1}(m)}{m^{s-1}\zeta(s)}.$$

Ramanujan considered the so-called elliptic modular function

$$q \prod_{n \geq 1} (1 - q^n)^{24} =: \sum_{n \geq 1} \tau(n)q^n$$

among many others. He conjectured that τ is multiplicative and that

$$\sum_{n \geq 1} \frac{\tau(n)}{n^s} = \prod_{p} \frac{1}{(1 - \tau(p)p^{-s} + p^{11-2s})}.$$

This Euler product of Ramanujan is remarkably different from the Dirichlet L-functions in two ways: its source is a modular form and, further, the Eulerian factors are *quadratic*. Ramanujan conjectured further that

$$|\tau(p)| \leq 2p^{\frac{11}{2}} \qquad \text{for each prime } p$$

which is equivalent to the hypothesis that the zeros of the polynomial

$$1 - \tau(p)T + p^{11}T^2$$

have both absolute value $p^{\frac{11}{2}}$. In this form they were known as *Ramanujan conjectures*. In Ramanujan's paper there is absolutely no hint about what led him to his conjectures. Given what we know about him, his education, and his style of writing mathematics, this is hardly surprising.

Shortly after Ramanujan's paper appeared, Mordell [19b] proved the Euler product formula above. Then Hecke pushed the ideas of Mordell further and considered *arbitrary* modular forms, not just the one that Ramanujan and Mordell considered. In Hecke's theory the modular form

$$\sum_{n=1}^{\infty} c(n)q^n$$

gets associated to the Dirichlet series

$$C(s) = \sum_{n=1}^{\infty} \frac{c(n)}{n^s};$$

the correspondence is not only formal but also analytical, the Dirichlet series being the *Mellin transform* of the modular form. Hecke extended Mordell's work and introduced a family of operators, nowadays called *Hecke operators*, which act on the space of modular forms [17]. If the modular form is of weight k and an *eigenform* of the Hecke operators, the function c is multiplicative and one has, exactly as in the case of the τ function,

$$C(s) = \sum_{n=1}^{\infty} \frac{c(n)}{n^s} = \prod_p \frac{1}{1 - c(p)p^{-s} + p^{k-2s}}.$$

He showed further that the Dirichlet series $C(s)$ has an analytic continuation and a functional equation that relates $C(s)$ to $C(k - s)$. Thus the class of Euler products had been widened enormously; modular forms had been discovered as a source of Euler products. But the earlier themes of analytic continuation and functional equation were still prominently in the mix.

In 1952 Gel'fand and Fomin wrote a paper [20] in which they pointed out that modular forms (actually cusp forms) of weight k correspond in a natural manner to the irreducible *unitary* representations of the group $\mathrm{SL}(2, \mathbf{R})$ that occur in Hilbert spaces naturally associated to the Poincaré upper half plane $\mathcal{H}$, which, in this context, is to be identified with

$$\mathrm{SO}(2, \mathbf{R}) \backslash \mathrm{SL}(2, \mathbf{R}) / \mathrm{SL}(2, \mathbf{Z}).$$

More precisely, they considered the natural unitary representation π_k of $\mathrm{SL}(2, \mathbf{R})$ on the Hilbert spaces $L_k^2(\mathcal{H})$ of sections of the line bundle on $\mathcal{H}$ defined by the character

$$\chi_k : \begin{pmatrix} \cos\theta & \sin\theta \\ -\sin\theta & \cos\theta \end{pmatrix} \longmapsto e^{ik\theta}$$

of $\mathrm{SO}(2, \mathbf{R})$ (modular forms of weight k) and proved that modular forms of weight k (up to scalar multiples) are in bijection with certain irreducible subrepresentations of π_k. Remarkably, and this was the essence of the discovery of Hecke (and Mordell), the groups $\mathrm{SL}(2, \mathbf{Q}_p)$ were also able to act on the holomorphic modular forms provided one enlarges the class of modular forms to include forms modular with respect to the principal congruence subgroups. The celebrated Hecke operators are nothing but a way of describing this action. Even nonholomorphic modular forms, such as the ones considered by Maass, could be brought into this circle of ideas. The various representations of the groups $\mathrm{SL}(2, \mathbf{R})$ and $\mathrm{SL}(2, \mathbf{Q}_p)$ can be combined into a single irreducible representation of the adelic group $\mathrm{SL}(2, \mathbf{A})$ where $\mathbf{A}$ is the ring of adeles over $\mathbf{Q}$. The weight condition can be interpreted in terms of the action of scalars. In this manner, a modular form appeared as a way of describing a *multiplicative object*—namely a certain representation, unitary, irreducible, and infinite dimensional of GL(2, **A**)—occurring in function spaces over

$$GL(2, \mathbf{A})/GL(2, \mathbf{Q}).$$

It is natural to call these the *automorphic* representations. Thus one could summarize the Hecke theory by saying that these automorphic representations of GL(2, **A**) have Euler products attached to them and that these Euler products have analytic continuations and functional equations.

Sometime in the 1960's Langlands entered the picture. In various letters and lectures he introduced the representation theoretic point of view and sketched the outlines of a broad theory of automorphic representations with astonishing implications that could be regarded as the emergence of a *nonabelian class field theory*. The content of these various works came to be known as the *Langlands conjectures* and eventually, as the ideas became more mature, as the *Langlands program*. In a monumental work [21a] published in 1970, Langlands, in collaboration with Jacquet (a student of Godement), worked out some of the consequences of the representation theoretic view point, not only over **Q**, but over any number field and even function fields. This study had several objectives, one of which was to construct a theory of irreducible representations of the adelic group GL(2, **A**(K)) that occur in the space of functions on the homogeneous space

$$GL(2, \mathbf{A}(K))/GL(2, K).$$

As a part of this unified point of view, the Grössencharakters of Hecke associated to a number field K were viewed as automorphic representations associated to GL(1) or, equivalently, as the representations of GL(1, **A**) occurring in the space of functions on the idele class group

$$\mathbf{C}(K) = GL(1, \mathbf{A}(K))/GL(1, K).$$

The irreducible representations of GL(2, **A**(K)) are of the form $\otimes_v' \pi_v$ where the product is over the places v of K, the π_v are irreducible representations of GL(2, **Q**$_v$), and the tensor product of the representations and their spaces are restricted in the following way: for almost all v, the space $\mathcal{F}_v$ of π_v contains a vector θ_v fixed by GL(2, $\mathfrak{r}_v$) ($\mathfrak{r}_v$ is the ring of integers in the completion K_v), the tensor product of the $\mathcal{F}_v$ being the linear span of all tensors whose components are equal to θ_v for almost all v. Jacquet and Langlands associated Euler products to such irreducible representations and proved their analytic continuation and functional equations. This work lifted the theory of Euler products to an entirely new level and provided a great unifying framework that included all previous L-functions associated to analytic objects. In addition, this work discussed the relation between automorphic representations of *different groups* and placed the work of Eichler-Shimura in a more precise and profound framework. Finally, this work suggested the remarkable fact that the automorphic L-functions contain among themselves the nonabelian Artin L-functions attached to Galois representations. Since the coincidence of the Artin *abelian* L-functions with the Dirichlet (or Hecke) L-functions lies at the very core of class field theory, this suggestion and the evidence presented for it in [21a] and other related works must be regarded as the beginnings of nonabelian class field theory.

To try to disentangle the various threads leading to these climaxes is very worthwhile, not just for finding out who did what and when, but to trace the evolution of the ideas and thereby understand better the fundamental issues. Anyone who attempts to get an understanding of these developments will probably be discouraged by the tremendous technical apparatus that clogs current expositions of these ideas. In what follows I have attempted a brief summary of some of these themes. The reader desirous of more should consult [7], [21a] and, above all, the material in the Langlands archive [21b] that contains all the works of Langlands, especially the letters and talks that led to the formulation of the original conjectures.

Classical modular forms and the Hecke Euler products associated to them. We give a brief discussion; the reader should consult [22a] for a concise but beautiful account of the basic theory and [22b] for a thorough and more profound discussion of the entire theory of automorphic forms. For both the classical theory and its representation theoretic incarnation, see [7].

Modular forms are functions on the (moduli) space of the isomorphism classes of elliptic curves. Elliptic curves are defined by lattices in the complex plane ($L \leftrightarrow \mathbf{C}/L$), and two such elliptic curves are isomorphic if and only if the lattices defining them are homothetic. One can thus regard modular forms as functions on the space $\mathcal{L}$ of lattices in $\mathbf{C}$. Let M be the set of pairs (ω_1, ω_2) of elements of $\mathbf{C}^2$ with $\Im(\omega_1/\omega_2) > 0$. The map taking $\omega = (\omega_1, \omega_2)$ to the lattice $\mathbf{Z}\omega_1 \oplus \mathbf{Z}\omega_2$ is surjective, and two such pairs ω, η define the same lattice if and only if

$$(\omega_1, \omega_2) = (\eta_1, \eta_2) \begin{pmatrix} a & c \\ b & d \end{pmatrix}, \qquad \begin{pmatrix} a & c \\ b & d \end{pmatrix} \in \mathrm{SL}(2, \mathbf{Z}).$$

Write

$$\Gamma = \mathrm{SL}(2, \mathbf{Z}).$$

Thus $\mathcal{L}$ can be identified with M/Γ. Note that $\omega_2^{-1} L = \mathbf{Z}\tau \oplus \mathbf{Z}$ where $\tau = \omega_1/\omega_2$, giving a map of M onto the Poincaré upper half plane $\mathcal{H}$ of all points τ with $\Im(\tau) > 0$. The action of Γ on M goes over to the fractional linear action

$$\tau \longmapsto \frac{a\tau + b}{c\tau + d}$$

of Γ on $\mathcal{H}$ via the map

$$(\omega_1, \omega_2) \longmapsto \tau = \omega_1/\omega_2.$$

Furthermore $\mathbf{C}^\times$ acts from the left on $\mathcal{L}$ by $L \mapsto \lambda L$, and it is clear that $\mathcal{L}/\mathbf{C}^\times$, which is naturally in bijection with the set of isomorphism classes of elliptic curves, can be identified with $\mathcal{H}/\Gamma$. A function F on the set of lattices is said to be of weight k if

$$F(\lambda L) = \lambda^{-k} F(L) \qquad (L \in \mathcal{L}).$$

Let L_τ be the lattice generated by $\{\tau, 1\}$. If

$$F(\tau) = F(L_\tau),$$

then the lattice functions F of weight k are in bijection with functions F on $\mathcal{H}$ satisfying the *modularity condition*

$$F\left(\frac{a\tau + b}{c\tau + d}\right)(c\tau + d)^{-k} = F(\tau) \qquad \left(\begin{pmatrix} a & b \\ c & d \end{pmatrix} \in \Gamma\right).$$

Now Γ is generated by

$$\begin{pmatrix} 1 & 1 \\ 0 & 1 \end{pmatrix} \quad \text{and} \quad \begin{pmatrix} 0 & -1 \\ 1 & 0 \end{pmatrix},$$

and so the modularity condition is equivalent to

(i) $F(\tau + 1) = F(\tau)$,

(ii) $F(-1/\tau) = \tau^k F(\tau)$.

Suppose now F is *holomorphic* in $\mathcal{H}$. Let $q = e^{2\pi i \tau}$. Then the map $\tau \mapsto q$ maps $\mathcal{H}$ onto the punctured unit disk $0 < |q| < 1$, and the condition (ii) above implies that $F(q) = F^\sim(q)$ for a function having a Laurent series expansion

$$F^\sim(q) = \sum_{n=0}^{\infty} a_n q^n.$$

We say that $F^\sim$ or F is *holomorphic at* ∞ if

$$a_n = 0 \text{ for } n < 0.$$

In this case we say that F is a *modular form*. If further $a_0 = 0$, we say that F is a *cusp form*. The reference here to a cusp comes from the fact that $\mathcal{H}/\Gamma$ can be made into a smooth compact Riemann surface by adding one point (cusp) at ∞. Note that

$$a_0 = \oint_{|q|=1} \frac{F^\sim(q)}{q} dq = \int_{\mathbf{R}/\mathbf{Z}} F(\tau) d\tau.$$

It is easy to see that

$$a_0 = 0 \Leftrightarrow \int_0^1 F(\xi + \tau) d\tau = 0 \quad (\xi \in \mathcal{H}).$$

With respect to the usual fundamental domain

$$D = \{\tau \in \mathcal{H} \mid |\Re(\tau)| \leq 1/2, |\tau| \geq 1\},$$

the cusp is at ∞. The measure $\frac{dx\,dy}{y^2}$ is invariant under the action of $\mathrm{SL}(2, \mathbf{R})$ on $\mathcal{H}$.

The classical modular forms are the *Eisenstein series* occurring, in the first place, as coefficients in the equation for the Weierstrass $\wp$-function:

$$G_k(L) = \sum_{0 \neq \omega \in L} \omega^{-2k} (k = 2, 3, 4, \dots), \quad G_k(\tau) = \sum_{(0,0) \neq (c,d) \in \mathbf{Z}^2} (c\tau + d)^{-2k}.$$

The G_k have weight $2k$. The Fourier series for G_k can be computed to be

$$G_k(\tau) = 2\zeta 2(k) + \frac{(2\pi i)^{2k}}{(2k-1)!} \sum_{n \geq 1} \sigma_{2k-1}(n) q^n.$$

The normalized Eisenstein series are

$$E_k(\tau) = \frac{1}{2\zeta(2k)} G_k(\tau)$$

so that

$$E_2 = 1 + 240(q + \sigma_3(2)q^2 + \dots + \sigma_3(n)q^n + \dots)$$

$$E_3 = 1 - 540(q + \sigma_5(2)q^2 + \dots + \sigma_5(n)q^n + \dots).$$

The (holomorphic) modular forms form an algebra graded by the weights and generated by E_2 and E_3, these two being algebraically independent. In particular only even weights occur. Let

$$\Delta = \frac{1}{1728}(E_2^3 - E_3^2) = q + \dots.$$

Then Δ is a cusp form of weight 12. It is a famous result of Jacobi that

$$\Delta(q) = q \prod_{n \geq 1} (1 - q^n)^{24}$$

so that

$$\Delta(q) = \sum_{n \geq 1} \tau(n) q^n.$$

The coefficients $\sigma_{k-1}(n)$ are multiplicative, and so if we associate the Dirichlet series

$$\Phi_k(s) = \sum_{n \geq 1} \frac{\sigma_{k-1}}{n^s}$$

to G_k, we see that

$$\Phi_k(s) = \zeta(s)\zeta(s - k + 1) = \prod_p \frac{1}{(1 - p^{-s})(1 - p^{k-1-s})}).$$

Ramanujan's idea was that

$$\sum_{n \geq 1} \frac{\tau(n)}{n^s}$$

was likewise also an Euler product, of degree 2.

Let $\mathcal{M}_k$ (resp. $\mathcal{S}_k$) be the space of modular (resp. cusp) forms of weight k. Once we know that E_2 and E_3 generate the graded algebra of modular forms, it is easy to compute the dimensions $\dim(\mathcal{M}_k)$. Of course $\dim(\mathcal{S}_k) = \dim(\mathcal{M}_k) - 1$. The space of modular forms of weight 12 is 2, and so the space of cusp forms has dimension 1, hence spanned by Δ. Mordell considered for each prime p the function

$$F(\tau) = p^{11} \left\{ \sum_{j=0}^{p-1} \Delta\left(\frac{\tau + j}{p}\right) p^{-k} + \Delta(p\tau) \right\}$$

and checked that F is a cusp form of weight 12. Hence F is a multiple of Δ, and comparison of Fourier coefficients gave the recursion formula for the $\tau(p^r)$. The multiplicativity also follows from a similar argument.

Hecke's theory does what Mordell did to *arbitrary* modular forms. The abstract *Hecke operators* $\mathcal{T}(n)(n \geq 1)$ are defined by *correspondences* $H(n)$ on the set $\mathcal{L}$ of lattices given by

$$H(n) = \{(L, L') \mid L' \subset L, [L : L'] = n\}.$$

If $\mathcal{G}(\mathcal{L})$ is the free abelian group generated by the elements of $\mathcal{L}$, each $H(n)$ gives rise to an operator on $\mathcal{G}(\mathcal{L})$ by

$$\mathcal{T}(n)L = \sum_{(L,L') \in H(n)} L'.$$

Notice that all the L' in the above sum are such that $nL \subset L' \subset L$ and hence are the preimages of subgroups of the finite subgroup L/nL of n^2 elements. Hence there are only finitely many of the L'. It can be verified that the $\mathcal{T}(n)$ satisfy the following:

(i) $\mathcal{T}(1) = $ id.
(ii) The $\mathcal{T}(n)$ are multiplicative:

$$\mathcal{T}(mn) = \mathcal{T}(m)\mathcal{T}(n) \qquad ((m, n) = 1).$$

(iii) If $\mathcal{T}(d,d)$ is the map $L \mapsto dL$ of $\mathcal{L}$, then

$$\mathcal{T}(d,d)\mathcal{T}(m) = \mathcal{T}(m)\mathcal{T}(d,d).$$

(iv) We have

$$\mathcal{T}(p^r)\mathcal{T}(p) = \mathcal{T}(p^{r+1}) + p\mathcal{T}(p,p)\mathcal{T}(p^{r-1}) \qquad (r \geq 1).$$

Note that if $(m,n) = 1$ and $[L : L''] = mn$, there is a unique L' with $[L : L'] = m$ such that $L \supset L' \supset L''$, leading to the multiplicativity of the $\mathcal{T}(n)$. It also follows from (iv) above that the $\mathcal{T}(p)$ and $\mathcal{T}(d,d)$ generate a commutative algebra that contains all the $\mathcal{T}(n)$. The actual Hecke operators on the space of functions on $\mathcal{L}$ of weight k are given by

$$(T(n)F)(L) = n^{k-1} \sum_{(L,L')\in H(n)} F(L').$$

Analytically, if $\Gamma = \mathrm{SL}(2,\mathbf{Z})$ and Γ_n is the set of matrices with integer entries and determinant n, then the L' with $[L : L'] = n$ are the gL with $g \in \Gamma_n$, and so the sum above is over Γ_n/Γ. A set of representatives of this is the set

$$\begin{pmatrix} a & 0 \\ b & d \end{pmatrix}, \qquad (a, d > 0, ad = n, 0 \leq b < d, a, b, d \in \mathbf{Z}).$$

Going over to the corresponding functions on the upper half plane we get

$$(T(n)F)(\tau) = n^{k-1} \sum_{a,b,d} d^{-k} f\left(\frac{a\tau + b}{d}\right)$$

where the sum is over a, b, d restricted as above.

   With the $T(n)$ defined this way, the properties of the $\mathcal{T}(n)$ lead to corresponding properties of the $T(n)$. The properties of the $T(n)$ are equivalent to the formal relation

$$\sum \frac{T(n)}{n^s} = \prod_p \frac{1}{I - T(p)p^{-s} + p^{k-1-2s}I}.$$

The $T(n)$ leave the space of modular (resp. cusp) forms stable. If we now assume, with Hecke, that a (nonzero) modular form F is an *eigenform* for all the $T(n)$, say

$$T(n)F = \lambda(n)F \qquad (\lambda(1) = 1),$$

then we get

$$a_n = a_1\lambda(n), \qquad a_1 \neq 0$$

where the a_n are the Fourier coefficients of f. If we normalize f so that $a_1 = 1$, then the a_n are multiplicative and

$$\sum_{n\geq 1} \frac{a_n}{n^s} = \prod_p \frac{1}{1 - a_p p^{-s} + p^{k-1-2s}}.$$

Notice that the modular form is determined by the $\lambda(n)$. For $F = \Delta$ we get to Ramanujan and Mordell. Indeed, since the space of modular forms of weight 12 is of dimension 2, being the span of E_2^3 and E_3^2, the space of cusp forms of weight 12 has dimension 1 and so must be spanned by Δ. Then Δ has to be an eigenform for the $T(n)$. This was in essence Mordell's argument. Actually Hecke proved that for $F \in \mathcal{M}_k$ the following are equivalent:

(a) F is an eigenform for the $T(n)$.
(b) If $F^{\sim}(q) = \sum_{n\geq 0} a_n q^n$, then the $a_n (n \geq 1)$ are multiplicative.

(c) The Dirichlet series $\sum_{n\geq 1} a_n n^{-s}$ has an Euler product.

Hecke could not prove that the eigenforms for the $T(n)$ span the space of cusp forms. This was done later by Petersson, a student of Hecke, who introduced the scalar product

$$(f,g) = \int\int_D y^{2k} f\bar{g} \frac{dxdy}{y^2}$$

(D being the fundamental domain for the action of the modular group) on the space of cusp forms of weight $2k$ and showed that the $T(n)$ are self adjoint with respect to it. Notice that for a cusp form f we have, for some $\alpha > 0$,

$$f(\sigma + i\tau) = O(e^{-\alpha\tau}) \qquad (|\sigma| \leq 1/2, \tau \to \infty)$$

and so the above integrals exist. The algebra generated by the $T(n)$ has thus a completely reducible action on the space of cusp forms, and being commutative, it follows that the space is spanned by the eigenforms. Moreover the eigenvalues of the $T(n)$ are real so that for the eigenforms which are cusp forms normalized by taking $a_1 = 1$, all the coefficients a_n are *real*. The Dirichlet series associated to the cusp forms either formally or by Mellin transform have Euler products converging on some right half plane. The Mellin transform interpretation allowed Hecke to generalize Riemann's method and prove for the corresponding Dirichlet series both the meromorphic continuation and the functional equation. Let $\mathcal{D}_k$ be the space of Dirichlet series

$$\varphi(s) = \sum_{n\geq 1} \frac{a_n}{n^s}$$

with the following properties:

(i) $\varphi(s)$ converges somewhere.

(ii) If

$$\Phi(s) = (2\pi)^{-s}\Gamma(s)\varphi(s),$$

then Φ has a meromorphic continuation to the whole s-plane with simple poles at $s = 0, k$ and no singularity anywhere else; let $a_0 = -\text{Res }_{s=0}\Phi(s)$.

(iii) The function

$$\Phi(s) + \frac{a_0}{s} + (-1)^{k/2}\frac{a_0}{k-s}$$

is of class EBV, i.e., entire and bounded on vertical strips.

(iv) Φ satisfies the functional equation

$$\Phi(s) = (-1)^{k/2}\Phi(k-s).$$

Then Hecke's results can be summarized as follows.

THEOREM (Hecke). *Let* $\mathcal{M}_k$ *be the space of modular forms of weight k and* $\mathcal{S}_k$ *the subspace of cusp forms. Then we have an isomorphism* $\mathcal{M}_k \simeq \mathcal{D}_k$ *given formally by*

$$f = \sum_{n\geq 0} a_n q^n \leftrightarrow \varphi = \sum_{n\geq 1} \frac{a_n}{n^s}$$

and analytically by the Mellin transform

$$\Phi(s) = \int_0^\infty f(iy)y^s \frac{dy}{y} \leftrightarrow f(iy) = \frac{1}{2\pi i}\int_{\sigma-i\infty}^{\sigma+i\infty} \Phi(s)y^{-s}ds.$$

Here in the first integral $\Re(s) >> 0$, *while in the second integral* $\sigma >> 0$.

The theory of Hecke can be extended in at least two directions. The first is to bring in forms which are modular with respect to subgroups of $\Gamma = \mathrm{SL}(2, \mathbf{Z})$, for example with respect to the principal congruence subgroups $\Gamma(N)$ defined by

$$\Gamma(N) = \{\gamma \in \Gamma \mid \gamma \equiv I \bmod N\}.$$

Hecke himself took this up. To get the description of these modular forms of higher level through their associated Dirichlet series, it became necessary to consider, with Weil, along with a modular form

$$\sum_{n \geq 1} a_n q^n,$$

all the *twisted* forms

$$\sum_{n \geq 1} a_n \chi(n) q^n$$

where χ is a Dirichlet character. By assuming analytic continuations and functional equations for all the Dirichlets series

$$\sum \frac{a_n \chi(n)}{n^s}$$

(or at least for a sufficiently large number of them) Weil was able to show that the original Dirichlet series is the Mellin transform of a modular form of some level N. The second generalization is to give up the holomorphicity and replace it by the condition that the forms are eigenforms with respect to the Laplace-Beltrami operator. This was done by Maass. Both of these generalizations will be covered in the adelic view which allows representation theory to be brought in. We shall turn to this next. But before doing that we mention that the original Ramanujan conjecture on the eigenvalues of the Eulerian factors can be generalized to the *Ramanujan-Petersson conjecture*: for a cusp form

$$F(q) = \sum a_n q^n \qquad (a_1 = 1)$$

of weight k we have

$$|a_p| \leq 2p^{(k-1)/2}.$$

This was finally proved by Deligne as a consequence of the Weil conjectures (see the nice exposition of Katz in [22c]).

The viewpoint of representation theory of the adelic group. We shall follow Deligne [23]. See also [7].

Let $e_i (i = 1, 2)$ be the standard basis vectors of $\mathbf{R}^2$. We write G for the set of all $\mathbf{R}$-isomorphisms of $\mathbf{R}^2$ with $\mathbf{C}$. For any $g \in G$ write

$$g(e_1) = \omega_1, \quad g(e_2) = \omega_2$$

so that $g(\mathbf{Z}^2)$ is the lattice with $\mathbf{Z}$-basis $\{\omega_1, \omega_2\}$. G is thus the set of *marked lattices*, namely, lattices with a given $\mathbf{Z}$-basis. We use ω_1, ω_2 as coordinates on G which allows us to view G as a complex manifold, namely the open set of pairs (ω_1, ω_2) in $\mathbf{C}^2$ such that $\Im(\omega_1/\omega_2) \neq 0$. From the definition of G there is an action of $\mathrm{GL}(\mathbf{R}^2)$ from the right and an action of $\mathrm{GL}_{\mathbf{R}}(\mathbf{C})$, the group of $\mathbf{R}$-linear automorphisms of $\mathbf{C}$ from the left. The right action is

$$(\omega_1, \omega_2) \longrightarrow (\omega_1, \omega_2) \begin{pmatrix} a & c \\ b & d \end{pmatrix}.$$

This action is holomorphic. The left action is

$$x(\omega_1, \omega_2) = (x(\omega_1), x(\omega_2)).$$

This is *not* holomorphic, but if λ is multiplication by $\lambda \in \mathbf{C}^\times$, then the corresponding action is holomorphic. We can identify G with $\mathrm{GL}(2, \mathbf{R})$ by identifying $\mathbf{C}$ with $\mathbf{R}^2$ via the ordered basis $(i, 1)$. Then the right action of $\mathrm{GL}(2, \mathbf{R})$ on G is right translation. The same basis for $\mathbf{C}$ allows us to identify $\mathrm{GL}_\mathbf{R}(\mathbf{C})$ with $\mathrm{GL}(2, \mathbf{R})$, and the left action of $\mathrm{GL}_\mathbf{R}(\mathbf{C})$ on G then becomes left translation. The space $\mathcal{L}$ of lattices gets identified with $G/\mathrm{GL}(2, \mathbf{Z})$, and so functions on $\mathcal{L}$ may be viewed as functions on G left invariant for $\mathrm{GL}(2, \mathbf{Z})$. On G we consider the space $\mathcal{F}_k$ of functions f with the properties:

(a) f is holomorphic.
(b) $f(\lambda g) = \lambda^{-k} f(g)$ $\quad (g \in G, \lambda \in \mathbf{C}^\times)$.
(c) $f(g\gamma) = f(g)$ $\quad (g \in G, \gamma \in \mathrm{GL}(2, \mathbf{Z}))$.
(d) f has moderate growth on G.

The condition (d) means, roughly speaking, that $f(g)$ does not grow faster than a polynomial in the entries of g and $\det(g)^{-1}$. This is a very mild condition. The map that takes an f as above with the function $f^\sim$ on the Poincaré upper half plane with coordinate $\tau(\Im(\tau) > 0)$ defined by

$$f^\sim(\tau) = f(\tau, 1)$$

establishes a linear isomorphism of $\mathcal{F}_k$ with the traditional space of functions $f^\sim$ such that

(i) $f^\sim$ is holomorphic.
(ii) $f^\sim\left(\frac{a\tau+b}{c\tau+d}\right) = (c\tau + d)^k f(\tau)$.
(iii) For some $N = N(f^\sim) > 0$, $|f(\tau)| \leq A\Im(\tau)^N$ as $\Im(\tau) \to \infty$, uniformly for $\Re(\tau)$ in compact sets.

That the growth condition (iii) on $f^\sim$ implies the growth condition (d) on f uses the fundamental domain for the modular group for which $\Re(\tau)$ is bounded (reduction theory).

We mentioned that the left action on G by $\mathrm{GL}_\mathbf{R}(\mathbf{C})$ is not holomorphic, except for the action by $\mathbf{C}^\times$. If we compute the infinitesimal action from the left we get vector fields $\ell(Z)$ for any element of Lie $(\mathrm{GL}_\mathbf{R}(\mathbf{C}))$. Let the basis for Lie $\mathrm{GL}_\mathbf{R}(\mathbf{C})$ be

$$H = \begin{pmatrix} 0 & -i \\ i & 0 \end{pmatrix} \quad X = (1/2) \begin{pmatrix} -1 & -i \\ -i & 1 \end{pmatrix}, \quad Y = \begin{pmatrix} -1 & i \\ i & 1 \end{pmatrix}.$$

Then the condition of holomorphy becomes

$$\ell(Y)f^\sim = 0$$

and the weight condition becomes

$$\ell(H)f = kf.$$

The last two conditions show that the element $f^\sim$ generates, under the action of Lie $(\mathrm{GL})_\mathbf{R}(\mathbf{C}))$, an *irreducible weight module*, namely the one with *lowest weight* k, $f^\sim$ being the lowest weight vector. From the theory of representations [24] it follows that a suitable completion of this space carries a representation of $\mathrm{SL_R}(\mathbf{C}) \simeq$ $\mathrm{SL}(2, \mathbf{R})$. Adding to this the action by the elements $\lambda \in \mathbf{C}^\times$ and the element

$$\begin{pmatrix} 0 & 1 \\ 1 & 0 \end{pmatrix}$$

we get an irreducible representation of $GL_{\mathbf{R}}(\mathbf{C}) \simeq \mathrm{GL}(2, \mathbf{R})$. If $f^\sim$ is a *cusp form*, it is *rapidly decreasing* when $\Im(\tau) \to \infty$, and the representation space will be contained in the Hilbert space with the Petersson scalar product, which can be identified with a subspace of the regular representation of $\mathrm{SL}(2, \mathbf{R})$ or the regular representation of $\mathrm{GL}(2, \mathbf{R})$ modulo the action of the center by a character. In this manner we have a bijection of holomorphic modular forms up to constants with certain irreducible representations of $\mathrm{GL}(2, \mathbf{R})$, the cusp forms corresponding to irreducible representations occurring in the regular representation (in the above sense) of $\mathrm{GL}(2, \mathbf{R})$. This was the original observation in [20].

To go beyond the classical theory and bring in the Hecke operators we have to show that the groups $\mathrm{GL}(2, \mathbf{Q}_p)$ also act on modular forms. The key point is that if f is a modular form and γ is an element of $\mathrm{GL}(2, \mathbf{Q})^+$ (+ means positive determinant), the transform of f by γ will no longer be modular but only modular with respect to *some* $\Gamma(N)$. However it turns out that if we consider the space V_k of forms of weight k modular with respect to *some* $\Gamma(N)$ (of level N), then $\mathrm{GL}(2, \mathbf{Q}^+)$ has an action on V_k and each $f \in V_k$ is fixed by some $\Gamma(N)$. If we *complete* $\mathrm{GL}(2, \mathbf{Q})$ with respect to the topology on it for which the $\Gamma(N)$ are open neighborhoods of the identity, we see that $\mathrm{GL}(2, \mathbf{A}_f)$ acts on V_k (recall that $\mathbf{A}_f$ is the ring of finite adeles) and the action has the property that each element is fixed by an *open* subgroup of $\mathrm{GL}(2, \mathbf{A}_f)$. The Hecke operators arise by the (convolution) action of elements in the Schwartz-Bruhat space of $\mathrm{GL}(2, \mathbf{A}_f)$, especially those that are constant on the double cosets of some open compact subgroup of $\mathrm{GL}(2, \mathbf{A}_f)$. Thus the entire adelic group acts on the holomorphic forms (level unspecified) and generates irreducible spaces of functions, and the entire classical theory with its generalization by Hecke is captured in this picture. It is in this manner that one may replace a modular form by an irreducible adelic representation that occurs, in a specified sense, in the space of functions on the adelic group.

The general notion of an automorphic form and/or representation in the adelic context can now be introduced. Let $G = \mathrm{GL}(2)$ so that for any commutative ring R with unit, $G(R) = \mathrm{GL}(2, R)$. Write $\mathbf{A}$ for the adele ring of $\mathbf{Q}$, $\mathbf{A}_f$ for the subring ring of finite adeles, this being the restricted direct product of the $\mathbf{Q}_p$. The automorphic forms are functions f on $G(\mathbf{A})$ which satisfy the following conditions:

(a) f is right invariant under $G(\mathbf{Q})$.

(b) Identifying the scalars in $G(\mathbf{A})$ with $\mathbf{J}(\mathbf{Q})$, there is a quasicharacter ω on on $\mathbf{J}(\mathbf{Q})$ trivial on $\mathbf{Q}^\times$ such that

$$\ell(x'x) = \omega(x')f \qquad (x' \in \mathbf{J}(\mathbf{Q}), x \in G(\mathbf{A})),$$

where ℓ is the action by left translations on $G(\mathbf{A})$.

(c) There is a compact open subgroup K_f of $G(\mathbf{A}_f)$ such that

$$\ell(x)f = f \qquad (x \in K_f).$$

(d) (finiteness condition) The transforms of f by left action of elements of $\mathrm{SO}(2, \mathbf{R})$ and the Casimir operator span a finite-dimensional space.

(e) f has moderate growth at infinity.

The condition (d) is a generalization of the holomorphy for classical modular forms and allows us to bring the modular forms (for example the Eisenstein series) of Maass into this picture. f is called a *cusp form* if its constant Fourier coefficient is 0, i.e.,

$$\int_{\mathbf{J}/\mathbf{Q}} f(xt(a))da = 0 \qquad \left(x \in G(\mathbf{A}), t(u) = \begin{pmatrix} 1 & u \\ 0 & 1 \end{pmatrix} \right).$$

By the action of $G(\mathbf{A}_f)$, $\mathrm{SO}(2, \mathbf{R})$, and the universal enveloping algebra of $\mathrm{SL}(2, \mathbf{R})$, such an f generates a subspace of the space of automorphic forms which is irreducible under these actions. These subspaces carry the automorphic representations. The action at the infinite place is not that of $\mathrm{SL}(2, \mathbf{R})$, but rather of the universal enveloping algebra of its Lie algebra on a module consisting of $\mathrm{SO}(2, \mathbf{R})$-finite vectors. However the theory of representations of Harish-Chandra allows one to pass freely between the representations of the enveloping algebra and those of the group itself. Thus, when we speak of the adelic action we have in mind an algebra of the form $\mathcal{U} \otimes \mathcal{D}$ where $\mathcal{U}$ is the enveloping algebra at the infinite place and $\mathcal{H}_f$ is the Hecke algebra (under convolution) of the Schwartz-Bruhat space of $\mathrm{GL}(2, \mathbf{A}_f)$. The transition between representations of $\mathcal{U} \otimes \mathcal{H}_f$ and the representations of $\mathrm{GL}(2, \mathbf{A})$ is straightforward modulo technical details. In this way an automorphic form gives rise to an irreducible representation of $G(\mathbf{A})$ or of $\mathcal{U} \otimes \mathcal{H}_f$. Denote this representation by π. Then we can write, in a certain sense, $\pi = \otimes' \pi_v$ where the notation denotes restricted tensor product in a certain sense. This is the representation theoretic point of view we talked about. Up to tensoring by a quasicharacter this representation is unitary and infinite dimensional.

In the classification of representations of $\mathrm{GL}(2, K)$ where K is a local field we have the *discrete series* and the *principal series* representations. It is possible to show that the Ramanujan-Petersson conjecture is equivalent to the statement that if π is a representation of $\mathrm{GL}(2, \mathbf{A})$ in the space of cusp forms and we write $\pi = \otimes' \pi_v$, then all the π_p belong to the principal series. It can also be shown that the space of holomorphic cusp forms of a given weight decomposes as a direct sum of irreducible modules under $\mathrm{GL}(2, \mathbf{A}_f)$, no two being equivalent (multiplicity free), thus generalizing the classical result of Hecke that the eigenvalues of the Hecke operators determine the cusp form up to a scalar. The proof depends on the theory of *new forms* developed by Atkins-Lehner.

In [21a] Jacquet and Langlands treated the theory of L-functions associated to automorphic representations of $\mathrm{GL}(2)$, their analytic continuations and functional equations. The Weil twists appeared in a natural manner in their theory. They did this by directly defining the Eulerian factors at the local level, and so their theory depended very much on a complete understanding of the representation theory of $\mathrm{GL}(2, K)$ where K is a local field. For $K = \mathbf{R}$ or $\mathbf{C}$ this is essentially a special case of the Harish-Chandra theory, but for K nonarchimedean it made use of the corresponding theory developed by Mautner, Gelfand-Graev, Kirillov, Bruhat and others.

The central question is of course the one posed already in [21a]: what is the arithmetic meaning of these automorphic L-functions? The preface of [21a] suggested that the Artin L-functions defined by irreducible representations of degree 2 of $\mathrm{Gal}(\bar{\mathbf{Q}}/\mathbf{Q})$ occur among the automorphic L-functions. This is a direct generalization of the situation for $\mathrm{GL}(1)$ where class field theory supplies the complete and affirmative answer through the Artin isomorphism of the Galois group with the profinite completion of the idele class group. Eventually Langlands formulated this question as a principle of *functoriality*; this principle has been a major driving force of the entire theory since then. But to get to this stage of the evolution of the subject one must first understand the simplest case, namely GL(1). This is class field theory or the theory of abelian extensions of global fields. We turn to this next.

6.4. Abelian extensions and class field theory

One of the simplest formulae we learn in trigonometry is

$$\sin 18° = \frac{\sqrt{5} - 1}{4}.$$

One consequence of this is that the field $\mathbf{Q}(\sqrt{5})$ is contained in the cyclotomic field of 20th roots of unity. We have already seen that this is a very general fact and that one of the proofs of quadratic reciprocity comes about by imbedding the quadratic field $\mathbf{Q}(\sqrt{d})$ in a cyclotomic field. Also the intervention of the cyclotomic fields $K_N = \mathbf{Q}(e^{2\pi i/N})$ make the Dirichlet theory of L-functions associated to Dirichlet characters very transparent, the reason being that the laws of prime factorization of the rational primes in a cyclotomic field are especially simple. All these facts must have suggested to Kronecker that the question of examining the role of the cyclotomic fields from the point of view of abelian extensions is very fundamental. Certainly, as their Galois groups are the groups G_N, they are rich enough to have *any* finite abelian group among their quotients. This is just an algebraic fact, but Kronecker discovered its geometric counterpart when he found the fundamental result that the cyclotomic extensions *contain* all abelian extensions. Weber later supplied a proof, which is known as the theorem of Kronecker-Weber.

THEOREM (Kronecker-Weber). *Every abelian extension of $\mathbf{Q}$ is contained in a cyclotomic extension. More precisely, if K is a finite Galois extension of $\mathbf{Q}$ with an abelian Galois group, there is a positive integer N such that $K \subset \mathbf{Q}(e^{2i\pi/N})$.*

It would have been highly surprising if it did not occur to Kronecker that one should examine the abelian extensions of fields other than $\mathbf{Q}$. With remarkable insight Kronecker conceived the idea that one should determine in a similar explicit fashion all the abelian extensions of the *imaginary quadratic fields* $\mathbf{Q}(\sqrt{-D})$ where D is an integer which is both > 0 and square free. Since the roots of unity which generate the abelian extensions are the values of the exponential function $e^{2\pi i z}$ at the rational numbers $z = \frac{m}{n}$, it was natural for Kronecker to search for the analogues of the exponential function in the case when the ground field is an imaginary quadratic field. By a miracle of insight he succeeded in divining that the role of the exponential function will be played by the elliptic modular function associated to the quadratic imaginary field. In a famous letter to Dedekind written on March 15, 1880 [25], Bd. V, pp. 455-457, he describes what he did with this problem. Here is the beginning of his letter:

> ... *My best thanks for your friendly note of the 12th of this month. I would like to take this opportunity to tell you that I believe I have overcome today the last of my difficulties which have been preventing the conclusion of an investigation with which I have been occupied again in the last months. The subject is*

> *my dearest dream of youth, namely the proof that the abelian
> equations with square roots of natural numbers as coefficients
> are given by transformation equations of elliptic functions with
> singular moduli, just as the abelian equations with integer coeffi-
> cients are given by the cyclotomic equations. I believe I have now
> completely succeeded in giving such a proof, and I want to direct
> all my efforts in writing it out. I hope that no new obstacles will
> arise. It is not only a valuable result as it seems to me, but the
> insights which I have had on the way have greatly satisfied my
> curiosity. I have also had the joy of satisfying the mathematical
> heart of my friend Kummer with my remarks concerning these
> problems because it appears now that there are some hopes for
> solving questions that are very dear to him . . . .*

Kronecker concludes his letter thus:

> *. . . I hope that you too, highly honored colleague, take interest
> in this insight and the nature of these equations, even though
> this presentation may appear at present somewhat strange. The
> hope to treat the central problem to define for the general complex
> numbers the analog of the singular moduli, I shall probably have
> to postpone it. I have to clarify and write up the results obtained
> up to now. . . .*

This letter, especially the phrase *meinen liebsten Jugendtraum*, meaning *the dearest dream of my youth*, has since become famous, and the Kronecker problem is generally referred to as the Jugendtraum. Simply put, it asks for the explicit determination of the abelian extensions of an arbitrary imaginary quadratic field. The integers of this quadratic field form a lattice in the complex plane and so determine a field of elliptic functions. The Kronecker program is to show that the abelian extensions are generated by the J-function and other related functions of this function field. Kronecker did not complete this program; it was completed by his successors Weber, Feuter, Takagi, and many others.

Let me now give a brief description of one of the basic results of this program [25], [22b]. For the lattice $L \subset \mathbf{C}$ we have the functions

$$g_2(L) = 60 \sum_{\omega \in L \backslash 0} \frac{1}{\omega^4}, \quad g_3(L) = 140 \sum_{\omega \in L \backslash 0} \frac{1}{\omega^6}.$$

Define

$$j(L) = 1728 \frac{g_2^3}{g_2^3 - 27g_3^2}.$$

Then j, called the j-invariant of the lattice L, is invariant under homotheties. Suppose now that we have an imaginary quadratic field $K = \mathbf{Q}(\sqrt{-D})$. The ring of integers of this field is a lattice in $\mathbf{C}$, and in fact this is the case for any ideal $\mathfrak{a}$ of the ring of integers. If j is the associated j-invariant, $j(\mathfrak{a})$ depends only on the ideal class of $\mathfrak{a}$. Thus, as the group of ideal classes is finite, we obtain a finite set of numbers $j(\mathfrak{a}_1), \ldots, j(\mathfrak{a}_h)$ where $\mathfrak{a}_1, \ldots, \mathfrak{a}_h$ is a complete set of representatives of ideal classes; here h is the class number of K. One of the main theorems of this subject is the following.

THEOREM. *The $j(\mathfrak{a}_i)(1 \le i \le h)$ are algebraic integers and form a complete set of conjugates over K. The field obtained by adjoining any one of them to $\mathbf{Q}(\sqrt{-D})$*

contains all of them and is the Hilbert class field of K, i.e., the maximal unramified abelian extension of K.

It is absolutely remarkable that the function j, which is highly transcendental, gives algebraic numbers (let alone algebraic integers) when evaluated at any ideal. That these numbers generate abelian equations is a further miracle. Finally, that Kronecker could have dreamt of this when he was young is the biggest miracle of all!

The corresponding theorems for ramified abelian extensions involve additional transcendental functions. The interested reader can look these up from a variety of sources [22b], [26b], [26c].

Weber of course knew of Kronecker's ideas and had himself pushed the Kronecker program to a considerable extent. Perhaps influenced by the last sentence of Kronecker's letter as well as his own work, it was thus natural for him to start thinking whether the construction of abelian extensions of *arbitrary* number fields could be achieved, perhaps not in the same detail as for **Q** and the quadratic imaginary fields, but still in terms of objects created from the ground field. However it was Hilbert who really formulated the goals and conjectured the contours of the general theory of abelian extensions of arbitrary number fields. Starting with his monumental work [27a] in 1897 and continuing with his own papers, he outlined a remarkably precise blueprint for the entire theory. It is perhaps this experience that led him to raise in his famous Paris address the problem (12th) of *explicit construction* of transcendental functions associated to a number field k whose (special) values will generate the abelian extensions of k [27b]. Here are some excerpts from his section entitled *Extension of Kronecker's theorem on Abelian fields to any algebraic realm of rationality*:

> ... *Finally, the extension of Kronecker's theorem to the case that*, in place of the realm of rational numbers or of the imaginary quadratic field, any algebraic field whatever is laid down as realm of rationality, *seems to me of the greatest importance. I regard this problem as one of the most profound and far-reaching in the theory of numbers and of functions....*

> ... *It will be seen that in the problem just sketched the three fundamental branches of mathematics, number theory, algebra, and function theory, come into closest touch with one another, and I am certain that the theory of analytical functions of several variables in particular would be notably enriched if one should succeed* in finding and discussing those functions which play the part for any algebraic number field corresponding to that of the exponential function in the field of rational numbers and of the elliptic modular functions in the imaginary quadratic field.

This particular section of Hilbert's address is well worth reading carefully, since it goes into detail regarding the analogy between algebraic number fields and function fields, namely the theory of compact Riemann surfaces. This line of thought has emerged full-fledged in current developments as the *geometric Langlands program*. Here are Hilbert's thoughts on this analogy:

> ... *Abel's theorem in the theory of algebraic functions expresses, as is well known, the necessary and sufficient condition that the*

points in question are the zero points of an algebraic function be-
longing to the surface. The exact analogue of Abel's theorem, in
the theory of number fields of class h = 2 is the law of quadratic
reciprocity:

$$\left(\frac{a}{j}\right) = +1$$

which asserts that an ideal j is then and only then a principal
ideal of the number field when the quadratic residue of the number
a with respect to the ideal j is positive....

Hilbert, with his profound feeling for form and generality, was also aiming for a theory of abelian extensions erected over any number field. Such a theory should be viewed as being in counterpoint to the problem of explicit construction of abelian extensions, a la Kronecker, of these number fields. (Actually most of the current expositions and proofs of the Kronecker program already assume class field theory.) Such a theory of abelian extensions of arbitrary number fields is a structure that was created by the combined efforts of a galaxy of great number theorists of the 19th and 20th centuries. This is known as class field theory, and we now turn to a brief description of it. We shall give just the highlights, as any detailed discussion will take too much space and time. For a history of the development of class field theory, the reader should look at the article of Hasse [28]. For treatments of class field theory one may look up [16a], [29a], [29b], [29c].

Class field theory in classical language. Class field theory really began perhaps with the work of Weber attempting to generalize the main results of Dirichlet on primes in residue classes to the context of arbitrary number fields. We have already seen that the argument of Dirichlet for proving that $L(1 : \chi) \neq 0$ when $\chi \neq \chi_0$ becomes very simple if we know the basic theory of the Dedekind zeta function and the formula that expresses the Dedekind zeta function of $K_N = \mathbf{Q}(e^{2\pi i/N})$ as a product of all the Dirichlet L-functions $L(s : \chi)$. Actually it is enough to know much less, in fact to know only that the rational prime p splits completely in K_N if and only if $p \equiv 1 \bmod N$. Since this fact is at the heart of Weber's definition of a class field and since it springs from Dirichlet's arguments depending on the Euler product representations of the L-functions, it is appropriate for us to consider it in some detail.

A set Ω of primes of a number field K is said to have *Dirichlet density* α if

$$\lim_{\sigma \to 1+0} \frac{\sum_{\mathfrak{p} \in \Omega} N\mathfrak{p}^{-\sigma}}{\log \frac{1}{\sigma-1}} = \alpha.$$

Note that the existence of the limit is part of the definition, as not every set has a well defined Dirichlet density. Clearly Dirichlet density is a finitely additive probability measure on the class of sets for which it has a meaning. Suppose now that L/K is a finite Galois extension. Let $\mathfrak{S}(L/K)$ be the set of primes $\mathfrak{p}$ in K that are unramified in L and split completely in L, and $\mathfrak{S}'(L/K)$ the set of primes of L above those in $\mathfrak{S}(L/K)$. Then

$$\log \zeta_L(\sigma) \approx \log \frac{1}{\sigma - 1} \approx \sum_{\mathfrak{P}} N\mathfrak{P}^{-\sigma} \approx \sum_{\mathfrak{P} \in \mathfrak{S}'(L/k)} N\mathfrak{P}^{-\sigma}$$

since for $\mathfrak{P} \notin \mathfrak{S}'(L/K)$ and unramified over $\mathfrak{p}$ we have $N\mathfrak{P}^{-\sigma} = N\mathfrak{p}^{-r\sigma}$ for some $r \geq 2$ and we have the familiar fact that

$$\sum_{\mathfrak{p}} N\mathfrak{p}^{-r\sigma} \approx 0 \qquad (r \geq 2).$$

On the other hand,

$$\sum_{\mathfrak{P} \in \mathfrak{S}'(L/K)} N\mathfrak{P}^{-\sigma} = [L:K] \sum \sum_{\mathfrak{p}} N\mathfrak{p}^{-\sigma}.$$

Since the left side is $\approx \log(1/\sigma - 1)$ we see that $\mathfrak{S}(L/K)$ has Dirichlet density $[L:K]^{-1}$ and that $\mathfrak{S}'(L/K)$ has Dirichlet density 1.

Let us now write for sets A, B of primes in K, $A \prec B$ if the set $A \setminus B$ has Dirichlet density 0; if $A \prec B$ and $B \prec A$, we write $A \approx B$. If E, L are two Galois extensions of K with $E \subset L$, then $\mathfrak{S}(L/K) \subset \mathfrak{S}(E/K)$. We now wish to show that conversely, if $\mathfrak{S}(L/K) \prec \mathfrak{S}(E/K)$, then $E \subset L$. To see this, note that if $\mathfrak{P} \in \mathfrak{S}'(L/K)$ and $\mathfrak{p}$ is the prime of K below $\mathfrak{P}$, then $\mathfrak{P}$ splits completely in the compositum LE if $\mathfrak{p}$ does so in E. So, if $\mathfrak{S}_1(L/K)$ is the set of $\mathfrak{P}$ in $\mathfrak{S}'(L/K)$ for which $\mathfrak{p} \notin (E/K)$, all primes $\mathfrak{P}$ in $\mathfrak{S}'(L/K) \setminus \mathfrak{S}_1(L/K)$ split completely in LE. On the other hand, the set of primes $\mathfrak{p}$ in K that are below elements of $\mathfrak{S}_1(L/K)$, being contained in $\mathfrak{S}(L/K) \setminus \mathfrak{S}(E/K)$, has Dirichlet density 0, since $\mathfrak{S}(L/K) \prec \mathfrak{S}(E/K)$. Hence $\mathfrak{S}_1(L/K)$ has Dirichlet density 0, so that $\mathfrak{S}(L/K) \setminus \mathfrak{S}_1(L/K)$ has Dirichlet density 1. This means that $\mathfrak{S}(LE/L)$ has Dirichlet density 1. By what we showed above, this density is $[LE:L]^{-1}$, and so we have $[LE:L] = 1$, i.e., $LE = L$. Hence $E \subset L$ as we claimed.

It follows from this that for two Galois extensions $E/K, L/K$,

$$E = L \Leftrightarrow \mathfrak{S}(L/K) \approx \mathfrak{S}(E/K).$$

In other words, the knowledge of the set of primes that split completely in L/K determines L uniquely.

We now return to the context of the Dirichlet theorem. Suppose now that we assume only that the rational primes $p \equiv 1 \bmod N$ are the primes that split completely in the cyclotomic field K_N. Let $m(\chi)$ be the order of $L(s:\chi)$ at $s = 1$, $\chi \neq \chi_0$. The Dirichlet argument (which we have gone over in section 2) gives, for C a residue class mod N,

$$\sum_{p \in C} p^{-\sigma} \approx \varphi(N)^{-1}(1 - \sum_{\chi \neq \chi_0} m(\chi)) \log \frac{1}{\sigma - 1}.$$

Taking C to be the residue class of 1 gives the result that the primes $p \equiv 1 \bmod N$ have Dirichlet density

$$\varphi(N)^{-1}(1 - \sum_{\chi \neq \chi + 0} m(\chi)).$$

If we now know that these are the primes that split completely in K_N, we know that they have Dirichlet density $\varphi(N)^{-1}$. Hence we must have

$$\sum_{\chi \neq \chi_0} m(\chi) = 0.$$

Before going to the existence problem of Weber let me describe how the groups G_N can be generalized to the context of an arbitrary number field. More precisely, given any integral ideal $\mathfrak{n}$ in K, one can define a congruence group, called an *ideal class group mod* $\mathfrak{n}$. To introduce this let us first observe that the Dirichlet groups G_N can be defined in another way that makes their generalization to a number field straightforward. Let $I(N)$ be the multiplicative group of rational numbers > 0 generated by the rational primes not dividing N. This is the multiplicative group of numbers of the form a/b where $a, b > 0$ are integers mutually prime and prime to N. The residue class map $x \mapsto [x]$ that takes an integer x to its residue class $[x]$ mod N extends to the homomorphism of $I(N)$ *onto* G_N that takes a/b to $[a][b]^{-1}$; its kernel is the group $\Gamma(N)$ of elements a/b with $a \equiv b \pmod{N}$ which is equivalent to

$$\mathrm{ord}_p \left(\frac{a}{b} - 1 \right) \geq N_p$$

for all primes p dividing N, N_p being the power of p appearing in the unique factorization of N. If K now is a number field and $\mathfrak{n}$ is an integral ideal with

$$\mathfrak{n} = \prod_v \mathfrak{p}_v^{n(v)}$$

as its unique factorization, we introduce the group $I(\mathfrak{n})$ of fractional ideals prime to $\mathfrak{n}$ and its subgroup $B(\mathfrak{n})$ defined by

$$B(\mathfrak{n}) = \left\{ (\xi) \mid \xi \in K^\times, \xi >> 0, \xi \equiv 1 \pmod{\mathfrak{n}} \right\}.$$

Here $\xi >> 0$ means that ξ maps to a positive number in each imbedding $K \hookrightarrow \mathbf{R}$, and $\xi \equiv 1 \pmod{\mathfrak{n}}$ means that for each finite place v for which the prime ideal $\mathfrak{p}_v$ divides $\mathfrak{n}$,

$$\mathrm{ord}_v(\xi - 1) \geq n(v).$$

(We write ord_v for the valuation defined by v.) The group

$$C(\mathfrak{n}) = I(\mathfrak{n})/B(\mathfrak{n})$$

is finite and is the *ideal class group mod* $\mathfrak{n}$. A *congruence group* is any group Γ such that

$$B(\mathfrak{n}) \subset \Gamma \subset I(\mathfrak{n}).$$

It must be noted that the groups $I(\mathfrak{n})/B(\mathfrak{n})$ form an inverse system (the larger $\mathfrak{n}$ is, the higher this quotient is), and one must really go to the inverse limit for a clean theory. Classically this was done by defining an equivalence and considering the equivalence classes. Note that for $\mathfrak{n} = 1$ we get the wider class group; the usual class group is a quotient of this.

Weber's fundamental definition can now be formulated as follows: suppose that we are given a congruence group Γ,

$$B(\mathfrak{n}) \subset \Gamma \subset I(\mathfrak{n}).$$

A finite Galois extension L_Γ/K is said to be a *class field* for K if

$$\mathfrak{G}(L/K) \approx \Gamma.$$

More precisely, L_Γ/K is a class field if and only if the primes of K that split completely in L_Γ are those in Γ up to a set of Dirichlet density 0. By what we have seen now, L_Γ is unique if it exists. Weber's fundamental existence problem was to show that such a class field exists always for any congruence group. With the cyclotomic extensions and the extensions of quadratic imaginary fields before him, Weber was convinced of the affirmative solution to the existence problem. Clearly, once the class field is shown to exist, the extension of Dirichlet's theorem to the present situation, namely, that there are infinitely many primes in each congruence class defined by elements of $I(\mathfrak{n})/B(\mathfrak{n})$, is fairly straightforward, as is also the fact, which comes out of the method of proof used, that the primes of K are equidistributed in the various classes.

Weber did not prove the existence of the class fields, although he proved by similar arguments that

$$[I(\mathfrak{n}) : \Gamma] \leq [L_\Gamma : K].$$

Hilbert, who knew of Weber's work and studied the special case of the so-called Hilbert class fields, which are the class fields for the divisor 1, offered good evidence for the structure he had divined. In this case $I(1)$ is the group of *all* fractional ideals, and $B(1)$ the subgroup of principal ideals generated by totally positive elements. The Hilbert class field is unramified everywhere, and its Galois group is the class group in the wider sense. From this experience and his work on relative quadratic fields he formulated the general results of abelian extensions. It was Takagi who established the full theory. In Takagi's theory the class fields exist and are abelian, and the correspondence between congruence groups and abelian extensions is bijective and possesses all the natural properties that it ought to possess, including completeness, namely that every abelian extension arises as a class field relative to some congruence group. Thus the tower structure of abelian extensions of K is captured completely by the tower structure of congruence groups. Moreover, with notation as above,

$$\mathrm{Gal}(L_\Gamma/K) \simeq I(\mathfrak{n})/\Gamma.$$

The isomorphism above was only formal in Takagi's work. It was not until Artin discovered the *canonical isomorphism* between the ideal class groups and the Galois groups through the *Artin symbol* that the theory was complete. We shall now discuss briefly this idea of Artin.

To describe the Artin isomorphism in this classical language, let us consider a Galois extension L/K, a prime $\mathfrak{p}$ of K, and a prime $\mathfrak{P}$ of L above $\mathfrak{p}$. If $\mathfrak{P}$ is unramified above $\mathfrak{p}$, the local Galois group, which is the subgroup of $\mathrm{Gal}(L/K)$ fixing $\mathfrak{P}$, is isomorphic canonically to the residue Galois extension defined by $\mathfrak{P}$ and $\mathfrak{p}$. So there is an element

$$\sigma = F_{L/K,\mathfrak{P}} \in \mathrm{Gal}(L/K),$$

the *Frobenius substitution*, such that

$$x^\sigma \equiv x^q \quad \mathrm{mod}\ \mathfrak{P} \qquad (x \text{ integral in } L).$$

Here q is the cardinality of the residue field at $\mathfrak{p}$. When we vary $\mathfrak{P}$ above $\mathfrak{p}$, $F_{L/K,\mathfrak{P}}$ varies in a conjugacy class, which therefore depends only on $\mathfrak{p}$, which we denote by

$$F_{L/K,\mathfrak{p}} \text{ or } F_{L/K,v} \quad (\mathfrak{p} = \mathfrak{p}_v).$$

If L/K is abelian, the conjugacy class is just a single element and so the Frobenius substitution is independent of $\mathfrak{P}$ above $\mathfrak{p}$ and so is a canonical element of $\mathrm{Gal}(L/K)$ that depends only on $\mathfrak{p}$. In this case we call it the *Artin symbol* of $\mathfrak{p}$ and write it as

$$\left(\frac{L/K}{\mathfrak{p}}\right).$$

If L'/K is a subextension of L/K, i.e., $K \subset L' \subset L$, then

$$\left(\frac{L'/K}{\mathfrak{p}}\right) = \text{Restriction of } \left(\frac{L/K}{\mathfrak{p}}\right) \text{ to } L'.$$

Let us look at the simplest example, namely when $K = \mathbf{Q}$ and L/K is a quadratic extension of a simple type. Let $K = \mathbf{Q}$ and $L = \mathbf{Q}(\sqrt{\ell})$ where ℓ is an odd prime; then the Galois group is $\{\pm 1\}$. For any odd prime p not dividing ℓ, we have

$$\left(\frac{\mathbf{Q}(\sqrt{\ell})/\mathbf{Q}}{p} \right) = \left(\frac{\ell}{p} \right) \qquad \text{(Legendre symbol)}.$$

Now 2 is also unramified, and the Frobenius substitution is ± 1 according to whether p splits completely or stays prime in L. It is standard that this is decided mod 8 and that in fact we have

$$F_{L/K,2} = (-1)^{(\ell^2 - 1)/8}.$$

Returning to the general case, the Artin symbol being defined for all primes $\mathfrak{p}$ that do not ramify in L, we can extend it multiplicatively to a homomorphism, also called the Artin symbol, from the group I of fractional ideals prime to the discriminant of L/K to the group $\mathrm{Gal}(L/K)$. Artin proved that *the kernel Γ of this homomorphism is a congruence group and that it gives an isomorphism of I/Γ with* $\mathrm{Gal}(L/K)$:

$$\left(\frac{L/K}{\cdot} \right) : I/\Gamma \simeq \mathrm{Gal}(L/K).$$

We have called the element

$$\left(\frac{L/K}{\mathfrak{p}} \right)$$

the Artin symbol because in various special cases it reduces to the various symbols, as in the case of the quadratic fields where it is the Legendre symbol. Notice that

$$\left(\frac{\ell}{p} \right) = \left(\frac{p}{\ell} \right)$$

by quadratic reciprocity, which implies that the homomorphism of the rationals prime to p into ± 1 determined by the map

$$p \longmapsto \left(\frac{\ell}{p} \right)$$

is trivial on the subgroup of n such that $n \equiv 1 \bmod \ell$, which is a congruence group. The Artin symbol isomorphism is the most general formulation of the reciprocity laws that pervade the theory of abelian extensions and is the culmination of the development that started with the quadratic reciprocity and went through several incarnations, the most important being the Hilbert norm residue symbol.

Hilbert posed the problem, the 9th in his list [27b], of finding the most general reciprocity law for a given number field. The examples he had in mind were the quadratic law of Gauss not only for $\mathbf{Q}$ but also for $\mathbf{Q}(\sqrt{-1})$, the cubic law of Eisenstien for $\mathbf{Q}(\omega)$ where ω is a primitive cube root of unity, and the power residue laws of Kummer. Artin's isomorphism gave the complete solution to this question. When K contains the ℓth roots of unity, the extensions $K(a^{1/\ell})(a \in K^\times)$ are abelian, and one can show that for a generic prime $\mathfrak{p}$ of K, the Artin symbol

$$\left(\frac{K(a^{1/\ell})/K}{\mathfrak{p}} \right)$$

is essentially what is known as the ℓth-power residue symbol

$$\left(\frac{a}{\mathfrak{p}}\right)_\ell.$$

The Artin isomorphism tells us that this depends only on the congruence class of $\mathfrak{p}$ with respect to some modulus $\mathfrak{n}$ in K. It is because of this that Artin viewed his theorem as a reciprocity law for *any* K, thereby completing the solution of the Hilbert problem.

Here is what Tate had to say about Artin's discovery [27c]:

> ... *How did Artin guess his reciprocity law? He was not looking for it, not trying to solve a Hilbert problem. Neither was he, as would seem so natural to us today, seeking a canonical isomorphism, to make Takagi's theory more functorial.... He was led to the law in trying to show that a new kind of L-series which he introduced really was a generalization of the usual L-series....*

The usual L-series

$$L(s:\chi) = \prod_p (1 - \chi(\mathfrak{p})N\mathfrak{p}^{-s})^{-1}$$

were associated to characters χ of the ideal (idele) class groups. Artin's L-series were associated to representations of the Galois groups of extensions of K. In the present case the representation was one-dimensional, a character χ. Artin's definition was (we shall see this in more detail in the next section)

$$L_{\text{Artin}}(s:\chi) = \prod_{\mathfrak{p}} (1 - \chi(\sigma_{\mathfrak{p}})N\mathfrak{p}^{-s})^{-1}$$

where χ is a character of $\text{Gal}(L/K)$, L/K a finite abelian extension, and

$$\sigma_{\mathfrak{p}} = F_{L/K,\mathfrak{p}}$$

is the Frobenius substitution at $\mathfrak{p}$; the product is over all primes unramified in L. It is now clear that if Artin's result is true, we can write

$$\chi(\sigma_{\mathfrak{p}}) = \chi^\sim([\mathfrak{p}])$$

where $[\mathfrak{p}]$ is the residue class of $\mathfrak{p}$ modulo some congruence group Γ modulo some divisor $\mathfrak{n}$, and $\chi^\sim$ is a character of $I(\mathfrak{n})/\Gamma$. Then χ is essentially a Hecke character and, up to a finite number of Eulerian factors,

$$L_{\text{Artin}}(s:\chi) = L_{\text{Hecke}}(s:\chi^\sim).$$

This would prove that $L_{\text{Artin}}(s:\chi)$ has analytic continuation and functional equation.

EXAMPLE. Let us consider a quadratic extension $L = \mathbf{Q}(\sqrt{\ell})$ of $\mathbf{Q}$ where ℓ is an odd prime with $\ell \approx 1(\bmod\ 4)$. The Galois group $\text{Gal}(L/K) = \{\pm 1\}$. From the calculation above and quadratic reciprocity it is immediate that for all primes $p \neq \ell$ we have

$$F_{L/K,p} = \left(\frac{p}{\ell}\right)$$

so that

$$L_{\text{Artin}}\ (s:\chi) = L_{\text{Dirichlet}}(s:\chi^\sim)$$

where $\chi^{\sim}$ is the Dirichlet character

$$p \longmapsto \left(\frac{p}{\ell}\right).$$

Class field theory in the language of ideles. This classical description of the main results of class field theory is clearly very beautiful. I have not mentioned several important results, such as Hilbert's principal ideal theorem (which he fore-saw from his analogy with Abel's theorem) stating that all ideals of K become principal in the Hilbert class field. Moreover I have not mentioned the fundamental work of Hasse which separated the global aspects of class field theory from its local aspects. Hasse's work showed that local class field theory, namely the theory of abelian extensions of *local fields*, was worthy of study in its own right and, further, that global class field theory can then be built on it, thus finally clarifying the local-global interplay. His work also created the complete theory of central simple algebras over global fields, including the determination of the global Brauer groups in terms of the local Brauer groups through the product formula for the elements of the global symbols in terms of the local symbols. For quaternion algebras this reciprocity law of Hasse reduces to the Hilbert product formula for the Hilbert symbols. For a complete treatment see [16a], [29a], [29b], [29c], and for additional perspective, [10], [28].

For our purposes it is necessary to describe these results in the language of ideles and adeles. Clearly the role of the ideal class groups will be played by the idele class group. Let K be a number field and let $\mathbf{C}(K)$ be the idele class group. Let Φ_K be the filter of open subgroups of $\mathbf{C}(K)$. These may also be regarded as open subgroups of $\mathbf{J}(K)$, and they are *always of finite index in* $\mathbf{J}(K)$ *(or* $\mathbf{C}(K)$). We now introduce

$$\mathbf{C}(K)^{\sim} = \varprojlim_{H} \mathbf{C}(K)/H \qquad (H \in \Phi_K),$$

the profinite completion of $\mathbf{C}(K)$. Note that all of the characters of $\mathbf{C}(K)^{\sim}$ are of finite order; indeed, they form the torsion subgroup of the character group of $\mathbf{C}(K)$. Of course $\mathbf{C}(K)$ has characters that are not of finite order; the characters of $\mathbf{C}(K)$, of finite order or not, are the Grössencharakters of Hecke. The theory created in the classical language can now be summarized in the following statements:

I. The ascending tower of abelian extensions of K is in canonical order-reversing bijection with the descending tower of subgroups in Φ_K. If $\Omega \in \Phi_K$ and L_Ω is the corresponding abelian extension of K, we have

$$L_{\Omega_1 \cap \Omega_2} = L_{\Omega_1} L_{\Omega_2}, \qquad L_{\Omega_1 \Omega_2} = L_{\Omega_1} \cap L_{\Omega_2}.$$

Moreover, we have an isomorphism

$$\mathbf{C}(K)/\Omega \xrightarrow{\sim} \mathrm{Gal}(L_\Omega/K).$$

Let us consider K^{ab}, the compositum of all abelian extensions of K. Then K^{ab}/K is an extension of *infinite degree* and $\mathrm{Gal}(K^{\mathrm{ab}}/K)$ is the profinite group

$$\mathrm{Gal}(K^{\mathrm{ab}}/K) = \varprojlim_{L} \mathrm{Gal}(L/K)$$

where the inverse limit is over all finite abelian extensions L/K.

II. The correspondence in I is induced by the *Artin symbol* which is a canonical isomorphism of $\mathbf{C}(K)^{\sim}$ with $\mathrm{Gal}(K_{ab}/K)$:

$$\mathfrak{A} : \mathbf{C}(K)^{\sim} \xrightarrow{\sim} \mathrm{Gal}\,(K^{\mathrm{ab}}/K).$$

This is the *Artin reciprocity law*. In particular, for any $\Omega \in \Phi_K$ the Artin symbol induces the isomorphism

$$\mathbf{C}(K)/\Omega \xrightarrow{\sim} \mathrm{Gal}(L_\Omega/K).$$

The image of $\Omega \in \Phi_K$ by $\mathfrak{A}$ is an open subgroup of $\mathrm{Gal}(K^{ab}/K)$ whose fixed points give the extension L_Ω in I. To understand the relation between L_Ω and Ω better, let us consider for any finite Galois extension L/K the norm map

$$N_{L/K} : \mathbf{J}(L) \longrightarrow \mathbf{J}(L).$$

Since this maps takes $L^\times$ into $K^\times$ we have the map

$$N_{L/K} : \mathbf{C}(L) \longrightarrow \mathbf{C}(K).$$

We then have

III. For any finite Galois extension L/K, $N_{L/K}(\mathbf{C}(L))$ is an open subgroup $\Omega(L/K)$ of $\mathbf{C}(L)$. Then the abelian extension $L_{\Omega(L/K)}$ is given by

$$L_{\Omega(L/K)} = K^{ab} \cap L.$$

In particular, L_Ω is the unique abelian extension of K such that

$$N_{L_\Omega/K}(\mathbf{C}(L_\Omega)) = \Omega.$$

This makes it clear that the Artin symbol is a canonical way to describe the *norm residue symbol* in the general case. In the classical reciprocity laws of Gauss, Eisenstein, Kummer, and others, the symbol was always a root of unity and could be given an invariant meaning only after the Galois group had been identified with some group of roots of unity.

The classical Artin symbol also fits nicely into this picture with $\mathfrak{A}$. Let L/K be a finite abelian extension. Let $R = R(L/K)$ be the finite set of places of K consisting of all infinite places and all finite places that ramify in L, i.e., that divide the discriminant $\mathfrak{d}$. We then have the subgroup $I(\mathfrak{d})$ of fractional ideals of K generated by the primes whose places are outside of R. By Artin's theorem the Artin symbol is the homomorphism

$$I(\mathfrak{d}) \longrightarrow \mathrm{Gal}(L/K)$$

which is surjective and whose kernel is a congruence subgroup of $I(\mathfrak{d})$, i.e., contains $B(\mathfrak{d})$ introduced earlier. Let us now introduce

$$\mathbf{J}(K)^R = \{x = (x_v) \in \mathbf{J}(K) \mid x_v = 1 \; \forall v \in R\}.$$

Then $\mathbf{J}(K)^R$ is a closed subgroup of $\mathbf{J}(K)$ and the subgroup

$$K^\times \mathbf{J}(K)^R$$

is dense in $\mathbf{J}(K)$ (approximation theory).

IV. There is a unique isomorphism

$$\left(\frac{L/K}{\cdot}\right) : \mathbf{J}(K) \longrightarrow \mathrm{Gal}(L/K)$$

such that

(a) $\left(\frac{L/K}{\alpha}\right) = 1 \quad \forall \alpha \in K^{\times}$.

(b) We have

$$\left(\frac{L/K}{x}\right) = \left(\frac{L/K}{\mathrm{id}(x)}\right) \qquad (x \in \mathbf{J}(K)^R)$$

where $\mathrm{id}(x)$ is the fractional ideal corresponding to the idele x.

The open subgroups $\Omega \in \Phi_K$ thus play the role of the congruence groups, and so we refer to $L_\Omega (\Omega \in \Phi_K)$ as the *class field* corresponding to Ω. The splitting of primes is also very simple to describe in this language. For any place v of K let us identify $K_v^{\times}$ as a subgroup of $\mathbf{J}(K)$ by identifying the element $y \in K_v^{\times}$ with the idele all of whose components are 1 except the one corresponding to v, and that one is y. Recall that U_v is the group of units of K_v, i.e., the group of elements y with $|y|_v = 1$ if v is finite, and $K_v^{\times}$ if v is infinite.

V. For a place v of K and an open subgroup $\Omega \in \Phi_K$, we have

$$v \text{ is unramified in } L_\Omega \Leftrightarrow U_v \subset \Omega$$

and

$$v \text{ splits completely in } L_\Omega \Leftrightarrow K_v^{\times} \subset \Omega.$$

Unramified and complete splitting are the same for infinite v. For v imaginary this is always so, and for v real this means that the places of L above v are all real also. Finally, the equality of the Artin L-function with the Hecke L-function can be described very simply. Let

$$\chi : \mathrm{Gal}(K^{ab}/K) : \mathrm{GL}(1, \mathbf{C}) = \mathbf{C}^{\times}$$

be a character. We may view this as a character of $\mathrm{Gal}(L/K)$ where L is the fixed point field of the kernel of χ, the latter being an open subgroup (hence of finite index) in $\mathrm{Gal}(K^{ab}/K)$. Then

$$L_{\mathrm{Artin}}(s : \chi) = \prod_{v \text{ finite unramified}} \left(1 - \chi(F_{v,L/K}) N \mathfrak{p}_v^{-s}\right)^{-1}.$$

Let

$$\chi^{\sim} = \mathfrak{A}^{-1} \circ \chi$$

be the character of $\mathbf{C}(K)$ obtained by transporting χ via the Artin symbol isomorphism $\mathfrak{A}$. Then

$$L_{\mathrm{Artin}}(s : \chi) = L_{\mathrm{Hecke}}(s : \chi^{\sim}).$$

Of course $\chi^{\sim}$ is of finite order and so is a Dirichlet character.

6.5. Artin nonabelian L-functions

In the years 1923-30 Artin introduced a new kind of L-functions. These are Euler products attached to representations ρ of the Galois group of finite extensions L/K of a number field K, with the corresponding Eulerian factors of degree equal to the $\dim(\rho)$. *This was a big step because it turned attention from the Galois groups themselves to their representations, which is the key to all further work.* Let v be a finite place and let $\bar{v}$ be a place of L above v. We write $G_{\bar{v}}$ for the subgroup of $\mathrm{Gal}(L/K)$ that stabilizes $\bar{v}$. We have a natural map

$$G_{\bar{v}} \longrightarrow \mathrm{Gal}(k_{\bar{v}}/k_v)$$

where $k_{\bar{v}}$ (resp. k_v) is the residue field at $\bar{v}$ (resp. v). This map is surjective, and we write $I_{\bar{b}}$ for its kernel. We write $F_{\bar{v}}$ for any element of $G_{\bar{v}}$ above the Frobenius generator of $\mathrm{Gal}(k_{\bar{v}}/k_v)$. If $\bar{v}$ is unramified over v, then $I_{\bar{v}}$ is trivial and so $F_{\bar{v}}$ is uniquely determined. Almost all places of K have this property (for all $\bar{v}$ above them). As usual let $q_v = N\mathfrak{p}_v$ where $\mathfrak{p}_v$ is the prime ideal defined by v. Following Artin we shall write

$$L_v(s:\rho) = \frac{1}{\det\left(I - \rho(F_{\bar{v}}q_v^{-s})\right)}.$$

We are writing L_v because the determinant is independent of the choice of $\bar{v}$ above v since all such $\bar{v}$ are conjugate under $\mathrm{Gal}(L/K)$. If v is ramified, one has to be a little more careful. Let V be the space of ρ and let $V^{\bar{v}}$ be the subspace of V of vectors fixed by the action of $I_{\bar{v}}$. Then the restriction $\mathrm{Res}_{\bar{v}}F_{\bar{v}}$ of F_{barv} to $V^{\bar{v}}$ is uniquely defined and the Eulerian factor at v is defined by

$$L_v(s:\rho) = \frac{1}{\det\left(I - \rho(\mathrm{Res}_{\bar{v}}F_{\bar{v}})q_v^{-s}\right)}.$$

Once again it is easy to see that this is well defined independently of the choice of $\bar{v}$ above v.

Let

$$\lambda_{v,1}, \ldots \lambda_{v,d} \qquad (d = \deg(\rho))$$

be the eigenvalues of $\rho(F_v)$. Then the Euler product is just

$$\prod_{v \notin S} \prod_{1 \le j \le d} \frac{1}{(1 - \lambda_{v,j}q_v^{-s})}$$

and so, as $|\lambda_{v,j}| = 1$, converges in $\Re(s) > 1$. With these definitions

$$L(s:\rho) = \prod_{v \text{ finite}} L_v(s:\rho).$$

The definition can be completed by adding the gamma factors corresponding to the infinite places, and the resulting Euler product is denoted by $\Lambda(s:\rho)$.

The key properties of the Artin L-functions are as follows:

(i) $L(s:\rho \oplus \rho') = L(s:\rho)L(s:\rho')$.

(ii) If $M/L/K$ is a tower with M/K also Galois, then any representation ρ_L of $G(L/K)$ may be viewed as a representation ρ_M of $G(M/K)$ by lifting through the map $G(M/K) \longrightarrow G(L/K)$; then

$$L(s:\rho_M) = L(s:\rho_L).$$

(iii) (inductive property) Let $L/M/K$ be a tower, let ρ_M be a representation of $\mathrm{Gal}(L/M)$, and let $\mathrm{Ind}_{\mathrm{Gal}(L/M)}^{\mathrm{Gal}(L/K)}(\rho_M)$ be the representation of $G(L/K)$ induced by ρ_M. Then

$$L(s:\mathrm{Ind}_{\mathrm{Gal}(L/M)}^{\mathrm{Gal}(L/K)}(\rho_M)) = L(s:\rho_M).$$

Property (iii) is remarkable and shows that in the theory of the Artin L-function the representations that are induced from suitable subgroups play an important role. Now these induced representations are *big* in the sense that they are seldom irreducible even if ρ_M is. On the other hand, the additivity expressed by (i) allows

us to extend the definition of the Artin L-function to even *virtual* representations
of $G(L/K)$, namely, to *virtual* linear combinations

$$\rho = \sum_j m_j \rho_i \qquad (m_j \in \mathbf{Z})$$

where the ρ_j are ordinary representations of $G(L/K)$; we just set

$$L(s : \rho) = \prod_j L(s : \rho_j).$$

Artin conjectured that these L-functions extend meromorphically to the whole
s-plane and satisfy functional equations and, further, that if ρ does not contain the
trivial representation, then $L(s : \rho)$ is *entire*. Brauer proved the extension property
as a consequence of his purely group theoretic result: if G is a finite group, then
every representation of G can be written as a *virtual* integral linear combination of
representations of the form $\mathrm{Ind}_H^G(\chi)$ where the H are suitable subgroups of G and
the χ are *one-dimensional*:

$$\rho = \sum_j m_j \mathrm{Ind}_{H_j}^G(\chi_j) \qquad (m_j \in \mathbf{Z}, \dim(\chi_j) = 1).$$

By class field theory, the $L(s : \chi)$ for one-dimensional χ can be identified with the
Hecke L-functions for suitable abelian extensions L/M. Hence follow the meromorphic continuation and functional equation, but not the fact that for ordinary
irreducible *nontrivial* ρ, $L(s : \rho)$ is entire; the presence of negative integers in the
virtual linear combinations prevent one from deducing this result. This conjecture
of Artin remains open.

6.6. The Langlands program

Given the story so far it was entirely natural for people to start thinking about
extending the theory of abelian extensions to a theory of *all* Galois extensions,
abelian or not. The theory of the Artin L-functions should be regarded as the
first step in this direction. The decisive step was taken by Langlands, as we mentioned earlier, when he put forward evidence that the automorphic L-functions are
exhaustive enough to contain all the Artin L-functions. In his vision the Galois representations ρ of degree n should determine automorphic representations π_ρ such
that

$$L(s : \rho) = L(s : \pi_\rho)$$

where the L-function on the left is the Artin one, and the one on the right is the
L-function of an automorphic representation of $\mathrm{GL}(n)$, namely an irreducible representation of the adele group $\mathrm{GL}(n, \mathbf{A}(K))$ that occurs in the space of automorphic
forms of $\mathrm{GL}(n, \mathbf{A}(K))/\mathrm{GL}(n, K)$. For general n these L-functions were defined
and their properties established by Godement and Jacquet. In this correspondence
the irreducible Galois representation ρ should correspond to *cuspidal* automorphic
representations π_ρ, these being the representations occurring in the space of cusp
forms. Of course we should use the adelic definition of what is meant by cusp forms.

In working out this program it is clearly important to explore the structural
features on both sides of the picture (Galois and automorphic). Analysis along these
lines made it clear that the notion of the Galois representation should be enlarged.
This was done by enlarging the Galois group to the so-called Weil groups and the
Weil-Deligne groups. Moreover, in order to accommodate the Galois representations

coming from geometry, it became necessary to include ℓ-adic representations on the Galois side. Finally, the special role of $GL(n)$ has been understood and the fundamental results formulated in a context where all reductive groups enter the picture.

This program has created tremendous excitement and interest ever since it was proposed. A whole generation of mathematicians has been attracted by its sweep, its generality, and the fact that it brings together geometry, arithmetic, and analysis in a manner that is unique in the history of these subjects. Great progress has been made, by Langlands himself and his collaborators, in specific cases, for instance, for $GL(2)$. Already these results have had a great impact, such as in Wiles's proof of Fermat's last theorem. However the general program, even for $GL(n)$, seems too difficult at present. Nevertheless Laurent Lafforgue succeeded in proving the fundamental theorems of this program for *function fields* with $GL(n)$ as the group on the automorphic side. In other words Lafforgue's work established a bijective correspondence between irreducible ℓ-adic representations ρ $(\ell \neq p)$ of degree n of $\mathrm{Gal}(\bar{K}/K)$ (K is a function field of characteristic p and $\bar{K}$ its separable algebraic closure) and irreducible cuspidal automorphic representations π_ρ of $GL(n, \mathbf{A}(K))$, $\rho \leftrightarrow \pi_\rho$, such that

$$L(s:\rho) = L(s:\pi_\rho).$$

He also proved the Ramanujan-Petersson conjecture in this context, that for an automorphic representation π with a central character of finite order at every place v where π is unramified, i.e., π_v has a unit vector fixed by $GL(n, K_v)$, the Hecke eigenvalues are all of absolute value 1; i.e., π_v belongs to the unitary principal series. For more on the work of Lafforgue see [30a] and the references given there. These results were proved earlier in the 1970's by Drinfel'd when $n = 2$. On the other hand, Michael Harris and Richard Taylor, followed a little later by Henniart (who gave a simplified proof), settled the local Langlands conjectures involving $GL(n, K)$ for any local field [30b]. The interested reader should consult [7] for a systematic introduction to the themes that we have touched upon in this section, as well as for further references.

Notes and references

[1]   L. Euler, *Opera Omnia*, I-14, 216-.
[2]   Letters to Goldbach, August 5 and October 28, 1752, in P.-H. Fuss, *Correspondance Mathematique et Physique*, Volumes I, II, The Sources of Science, Johnson Reprint Corporation, 1968.
[3]   P. G. Lejeune Dirichlet, *Mathematische Werke*, Bds. I, II, Chelsea, 1969.
[4]   H. Davenport, *Multiplicative Number Theory*, Springer-Verlag, 1980, Second Edition (revised by H. L. Montgomery).
[5]   B. Riemann, *Ueber die Anzahl der Primzahlen unter einer gegebenen Grösse*, Collected Papers, Raghavan Narasimhan (Ed.), Springer-Verlag, 1990, 177-185.
[6]   See the references in [4].
[7]   *An Introduction to the Langlands Program*, J. Bernstein and S. Gelbart (Eds.), Birkhäuser, 2003.
[8]   E. E. Kummer, *Collected Papers*, A. Weil (Ed.), Vol. I, Springer-Verlag, 1975.
[9]   H. M. Edwards, *Fermat's Last Theorem*, Springer-Verlag, 1977.
[10] R. P. Langlands, *Class fields*, Lectures given at UCLA, 2003.
[11] Z. I. Borevich and I. R. Shafarevich, *Number Theory*, Academic Press, 1966.
[12] N. Bourbaki, *Algèbre Commutative*, Ch. 7, "Diviseurs", Hermann, 1965.
[13] H. Weyl, *Algebraic Theory of Numbers*, Annals of Mathematical Studies, Princeton University Press, 1940.

[14a] E. Artin, *Collected Papers*, Addison-Wesley, 1965.

[14b] R. Dedekind, *Gesammelte Mathematische Werke*, t. I, Braunschweig, 1932.

[14c] E. Artin, *Algebraic Numbers and Algebraic Functions*, Gordon and Breach, 1967.

[14d] C. Chevalley, *Introduction to the Theory of Algebraic Functions of One Variable*, A.M.S. Math. Surveys, VI, 1951.

[15]   C. Chevalley, *La théorie du corps de classes*, Ann. Math., 41 (1940), 394-418.

[16a] A. Weil, *Basic Number Theory*, Springer-Verlag, 1967.

[16b] M. Kneser, *Semisimple algebraic groups*, in *Algebraic Number Theory*, J. W. S. Cassels and A. Frölich (Eds.), Thompson Book Company, Inc., 1967, 250-267.

[17]   E. Hecke, *Mathematische Werke*, Vandenhoeck & Ruprecht, Göttingen, 1959.

[18a] J. T. Tate, *Fourier Analysis in Number Fields and Hecke's Zeta-Functions*, Thesis, Princeton, 1950. Reprinted in *Algebraic Number Theory*, J. W. S. Cassels and A. Frölich (Eds.), Thompson Book Company, Inc., 1967.

[18b] R. Valenza and D. Ramakrishnan, *Fourier Analysis on Number Fields*, Springer-Verlag, 1999.

[18c] D. Bump, *Automorphic Forms and Representations*, Cambridge University Press, 1999.

[18d] H. Cartan and R. Godement, *Théorie de la dualité et analyse harmonique dans les groupes abéliens localement compacts*, Annls. Scient. Ec. Norm. Sup. Paris, (3) 64 (1947), 79-99.

[19a] S. Ramanujan, *On certain arithmetical functions*, Collected Papers, Chelsea, 1962, 136-162.

[19b] L. J. Mordell, *On Mr. Ramanujan's Empirical Expansions of Modular Functions*, Proc. Camb. Phil. Soc., 19 (1917), 117-124.

[20]   I. M. Gel'fand and S. V. Fomin, *Geodesic flows on manifolds of constant negative curvature*, Usp. Mat. Nauk 7(1) (1952), 118-137 (in Russian); *Collected Papers of Izrail M. Gel'fand*, S. G. Gindikin et al. (Eds.), Volume II, Springer-Verlag, 1988.

[21a] H. Jacquet and R. P. Langlands, *Automorphic Forms on GL(2)*, Springer Lecture Notes in Mathematics, Springer-Verlag, 1970.

[21b] Click on the website
       www.sunsite.ubc.ca/DigitalMathArchive/Langlands/intro.html

[22a] J.-P. Serre, *A Course in Arithmetic*, Springer-Verlag, 1973.

[22b] G. Shimura, *Introduction to the Arithmetic Theory of Automorphic Functions*, Iwanami Shoten Publishers and Princeton University Press, 1971.

[22c] N. Katz, *An overview of Deligne's proof of the Riemann hypothesis for varieties over finite fields*, in *Mathematical Developments Arising from Hilbert Problems*, Proc. Symp. in Pure Math., Vol. XXVIII, AMS, 1976, 275-305.

[23]   P. Deligne, *Formes modulaires et representations de GL(2)*, in *Modular Functions of One Variable*, Springer Lecture Notes in Mathematics #349, Springer-Verlag, 1973, 55-105.

[24]   V. S. Varadarajan, *An Introduction to Harmonic Analysis on Semisimple Lie Groups*, Cambridge University Press, 1989.

[25]   L. Kronecker, *Mathematische Werke*, Chelsea, 1968.

[26a] J.-P. Serre, *Complex Multiplication*, in *Algebraic Number Theory*, J. W. S. Cassels and A. Frölich (Eds.), Thompson Book Company, Inc., 1967, 292-296.

[26b] D. A. Cox, *Primes of the Form $x^2 + ny^2$*, Wiley, 1989.

[26c] S. Lang, *Elliptic Functions*, Addison-Wesley, 1973.

[27a] D. Hilbert, *Gesammelte Abhandlungen*, Bd. I., 62-363, Chelsea, 1965. French translation: *Théorie des Corps de Nombres Algébriques*, Éditions Jacques Gabay, 1991.

[27b] D. Hilbert, *Mathematical Problems*, in *Mathematical Developments Arising from Hilbert Problems*, Proc. Symp. in Pure Math., Vol. XXVIII, AMS, 1976, 1-34.

[27c] J. T. Tate, *Problem 9: The general Reciprocity Law*, in *Mathematical Developments Arising from Hilbert Problems*, Proc. Symp. in Pure Math., Vol. XXVIII, AMS, 1976, 311-322.

[28]   H. Hasse, *History of class field theory*, in *Algebraic Number Theory*, J. W. S. Cassels and A. Frölich (Eds.), Thompson Book Company, Inc., 1967, 266-279.

[29a] J.-P. Serre, *Local Class Field Theory*, 129-162, in *Algebraic Number Theory*, J. W. S. Cassels and A. Frölich (Eds.), Thompson Book Company, Inc., 1967, 128-166.

[29b] J. T. Tate, *Global Class Field Theory*, 162-203, in *Algebraic Number Theory*, J. W. S. Cassels and A. Frölich (Eds.), Thompson Book Company, Inc., 1967.

[29c] S. Lang, *Algebraic Number Theory*, Addison-Wesley, 1968.

[30a] M. Rapoport, in *The mathematical work of the 2002 Fields medalists*, Notices of AMS, February 2003, 212-217.

[30b] Henri Carayol, *Preuve de la conjecture de Langlands locale pour GL(n): Travaux de Harris-Taylor et Henniart*, in *Seminaire Bourbaki*, Exposé 857, Vol. 1998/99, exposés 850-854, Astérisque, 266 (2000), 191-243.

Gallery

This is a selective portrait gallery containing thumbnail sketches of some of the mathematicians whose work has been touched upon in this book. A good source to consult for all historical mathematical and personal aspects of mathematics and mathematicians is the MacTutor History of Mathematics Archive website:

http://turnbull.mcs.st-and.ac.uk/~history

For obvious reasons I have given a little more emphasis to mathematicians of more recent times.

Niels Henrik Abel (1802-1829). Abel was born in Norway. His entire life was a struggle against poverty and for the recognition of his mathematical genius. His manuscript on algebraic functions was lost by the carelessness of Cauchy, and Gauss did not even open his letter on the insolubility by radicals of the general equation of the fifth degree. But with the passage of time his work has been recognized for its originality, depth, and beauty. In addition to his pioneering work on algebraic functions and their integrals, he was the first mathematician, along with Weierstrass, to insist on absolute rigor in analysis. In this sense he is one of the founding fathers, along with Dedekind, of modern real analysis. He became a friend of Crelle who had started what became known as Crelle's journal, and in the first volume in 1827 several papers of Abel were published.

He proved that a general equation of the fifth degree does not admit a solution in radicals and in fact determined precisely the conditions under which this happens. His work on integrals of algebraic functions was revolutionary. He was able to subsume a substantial part of Legendre's and Jacobi's work on elliptic integrals under his work on algebraic functions and their integrals, and this work was recognized by Legendre as far-reaching and profound. One must therefore regard him as a great predecessor of Riemann in this area. In analysis he clarified the foundations of the theory of power series and proved the theorem that if the series converges at a boundary point, the function defined by the power series approaches the sum of the series as the argument approaches the boundary point radially.

He died young and largely unrecognized, but his name will live forever because of the genius that informs his work.

Emil Artin (1898-1962). Emil Artin was born in Austria and died in Hamburg, Germany, where he worked for many years and ran a famous seminar. In 1937, when Europe, after the emergence of Nazism, became a difficult place to live for mathematicians (among others) who were either Jewish or had Jewish connections, Artin, whose wife was Jewish, came to the United States. He worked in the Universities of Notre Dame and of Indiana at Bloomington before going to Princeton in 1946, where he spent most of the rest of his academic career, returning to Hamburg in 1958.

His work penetrated deeply into many areas of algebra and algebraic geometry. He was the first to introduce zeta functions of algebraic function fields of one variable defined over a finite field. For some of these he verified the analogue of the Riemann hypothesis; this was generalized to elliptic curves by Hasse, then to all curves by Weil, and finally to all smooth projective varieties by Deligne. He made the conjecture that any integer which is different from ± 1 is the primitive root of the field of integers mod p for a set of primes p of positive density; it is still unproved today. His work with Schrier produced the solution to one of Hilbert's problems, namely the structure of real fields. His most famous and enduring contribution is of course the general reciprocity law for abelian extensions of algebraic number fields which appears as a product formula for the so-called Artin symbol. It encompasses all reciprocity laws discovered earlier: namely, the quadratic law of Gauss, the biquadratic law of Gauss, the cubic law of Eisenstein, Hilbert's theory of Hilbert symbols, and the power law of Kummer. This solved another of Hilbert's problems, the 9th. His work, along with that of Hilbert, Hasse, Emmy Noether, Takagi, and others created the edifice of class field theory, which is the complete theory of abelian extensions of number fields. His books *Galois Theory*, *Rings with Minimum Condition*, *Geometric Algebra*, and *Class Field Theory* have become classics. His *Collected Papers* were edited by Serge Lang and John Tate, both his students.

The Bernoullis. In the whole history of science there is perhaps not another example like the Bernoulli family. Three generations of Bernoullis made distinguished contributions to mathematics, astronomy, probability, and the physical sciences, and strongly influenced the course of these sciences by their work. This started with Jacob Bernoulli and his brother, Johann Bernoulli, and then the torch was carried on by their sons, Nicolaus II and Daniel, respectively, and then their sons. Remarkably, they chose the academic path for their lives in spite of the fact that commerce and economics were the preferred career choices in their times and there was considerable pressure on them from their own families not to pursue the academic profession. They were originally from Basel, Switzerland, but in the course of their lives took academic positions in various parts of Europe, including Russia.

When Jacob and Johann were beginning their careers, the papers of Leibniz on calculus had just appeared, and because of their novelty and obscure structure, it was difficult for people to understand them and move forward. Jacob and Johann started a really deep study of Leibniz's work, and their understanding led them to make fundamental contributions to the theory of differential equations and to problems of maxima and minima in which the unknown was a curve, such as the brachistochrone, the isochrone, the catenary, and so on. Even more important for our story is the fact that Johann Bernoulli was a mentor for Euler when Euler was growing up, and this fact profoundly influenced the topics that Euler worked on early in his career. Undoubtedly the deep insights that Jacob and Johann had concerning the problems of maxima and minima of curves must have played the important role of catalyst in Euler's mind when he developed the calculus of variations and unified all these disparate problems under a single umbrella.

Jacob Bernoulli (1654-1705). Any enumeration of the contributions of Jacob Bernoulli must include his work on the isochrone problem and the study of certain nonlinear differential equations (Bernoulli's equation), his work on the sums $\sum_n n^k$

which led to what Euler would later call Bernoulli polynomials and Bernoulli numbers, and his work on the lemniscate of Bernoulli that would figure so prominently in Gauss's work on elliptic integrals. But his most original and epoch-making work was his treatise on the theory of probability, *Ars Conjectandi*. It was (posthumously) published in Basel in 1713 by his nephew Nicolaus. Among other things it contained the famous law of large numbers. The significance of this book and the results contained in it for the development of the theory of probability cannot be overemphasized. The book also contained the probability of winning in certain games under suitable assumptions, like the probability of what is called a *chase* in *real tennis*. One may therefore say that Jacob Bernoulli was the first major exponent of the theory of games.

Johann Bernoulli (1667-1748). Johann was the younger brother of Jacob. He occupied the chair of mathematics in Basel University after his brother's death. He was interested in problems of maxima and minima involving curves and posed the problem of brachistochrone, which was solved by several people including himself, his brother Jacob, Leibniz, and Newton. He became a tutor to the marquis de l'Hopital. The latter published a book on infinitesimal analysis based on the lectures of Johann, but Johann's essential contribution to this book became public after the death in 1704 of the marquis. However his most important contribution to mathematics might have been the fact that he was the mentor to Euler when Euler was growing up in Basel. Certainly, before branching out on his own in St. Petersburg, the problems that Euler worked on, like the calculus of variations, were inspired by his association with Johann Bernoulli.

Nicolaus Bernoulli I (1687-1759). Nicolaus I was a nephew of both Jacob and Johann Bernoulli. He corresponded with Euler and was able to solve the problem of finding the value of $\sum_n (1/n^2)$ after Euler had done it. He was the one who organized the posthumous publication of his uncle Jacob's great treatise *Ars Conjectandi*. He, like his uncles and cousins, worked on problems of calculus, differential equations, especially the Riccati equation. He was the first to prove the equality of mixed partial derivatives.

Nicolaus Bernoulli II (1695-1726). Nicolaus II was one of three sons of Johann Bernoulli and accompanied his brother Daniel to St. Petersburg. Unfortunately he died within a year of going there, apparently due to acute appendicitis.

Daniel Bernoulli (1700-1782). Daniel Bernoulli was a son of Johann and forged for himself a most distinguished career in mathematics and physics. He made lasting contributions to mechanics, theory of elastic bodies, and above all, to hydrodynamics. He was a companion to Euler in the early years at St. Petersburg but left for Basel in 1734. His most enduring contributions are certainly to hydrodynamics and to the theory of the vibrating string. He discovered the wave equation of the string and the fundamental fact that any mode of the string is a linear superposition of the characteristic harmonic modes. Guided by his research in the theory of the vibrating string, he came close to the idea of expanding functions as a series of sines and cosines, but it was left to Fourier to take the decisive step in this matter. He corresponded extensively with Euler.

Émile Borel (1871-1956). Émile Borel created two great monuments to his creative power. The first was his work on the theory of measure of sets of points. His great insight was to recognize that an effective theory of measure needs *countable additivity* and not finite additivity, as had been assumed hitherto. His point of view was later carried out to its logical end by Henri Lebesgue, the father of modern integration theory. Borel's ideas led him to the definition of Borel sets, which is the foundation for all work in modern functional analysis. His second great contribution was his creation of a systematic theory of divergent series. He was the first to do so, and his work conferred real legitimacy on Euler's work on divergent series done a century earlier. The concept of Borel summability has revived again in recent years, first in the asymptotic expansions of quantum field theory and then in the theory of multisummability, which is at the heart of recent work on the Hilbert problem of limit cycles. The French school of Ramis, Ecalle, and others has carried out deep generalizations of Borel's ideas.

Claude Chevalley (1909-1984). Chevalley was born in Johannesburg in Transvaal, South Africa, but eventually came to France for his studies in mathematics. He studied with Picard, Emil Artin, and Hasse. He came to the United States and was a professor at Columbia University from 1949 to 1957, when he returned to the University of Paris.

Chevalley is one of the seminal figures of modern mathematics. His impact through his original work, his ideas, and his books has been monumental. He introduced the concept of *idèles* in terms of which he recast class field theory in a famous paper in 1940. His point of view has thereafter been the preferred one in all subsequent treatments and uses of this theory. He created the theory of algebraic groups and carried out the classification of simple algebraic groups in all characteristics by direct group theoretic methods, which were new even in characteristic zero (complex numbers) because they did not rely on the Lie algebra (they could not, since the exponential map is a problematic tool in positive characteristic). This classification led, in his hands, to the construction of new finite simple groups and had a tremendous impact on the classification of finite simple groups. He worked on the foundations of algebraic geometry and was one of the pioneers in the theory of local rings. He, along with Serre, was mainly instrumental in reformulating algebraic geometry in a language similar to that of the modern theory of differentiable manifolds, thereby bringing geometric intuition back to algebraic geometry, something that had been lacking in the purely algebraic views of Zariski and Weil. His work in this area paved the way for Grothendieck and his schemes.

His books, such as *Theory of Lie Groups*; *The Algebraic Theory of Spinors* (the first to formulate the theory of spinors over arbitrary fields); *Algebraic Functions of One Variable*, which recast the Dedekind-Weber theory over any field and advanced it in a major way; *The Construction and Study of Certain Important Algebras*; his notes *Fondements de la Géométrie Algébrique*; and his Paris seminars, *Classification des Groupes de Lie Algebriques* (Séminaire C. Chevalley, 1956–58, Vols. 1 and 2) on algebraic groups and their classification, have educated nearly everyone who works on these subjects. He was the youngest member of the Bourbaki group when he joined.

He was somewhat of an iconoclastic figure and had absolute standards of writing that sometimes made it difficult to read his books. The precision of his writings was legendary. Occasionally he got carried away, as when he wrote, in the introduction to his book *Fundamental Concepts of Algebra*:

Algebra is an exercise in the rectitude of thought of which it is futile to disguise the austerity.

The theory of algebraic groups is permeated from top to bottom with his ideas and the objects he created: Chevalley basis, Chevalley groups, Chevalley isomorphism, and so on. The project of editing and publishing his *Collected Works* has begun, and when finished, it will be a substantial history of the modern theory of algebraic groups and their representations.

Richard Dedekind (1831-1916). Dedekind was a contemporary of Riemann and Dirichlet in Göttingen and spent many years there before going to Zürich. His contributions in mathematics to number theory, to real analysis, to algebraic geometry and so on were almost always at the foundational level. His theory of *cuts* made the real number system a rigorous object and laid the foundation of the modern theory of real variables. His theory of ideals in algebraic number fields extended Kummer's theory of ideals in cyclotomic fields to *all* number fields. His famous paper with Weber on the algebraic theory of Riemann surfaces provided a completely algebraic development of Riemann's theory and paved the way for the modern theory of function fields in all characteristics. He was in addition a great expositor and teacher, and hence his papers and lectures played a big role in the spread of the basic ideas in the many fields in which he worked. In fact, his theory of ideals in number fields appeared as a supplement to the fourth edition of Dirichlet's lectures on number theory, which he edited. This book, although referred to by him always as Dirichlet's lectures, was entirely written by him after Dirichlet's death and included in an appendix his theory of ideals. It was a landmark in the theory of algebraic number fields, remaining as the fundamental book in the subject for decades. His discoveries include the Dedekind zeta function of a number field and Dedekind domains.

Pierre Deligne (1944-). Pierre Deligne must be regarded as one of the greatest mathematicians of our era. Born and educated in Brussels, he was a prodigy and became a student of Grothendieck in Bures-sur-Yvette, going on to become one of the greatest algebraic geometers of his generation. He solved the famous conjectures of Weil on the zeta functions of smooth projective varieties over finite fields, and, as a consequence, the Ramanujan-Petersson conjecture on the coefficients of cusp forms such as the Ramanujan τ-function. He worked out the solution to the Riemann-Hilbert problem from an entirely new perspective and on arbitrary compact Riemann surfaces, and extended his solution to include higher-dimensional varieties and the monodromies of flat regular singular connections on them.

Deligne has been in the forefront of research in all aspects of geometry, topology, and arithmetic: Hodge theory, algebraic cycles, intersection homology, the Langlands program and its geometric incarnation, and so on. In recent years he has become deeply interested in the interplay of quantum field theory and string theory with geometry, both conventional and super.

His writings are characterized by exceptional clarity couched in the most economical terms, a combination that is Mozartian.

Paul Adrien Maurice Dirac (1902-1984). History already regards Dirac as one of the greatest physicists of all time. His place in the physics pantheon rivals that of Newton and Einstein. His approach to physics was characterized by its mathematical elegance and clarity, and its central feature was that the mathematics guided the physics (to summarize it inadequately). It has since then become known as the *Dirac mode* of discovery. It led him to the introduction of noncommutative algebra in quantum mechanics, to magnetic monopoles, and above all, to the discovery of *antimatter*, which is regarded by many physicists as the greatest discovery in physics in the 20th century.

Among mathematicians he became famous for his discovery and use of the Dirac delta function, which is the distribution representing a measure with all its mass concentrated at a single point; and for his discovery of the relativistic equation of the electron, based on the spin representations of Clifford algebras. He was a pioneer in the theory of infinite-dimensional representations of Lie groups and was the teacher of Harish-Chandra, perhaps the greatest master of this theory in our times. Dirac discovered so to speak the conformal group and its representations. Unhappy with the highly nonrigorous method of *renormalization* used by physicists in quantum electrodynamics (and since then in all parts of the theory of elementary particles and their interactions), Dirac spent a lifetime looking at the foundations of electrodynamics on arbitrary manifolds. Modern string theory, with its emphasis on mathematics, is an attempt in the Dirac mode to quantize the theory of gravitation and hence unify the fundamental forces.

Dirac's dictum was that a fundamental theory should have beauty. In this he was of the same opinion as Herman Weyl. Dirac was famously taciturn, expressing himself with almost painful conciseness. Bohr once described him as the physicist with the purest soul.

Lejeune Dirichlet (1805-1859). Dirichlet's name was really Peter Gustav Dirichlet; his grandfather was from the Belgian town of Richlet, which is responsible for his name. He was a great number theorist and an analyst. He was especially famous for the clarity of his thinking and his capacity to view problems always from that point of view which rendered the whole structure transparent. In number theory he solved the problem of infinitude of primes in arithmetic progressions which had resisted solution by everyone before him. It was in the course of solving this problem that he introduced the so-called Dirichlet series $L(s:\chi)$ corresponding to characters of the group of residue classes for a given modulus. These had a product representation similar to Euler's representation of the zeta function. Dirichlet analyzed their behavior near $s = 1$ and used it to deduce his epoch-making results on the density of primes in residue classes. Such Dirichlet series in their original form and their modern reincarnations due to Artin, Hecke, Weil, and Langlands have dominated number theory ever since. Even today they are called L-series, retaining Dirichlet's notation. He proved that if χ is a quadratic character, $L(1:\chi)$ is a positive number that was later identified as (up to a normalization) the *class number* of the quadratic subfield of the cyclotomic field defined by the modulus in question. This is the famous *class number formula*. The class number, which is the number of ideal classes modulo principal ideals, is a deep invariant of the algebraic number field, which reflects its departure from being a field obeying the classical form of unique factorization. Its computation as a special value of the L-series by Dirichlet was a major discovery.

In addition to number theory Dirichlet made profound contributions to analysis and mathematical physics. He was in some sense a protégé of Fourier, having attended the latter's lectures in Paris, and he was the first person to establish precise conditions on a function that would ensure the convergence of its Fourier series. He proved that the Fourier series of a periodic function which had a finite number of maxima and minima in a period interval converged to the function at any continuity point. His proof of this result introduced the famous Dirichlet kernel. His work in real analysis led him to the modern concept of a function as a rule that assigns a value to every point in the domain of the function. His work on potential theory led to the famous Dirichlet problem of finding a harmonic function on a domain with given boundary values. He was the first person to study rotating bodies and their equilibria, and his work inspired Riemann's fundamental paper on the subject, which in turn has led to enormous work in modern times.

Dirichlet was a close friend of Jacobi and took care of Jacobi's children after Jacobi's death. He was married to Rebecca Mendelssohn, who was a sister of the great composer, and so was a member of the artistic milieu of his times. He was also a mentor of Riemann when the latter was in Berlin.

Vladimir Gershonovich Drinfel'd (1954-). Vladimir Drinfel'd was born in Kharkov in 1954. He did his doctorate under Yuri Manin at Moscow University. He has made profound contributions in number theory, quantum groups, and mathematical physics. In number theory he proved in the 1970's the Langlands conjectures (for the case of $GL(2)$) for the function field of a smooth projective algebraic curve over a finite field and in the process discovered new objects (shtuka). His ideas were important in the solution by Lafforgue in the late 1990's of the full set of conjectures for $GL(n)$ for all n. Drinfel'd was awarded the Fields Medal for this work and his work on quantum groups. He was also the author of a famous work with Manin, which, together with the work of Atiyah and Hitchin, described the structure of self-dual instantons. He is currently at the University of Chicago, USA.

Leopold Fejér (1880-1959). Fejér made many fundamental contributions to Fourier series, potential theory, and approximation theory, but he is most famous for his result asserting that the Fourier series of a periodic continuous function is uniformly summable $(C, 1)$ to the function. Dirichlet's result on pointwise convergence required a more profound limitation on the function, such as being of bounded variation, and so Fejér's result attracted a lot of attention. He was only 20 when he proved this result. His basic observation was that the kernel, called the *Fejér kernel*, which governs the $(C, 1)$-summation of the Fourier series of a function, is always *positive*, unlike the Dirichlet kernel, which is highly oscillatory.

Richard Feynman (1918-1988). Feynman was a whiz kid when he was young and grew up to be one of the greatest theoretical physicists of his era—profound, original, and iconoclastic. His doctoral thesis at Princeton introduced a completely novel way of developing quantum mechanics, the so-called *path integral method*. Its essence was the idea that unlike classical mechanics where the paths are determined by a variational principle, in quantum theory *all* paths have to be taken into account. Each path comes associated with a complex number called its probability amplitude, and the total probability amplitude for an event is the sum or the integral over all possible paths; the actual probability is then the absolute square of this probability amplitude. Although originally formulated for nonrelativistic quantum mechanics, this idea has been made into a fundamental principle in all of quantum theory, including the quantum theory of fields. It is very difficult to formulate it in rigorous mathematical terms, but the formula, generalized to the setting of quantum field theory and referred to as the *sum over histories*, has proved to have remarkable predictive power. People like Witten have used the intuition coming from quantum field theory and the path integral formalism to predict remarkable results in topology and algebraic geometry, many of which have been verified subsequently.

Feynman had great physical intuition, and guided by it, he made conceptual mathematical leaps that were truly startling. Rigor per se was not important to him and was certainly abandoned when it stood in the way of conceptual progress. In this he bears a great resemblance to Euler, who often went far beyond conventional limits in the pursuit of ideas.

Feynman contributed to many parts of theoretical physics, such as quantum electrodynamics, weak interactions, superfluidity and superconductivity, and so on. He was awarded the Nobel Prize in 1963 jointly with Schwinger and Tomanaga for his work on quantum electrodynamics. He never lost his wry, irreverent approach to science and life. He delivered a famous course of lectures to undergraduates at the California Institute of Technology. These lectures were later made into a three-volume series of books. They have educated literally thousands of physicists and mathematicians since their publication.

Jean Baptiste Joseph Fourier (1768-1830). Scholar, creative mathematician, statesman, diplomat, and administrator, Fourier left an indelible mark on the mathematics and scientific life of his times. He was the governor of Nile during Napolean's conquest of Egypt. He was the secretary of the French Academy of Sciences for many years. His diplomatic skills enabled him to survive the many changes in the political landscape of his times in France.

Fourier was the creator of the theory of Fourier series and integrals. While it can be said (with some justification) that the idea of expanding functions as series of sines and cosines goes back to Daniel Bernoulli and the formulae for the Fourier coefficients was already in Euler's work, the theory of *Fourier integrals* is entirely his own. He created the theory of Fourier series and integrals as a tool for the study of problems of heat conduction, and his work was the first to reveal the depth of the concept of the heat equation. In doing this Fourier found that he had to have a new conception of a function, wherein a function could be given by different analytical expressions in different parts of its domain of definition. This was not appreciated by even the greatest mathematicians of his time, like Lagrange, and led to heated controversies that delayed the publication of his epoch-making *Theorié de analytique de la chaleur* by more than a decade, the book finally making its appearance as a publication of the French Academy of Sciences after he became its secretary.

It was in the course of developing the theory of Fourier series that Fourier gradually reached his view of a function as a more general object than hitherto conceived. Very simple problems of heat conduction in which different parts of the body under investigation are kept at different temperatures convinced him of the necessity of liberating the concept of a function from its classical limitations.

This change in point of view reached its logical end with the ideas of Dirichlet and Riemann. Contrary to popular impression, Fourier was quite aware of the subtleties that were demanded by his theory, such as, for instance, the fact that the Fourier series and integrals of functions with discontinuities in general converged only conditionally.

But what made his work so fundamental was the *inversion formula* for the Fourier integral. It paved the way for the Fourier transform to become a fundamental and powerful tool in all problems of analysis, differential equations, communication theory, and quantum theory. It is impossible to imagine life, at least in the linear world, without the Fourier transform.

Ferdinand Georg Frobenius (1849-1917). Frobenius is the founder of the modern theory of group representations, which he developed for finite groups. All the major results of the theory were obtained by him in a series of papers of astonishing creativity: the general theory of characters, the determination of the irreducible characters as well as the construction of the corresponding representations (for special groups such as the permutation groups and the linear groups over finite fields), the concept of induced representations, the Frobenius reciprocity theorem, and so on. But his mathematical range was even wider. He introduced the so-called Frobenius method of finding solutions to regular singular differential equations and worked on the theory of theta functions and on what are now called Frobenius matrices. The transformations of algebraic varieties induced by the map $x \longmapsto x^p$ in algebraic geometry over the prime characteristic $p > 0$ are named after him. He discovered the conditions for the integrability of a system of first-order linear partial differential equations, a result that is fundamental in the theory of Lie groups and also in the theory of multivalued solutions of linear ordinary differential equations in the complex domain. He spent his initial years as a professor in Zürich before moving to Berlin, where he succeeded his teacher Weierstrass to the mathematics chair.

The influence of Frobenius in representation theory is pervasive, even today. No matter what category of groups one works with – Lie groups over real, local, or finite fields, over adele rings, and more generally, super Lie groups – the notions of induced representations and characters are indispensable ingredients for any sustained development of the theory of representations in that category. His great successors – Weyl, Gel'fand, Mackey, Mautner, Bruhat, Harish-Chandra, and others – made many of their discoveries while attempting to make sense of his ideas in their contexts.

Carl Friedrich Gauss (1777-1855). One of the greatest mathematicians of all time, Gauss spanned the period between Euler and Riemann. A prodigy at an early age, he was unsure whether to choose mathematics or philology as the discipline in which to forge a career and decided in favor of mathematics, fortunately for our science, when he discovered the possibility of the construction of the regular 17-gon by ruler and compass at the age of 19. He wrote down in a very terse form his results in a diary of this time of his life, a perusal of which should fill anyone with wonder.

Gauss's monumental *Disquisitiones Arithmeticae* created number theory as an independent discipline in mathematics and set the standard for centuries to come. Here he proved the quadratic reciprocity law, discovered by Euler and incompletely established by Legendre. He was to discover many different proofs of it during his lifetime. He proved the fundamental theorem of algebra, again furnishing many different proofs for it. He discovered the notion of curvature for two-dimensional surfaces, the relationship between this local concept and global aspects, the double periodicity of the lemniscatic functions, and was the first mathematician to think and work comfortably in the complex domain. He also was the first to formulate the prime number theorem.

Gauss had a universal conception of mathematics and made many contributions to what we may call applied mathematics today: geodesy, magnetism, the method of least squares, which he created for astronomical purposes, the Gaussian distribution, and so on. He was, along with Bolyai and Lobachevsky, the co-discoverer of non-Euclidean geometry, although he never published his ideas, his work becoming known only after his death. From the letters of Betti to Tardy* one now knows that Riemann's idea of dissecting a Riemann surface of arbitrary genus to make it simply connected and even the discovery of combinatorial topology and what we now call the Betti groups had their origins in discussions of Riemann with Gauss (and of Riemann with Betti).

It is essential to understand Gauss in order to appreciate what Riemann and Dirichlet and everyone else did after him.

Israel Moisevitch Gel'fand (1913-). One of the greatest mathematicians of the 20th century, Gel'fand has shaped modern analysis by his ideas, discoveries, and contributions almost more than anyone else. He went to Moscow when he was 16 and worked on various odd jobs to take care of himself. He did not attempt to get a formal degree at this time, since he was already tutoring graduate students from the Department of Mathematics of the Moscow University!

Gel'fand wrote a famous paper with A. N. Kolmogorov on the ideals of the ring of continuous functions on a compact Hausdorff space. This was the starting point of the idea that a space is determined by the ring of functions on it, a theme that is so profound that it dominated and still dominates everything in geometry even today. He created his celebrated theory of *normed rings* or *Banach algebras* soon after his thesis, in which he was able to pursue this idea and represent a commutative Banach algebra as an algebra of functions on its spectrum. He realized that this representation, which is akin to the Fourier transform (and in fact reduces to it in a special case) and called the *Gel'fand transform* nowadays, could have a kernel the elements of which would be nilpotent in the sense of functional analysis and would have vanishing transform. Thus his theory was the forerunner to the Grothendieck theory, which did the same thing but in the context of *all* commutative rings.

The Gel'fand-Grothendieck point of view has been generalized in two directions. In the first, the algebras become supercommutative and lead to the concept of *superspace* which is localizable and has local coordinates which are both commuting and anticommuting. In the second instance the commuting algebras of global rings are deformed, leading to the current theory of quantized varieties and quantized groups, although such geometries do not appear to be localizable. Supergeometry has been studied intensively by physicists since the 1970's, while noncommutative

---

*See A. Weil, *Riemann, Betti, and the birth of topology*, Archive for the History of Exact Sciences, 21 (1979), 91-96; 21 (1979/80), 387.

geometry has been pursued in more recent times as a possible framework for grand unification.

Gel'fand was a pioneer in the theory of *unitary* representations of Lie groups, necessary for the theory of quantum particles and fields and important in its own right. He saw clearly that the heart of the problem of studying the unitary representations of Lie groups was in the *semisimple* Lie groups of Cartan and Weyl. He worked with a succession of collaborators such as Neumark, Graev, Kazhdan, Kirillov, Berezin, and others, and his work made it possible for Harish-Chandra eventually to take the subject to its proper level. He already saw in 1954 that the theory of distributions would play a major role in representation theory, a vision that was fulfilled with Harish-Chandra's epoch-making work in the 1960's. His 1952 paper with Fomin set the stage for the representation theoretic view of modular forms.

It is impossible to describe even casually the various areas he has made contributions to: Sturm-Liouville problems, formal Hamiltonian structures, hypergeometric systems in many variables, integral geometry, number theory and automorphic forms, the mathematics of biological systems, to name just a few. He conducted a legendary seminar in Moscow University for decades till he left Russia to settle down in the United States. Almost all great Russian mathematicians of the modern era have been members of this seminar, and they all attest to his universal and inspiring presence in that seminar.

Roger Godement (1921-). Roger Godement did his doctorate under Henri Cartan in Paris. He was a pioneer in the theory of infinite-dimensional representations of Lie groups and in the representation theoretic aspects of the theory of automorphic forms. His work on the L-functions of automorphic representations of GL(2) and GL(n) and the group of units of simple central algebras with Jacquet (who was his student) was very influential in the formulation of the Langlands conjectures.

Christian Goldbach (1690-1764). A lifelong friend and patron of Euler, Goldbach played a very important role in the development of Euler's mathematical interests during the latter's first stay in St. Petersburg. During the course of his life an enormous number of letters were exchanged between them. These letters make fascinating reading and give a deep glimpse into the formation of Euler's ideas in various subjects.

Goldbach was himself a very talented mathematical amateur and is known today for a conjecture he made, namely that every even number greater than 2 is a sum of two odd primes. This conjecture remains unproved, although much progress has been made, especially by Hardy-Littlewood and Vinogradov.

Alexander Grothendieck (1913-). Born in Berlin, Grothendieck moved to France in 1941 and has remained there more or less ever since. His father was Russian. He did his doctoral thesis under L. Schwartz and was a member of the Bourbaki group for a few years.

In his doctoral thesis *Produits tensoriels topologiques et spaces nucléaires*, he demonstrated his ability to think in ways that were completely abstract but penetrating nevertheless. Then in the years 1959-70 he was the leader of a school in modern algebraic geometry at the IHES in Bures-sur-Yvette which revolutionized algebraic geometry and established the language, format, and context in which algebraic geometry would henceforth be studied. Ideas poured forth from him at an incredible pace during these years. The only thing more remarkable than this achievement was the prodigal generosity with which he shared these ideas with the members of his group. Even a mere enumeration of the concepts and ideas he has created strains belief and makes us wonder how a single mind conceived all of it in such a short span of time. His discovery of schemes and of toposes may be described as one of the most profound in the long history of the evolution of the idea of space.

It was by building on Grothendieck's framework and methods that Pierre Deligne found the solution to the Weil conjectures. All major advances in algebraic geometry since then have been made within the edifice of ideas that Grothendieck created. This framework that united geometry and arithmetic will almost certainly be the one in which the central conjectures of arithmetic and geometry will be proved. To modify Hermann Weyl's remark about Hilbert, no person of equal or even comparable stature in algebraic geometry has arisen in the 20th century.

He was strongly pacifist and resigned from the IHES over its policies of accepting grants from military institutions. He lives in semiretirement near Montpellier, France. For a moving and compelling portrait of Grothendieck, both as a man and as a mathematician, see * below.

Godfrey Harold Hardy (1877-1947). The English mathematician *par excellence* of the 20th century, Hardy was one of the greatest analysts of his time. His lifelong collaboration with Littlewood is perhaps the longest and most productive joint effort in the history of mathematics. Their discovery of the so-called circle method, which evolved out of his joint work with Ramanujan, led to the solution of some of the most difficult asymptotic questions in analytic number theory.

However Hardy will perhaps be most remembered for his discovery of Ramanujan and his subsequent collaboration with him. The story, one of the most famous in mathematics, has been told by Hardy himself in his famous book on Ramanujan.** He alluded to his association with Ramanujan as the single romantic incident in his life. His paper on the asymptotics of the number of partitions of a large positive number, written with Ramanujan, is perhaps the most beautiful of all of his work.

He wrote with exceptional clarity and was almost singlehandedly responsible for the modernization of the English mathematical curriculum as well as the famous Cambridge Tripos examinations. He wrote beautiful books, like the *Divergent Series* and *A Mathematician's Apology*. He had many eccentricities but made no effort to hide them. He was fond of cricket and used its personalities in amusing rankings of mathematicians. For example, when he said that a certain mathematician was of the Bradman class, it meant that he was in the highest class. He was a self-professed atheist, but his beliefs were his own, never intrusive, as evidenced by his happy collaboration with the deeply religious Ramanujan. In discussing these aspects, he was open and the ultimate perfect English gentleman.

Harish-Chandra (1923-1983). Without any doubt Harish-Chandra and Ramanujan are the greatest of the mathematicians that India has produced in modern times. Although Harish-Chandra lived almost all of his mature life in the United States,

---

*Allyn Jackson, *Comme Appelé du Néant–As if summoned from the void*: *The life of Alexandre Grothendieck*, in Notices of the AMS, Vol. 51, No. 9, 1038-1056; Vol. 51, No. 10, 1196-1212.
  **Ramanujan*, Chelsea, 1940.

his basic training was in India. He started out as a theoretical physicist but became a mathematician because he felt more at home in the rigor and internal esthetics of mathematics.

His major interest was the theory of representations of Lie groups. In this he was the greatest master since Hermann Weyl. Indeed he did to infinite-dimensional unitary representations of all semisimple Lie groups what Weyl did to the finite-dimensional representations of compact Lie groups. He extended in a novel and profound manner the concept of the character to apply to *infinite-dimensional* irreducible representations of a semisimple Lie group, in particular to infinite-dimensional irreducible unitary representations. Nowadays it is called the *Harish-Chandra character*. His theory of characters led him to a formula for the irreducible unitary characters which rivals Weyl's formula in beauty, simplicity, and impact. In later years he became deeply interested in the arithmetical aspects of representation theory and established, in joint work with Borel, the fundamental theorems of arithmetically defined homogeneous spaces.

His conception of representation theory was revolutionary for his time: the representations themselves were in the background and were studied with the help of analysis and differential equations on the group manifold. Along the way he singlehandedly created nonabelian Fourier analysis. If there is one feature that distinguishes nonabelian Fourier analysis from its abelian counterpart, it is the concurrent appearance of both discrete and continuous spectra in the harmonic analysis of functions and distributions on semisimple Lie groups and their homogeneous spaces, and Harish-Chandra discovered the central principles of this analysis. His work paved the way for the emergence of Langlands's great program. If one were asked to choose the three greatest figures in Fourier analysis, one must go with Fourier himself, Wiener, and Harish-Chandra.

He worked at Columbia University for a number of years before moving in the early 1960's to Princeton as a permanent member of the Institute for Advanced Study. He lived in Princeton for the rest of his life. He died in 1983, a few months before a conference in Princeton that was supposed to celebrate his 60th birthday. As in the case of Pauli, the birthday conference became a memorial conference. His *Collected Papers*, which have been published by Springer, would make anyone marvel disbelievingly at the prodigious powers and insight of this man who achieved greatness by a solo effort in an era where collaboration was the norm.[*]

Helmut Hasse (1898-1979). Hasse was a student of Hensel in Marburg, Germany, in 1920. His work was of fundamental importance in shaping the modern arithmetic of abelian extensions of algebraic number fields, hence of class field theory. He discovered the *Hasse Principle*, of which one example is that a quadratic form with rational coefficients represents 0 rationally if and only if it does so over the field of real numbers and over all p-adic fields. This *local-global* principle has been the cornerstone of all advances in number theory since then. He also proved fundamental results in the theory of central simple algebras over local fields, thus in the theory of the Brauer groups over such fields. For such algebras over a global field like the rationals he formulated the Hasse principle as a product formula describing

---

[*] *Harish-Chandra Collected Papers*, V. S. Varadarajan (Ed.), Vols. I-IV, Springer-Verlag, 1984. See also *Harish-Chandra, 1923-1983*, by R. P. Langlands in the Biographical Memoirs of Fellows of the Royal Society, 31 (1985), 199-225. In addition, see the review by Langlands of the *Collected Papers of Harish-Chandra*, Bull. London Math. Soc., 17 (1985), 175-179.

the global Brauer group in terms of the local Brauer groups, a result equivalent to the Artin reciprocity law. He was thus a great pioneer in abelian field theory.

Hasse was the successor to Hensel in Marburg and later of Weyl in Göttingen. His political attitudes during the Nazi era have generated controversy, but his devotion to mathematics has not been seriously questioned. He went to the Humboldt University in East Berlin in 1949 and had several books on number theory published after that. All in all, he must be regarded as one of the masters of modern number theory.

Erich Hecke (1887-1947). Hecke was a student of Hilbert and in his dissertation studied, at Hilbert's suggestion, modular forms in two variables. Modular forms became a lifelong concern for him, and he made profound contributions to the theory of Dirichlet series and modular forms in two remarkable ways. In the first, he enlarged the class of characters to which one could associate an Euler product and introduced for this purpose the *Grössencharakter*. Grössencharakters are the characters of the idele class group, and those of finite order may be identified with the generalizations of the Dirichlet characters to the context of an arbitrary number field. He proved the analytic continuation and functional equation for the Dirichlet series associated to these. The second contribution was even more original. He discovered, by a deep generalization of Mordell's method, that a classical modular form can be acted upon by what are now called *Hecke operators* and that if the form is an eigenform with respect to these, one could associate a Dirichlet series and Euler product to it. Again these would possess analytic continuation and satisfy a functional equation. Eventually these two types of Dirichlet series with Euler products would be unified under Langlands's scheme.

Hamburg, with Artin and Hecke, was a great place to be for a number theorist in those years. It is interesting that the objects that were so important to them, namely Artin *L*-functions and Hecke Euler products for modular forms, would be related eventually as parts of the same picture, namely the Langlands version of a higher-dimensional reciprocity law, but this was not obvious at that time, and so there was very little scientific resonance between them.

Kurt Hensel (1861-1941). Hensel was one of a great succession of number theorists from Germany. He was a student at Berlin and studied under the guidance of the Berlin triumvirate of Kummer, Kronecker, and Weierstrass. He was especially inspired by Kronecker's vision of algebraic number theory. It was at this time that Weierstrass had been stressing the concept of power series developments as a foundation for complex function theory. In Hensel's mind a profound unification of the Weierstrassian and Kroneckerian views took place, and he discovered during this process the idea of a field with valuation. In essence the Weierstrassian idea of local power series expansion became, in Hensel's hands, the *p*-adic expansion of a rational number, the local aspect coming from the fact that one is operating with a fixed prime. This analogy between function fields and number fields, developed by Hensel, would lie at the heart of Hilbert's view of class field theory and reciprocity laws. If the function field was taken over a *finite field*, the analogy became even closer, and the theories became parallel. This was Artin's theme, and in modern times it has bloomed into an algebraic geometric version of the Langlands program.

The line of succession
$$Kummer \rightarrow Kronecker \rightarrow Hensel \rightarrow Hasse \rightarrow Witt$$
is one of the most remarkable in the history of mathematics. It was the main force in the development of number theory in Germany in the 19th century.

David Hilbert (1862-1943). Hilbert was the greatest master of mathematics of his generation and a true universalist. He did his doctorate under Lindemann in Königsberg. His first appointment was in Königsberg, and after some years he moved to Göttingen, where he held the chair till he retired.

To trace the mathematical trajectory of Hilbert is to trace the development of mathematics in the 19th century. His mathematical achievements were summarized by Weyl in the obituary he wrote for Hilbert,[*] one that has become the standard for such writings. Hilbert achieved his universality by concentrating on one area for a length of time and then moving on to another area. These changes of interest are quite abrupt and hard to explain. Even the great obituary of Weyl does not give any clue to the true creative processes within Hilbert's mind.

His first achievement was in classical invariant theory, where he proved the central finiteness theorems: the finite generation of invariants for the classical groups. In the process he built the foundations of the modern theory of polynomial algebras and their ideals, from which he could develop algebraic geometry in a proper rigorous manner. His construction of the moduli space for group actions would later be put in its proper perspective by Mumford when the latter discovered what is now called the Hilbert-Mumford compactification of the orbit space. This was also the time in which Hilbert developed the theory of syzygies that has now become the homological theory of commutative algebras.

From invariant theory and polynomial algebras Hilbert went to number theory when he began writing an encyclopedia article for the German mathematical society on the status of number theory. This article was a landmark in the subject for decades and influenced arithmetic research profoundly. In it he unified the work of Euler, Gauss, Dirichlet, Kummer, Eisenstein, Kronecker, and others under the framework of abelian extensions of number fields and asked for an extension of the Kronecker Jugendtraum to the case of abelian extensions of an *arbitrary* number field. As an example he formulated the theory of quadratic extensions of arbitrary number fields. His analogy between function fields and number fields led him to formulate the quadratic reciprocity as a product formula for the *Hilbert symbol* and to his formulations of even more general reciprocity laws. Eventually Artin and Hasse would establish the general reciprocity laws in the Hilbertian format, namely as product formulae.

From number theory it was to foundations of geometry that Hilbert turned his attention. His contributions were collected in books on the subject that he published. In his characteristic way he clarified the axiomatic structure of euclidean and noneuclidean geometries.

His next interest was in the spectral theory of bounded self adjoint operators, where he introduced the most profound notion of a projection valued measure (same as the resolution of the identity) and lifted spectral theory beyond its prison of Fredholm and compact operators to include *all bounded self adjoint operators.*

---

[*] *David Hilbert and his mathematical work*, Bull. Amer. Math. Soc., 50 (1944), 612-654; Gesammelte Abhandlungen, Band IV, 130-172.

The spectral theory would then go on to reach its full flowering with the work of von Neumann on unbounded operators a few years later. In addition he studied the spectral theory (by the method of parametrix) of operators such as the Laplacian on compact manifolds like the sphere and began a theory that would prove to be one of the most fertile in 20th-century mathematics.

From spectral theory Hilbert went to physics. In physics his work was climaxed by his variational derivation of the Einstein equations of general relativity. In spite of the fact that he arrived at them before Einstein, he always expressed the view that the real achievement was that of Einstein and his profound physical understanding and that, he, Hilbert, merely supplied the mathematics. This attitude is at remarkable variance with many contemporary mathematicians who think that their job is to discover the functor in the sky that would solve all the problems of physics.

All of this took place before the First World War, and Hilbert's postwar concern was an attack on the very foundations of mathematics by Brouwer and his concepts of intuitionistic mathematics. Hilbert was, in the words of Weyl, the *champion of axiomatics* and rose to the defense of the free-flowing methods of Cantorian set theory as a foundation for all of mathematics. *No one will banish us from the paradise created by Cantor*, he said. Unfortunately he did not succeed in this effort. Indeed, we now know that he could not have succeeded. The story has become much more complicated with the discoveries of Gödel, von Neumann, and, in recent times, Cohen and others. In the words of Weyl, the boundary between what is true and what is provable has once again become vague.

At the Paris International Congress of Mathematicians in 1900 Hilbert delivered the opening address, entitled *Mathematical problems.* He formulated twenty-three problems in all fields of mathematics as deserving the attention of mathematicians. Each one of them has proved to be profound, and the solutions have typically greatly advanced our knowledge of the subject.*

No one could disagree with Weyl's remark on Hilbert that *no mathematician of equal stature has risen from our generation.*

Carl Gustav Jacob Jacobi (1804-1851). Jacobi was a formalist surpassed perhaps only by Euler, and later perhaps only by Ramanujan; an innovator of new methods and ideas in the theory of elliptic functions; a man who made profound advances in the calculus of variations, such as the theory of what are now called Jacobi fields; whose conception of analytical mechanics changed the course of the subject; and who began the study of curves of higher genus. One can go on and on and still not be finished with Jacobi's achievements as a mathematician. His was a universal mind to which was coupled a technical power of formidable proportions. Concepts like the Jacobian of a curve, the Jacobian of a transformation of coordinates, Jacobi sums, Jacobi symbol, Jacobi theta functions, Hamilton-Jacobi equations, Jacobi fields, Jacobi identity in a Lie algebra, and so on are evidence of the diversity and complexity of Jacobi's intellectual horizons.

He was a very close friend of Dirichlet. Dirichlet's wife was a sister of the great composer Felix Mendelssohn, and so Jacobi was a member of the artistic and scientific milieu of his time. He used to go to Dirichlet's house, and the two would

---

Mathematical Developments Arising from Hilbert's Problems, Proceedings of Symposia in Pure Mathematics, XXVIII, Amer. Math. Soc., Providence, RI, 1976.

sit for long periods of time, not uttering a single word. Jacobi used to say that these *mathematical silences* stimulated him!

He suffered from diabetes and had problems getting proper medical attention. He had great difficulty raising money to go to sunny Italy, where the better climate would have helped him. He died early, and his children were adopted by Dirichlet and brought up by him.

Hervé Jacquet (1939-). Jacquet was a student of Godement and came to Princeton in the late 1960's to work with Langlands on the representation theoretic side of automorphic forms and their L-functions.* It was during this stay that he wrote the monumental work *Automorphic Forms on* $\mathrm{GL}(2)$ with Langlands, in which they advanced the suggestion that the Artin L-functions are to be found among the automorphic L-functions. I quote from their introduction:

> *Two of the best known of Hecke's achievements are his theory of L-functions with grössencharakter, which are Dirichlet series which can be represented by Euler products, and his theory of the Euler products associated to automorphic forms on* $\mathrm{GL}(2)$. *Since a grössencharakter is an automorphic form on* $\mathrm{GL}(1)$, *one is tempted to ask if the Euler products associated to automorphic forms on* $\mathrm{GL}(2)$ *play a role in the theory of numbers similar to that played by the L-functions with grössencharakter. In particular do they bear the same relation to the Artin L-functions associated to two-dimensional representations of the Galois group as the Hecke L-functions bear to the Artin L-functions associated to one-dimensional representations? Although we cannot answer the question definitively one of the principal purposes of these notes is to provide some evidence that the answer is affirmative....*

After this he went on to make fundamental contributions to the subject. He has been one of the leaders of the subject for many decades.

Mark Kac (1914-1984). Mark Kac, an immigrant from Poland to the United States, was one of the great exponents of probability theory and statistical mechanics in modern times. His discovery, jointly with Erdös, of what is called the *invariance principle* was important because in the attempt to understand it the methods of probability in function space were developed. But he is most widely known for what is called the *Feynman-Kac formula.* In quantum theory, for instance of a single particle moving on a line, the dynamical operator that allows us to go from the state at time $t = 0$ to the state at time $t = T$ is an integral operator, and its kernel is the so-called *propagator.* Feynman had derived a heuristic formula

---

*It was in 1968 also that I was first introduced to the subject. I was at Princeton as a Sloan fellow and distinctly remember two conversations. The first, with Weil, was rather brief. He advised me, in view of my knowledge of the representation theory of semisimple groups, to study the work of Jacquet-Langlands and added with a twinkle that it is a heavy task. It was also during this meeting that he gave me reprints of his papers in the *Acta* on Siegel's work, signed with best wishes! The second was a long conversation with Haris-Chandra, conducted in the parking lot near Building C, when we were about to go home, in which he spoke admiringly of the work of Jacquet and Langlands and emphasized that "they think they have Artin's nonabelian L-functions." The two conversations made a deep impression on me. It was during this latter conversation that Harish-Chandra advised me to study the theory of elliptic curves.

for it as an integral over the path space and based his entire view of quantum theory on this formula. Unfortunately, the measure on path space, with respect to which the path integration has to be done, does not exist, and it was very difficult to get a rigorous understanding of what Feynman's formula meant and how it could be used. Kac changed the situation completely when he realized that by taking *imaginary time* he could work with *Wiener measure*, which is a legitimate object on path space, and then get back to actual time by analytic continuation. It was a great achievement, one of which he was justifiably proud. Later on, Kac's idea would be extended by Schwinger when he formulated quantum field theory in the euclidean (as opposed to the Minkowskian) setting, on a spacetime (*Schwinger spacetime*) where the time coordinate is imaginary. The rigorous construction of quantum field theories in modern times utilizes the euclidean framework and obtains results for Minkowski spacetime by analytic continuation from the euclidean picture.

Kac had a great personality: congenial, deeply understanding, and forever curious about the world around him. He made an impact on his scientific times that his family, friends, and admirers can never forget.

Andrei Nikolaevich Kolmogorov (1903-1987). The creator of modern probability theory, the teacher of some of the greatest Russian mathematicians including Gel'fand, a mathematician as interested in teaching as in discovering new results, a master of classical analysis and its modern offshoots like dynamical systems and ergodic theory, and one of the pioneers of complexity theory and theory of information and automata, Kolmogorov must be regarded as perhaps the greatest mathematician from Russia in the first half of the 20th century. His axiomatic construction of probability theory as a branch of measure theory, as contained in his masterful but terse 1933 monograph, is still the framework used in that subject. He was a prodigy, as is obvious from the fact that he discovered when he was barely 20 that there are L^1-functions whose Fourier series diverge everywhere. He discovered the relation between Markov processes and diffusion equations, created the theory of stationary stochastic processes, and was very influential in Gel'fand's initial work on unitary representations of noncompact Lie groups. In the years 1950-60 he had a second burst of creativity when he discovered the mathematical meaning of entropy of an abstract dynamical system, using it to settle open problems in the ergodic theory of such systems and in proving that a generic Hamiltonian system is stable around a fixed point, a result which initiated the so-called KAM theory of dynamical systems. His students form a veritable who's who of Russian mathematics.[*]

Leopold Kronecker (1823-1891). *God created the integers, and everything else was made by Man.* This famous remark attributed to Kronecker might have hindered modern appreciation of his genius and the impact of his ideas in number theory. Although he wrote his thesis under Dirichlet, he had earlier been a student of Kummer in the gymnasium and was a close friend of Kummer for the rest of his life. In later years he became a champion of algorithmic methods in mathematics and so was a bitter opponent of works in which the methods of set theory were used with great freedom, such as those of Cantor and Dedekind, and even Weierstrass. This led to many conflicts of personality while he was in Berlin. In modern

---

[*]For a deeply moving and personal view of Kolmogorov's life and times, see *Kolmogorov in Perspective*, History of Mathematics, Vol. 20, Amer. Math. Soc., 2000.

times, his philosophy would find resonance in the intuitionism of Brouwer and Weyl. With the development of computers the attractiveness of algorithmic methods has increased, and the Kroneckerian view has emerged again to the forefront.

In number theory his work is of extraordinary depth and significance even today. He extended (along with Dedekind) Kummer's theory of cyclotomic ideals to arbitrary algebraic number fields (and even to much more general fields). He formulated the *Jugendtraum*, which is one of the most beautiful parts of arithmetic, namely the explicit description of the abelian extensions of a quadratic number field, a program to which he made basic contributions. However it was completed only later by Weber, Takagi, and others. It was this work of Kronecker that led Hilbert to formulate his 12th problem, which asks for similar explicit constructions of the abelian extensions of arbitrary algebraic number fields.

Ernst Eduard Kummer (1810-1893). Kummer is one of the founding figures of modern number theory. He created the ideal theory of cyclotomic fields, an achievement without which algebraic number theory as we understand it now would not exist. One of his reasons was to restore unique factorization to numbers, and for this purpose he introduced his ideal numbers, which are the prime divisors or ideals in the modern view. He made fundamental contributions in the further development of number theory, such as reciprocity laws, ℓ-adic analysis, class numbers, and so on. But his genius was not confined to number theory alone. He made discoveries in geometry, such as the Kummer surface.

Kummer was a great teacher, and in his early career when he was teaching in a gymnasium he had Kronecker and Joachimsthal as his students. Later, when he went to Berlin, his stay was a period of great prestige and growth for the Berlin mathematical school, because Kummer, Kronecker, and Weierstrass were together as colleagues for many years. Eventually the relations between these three became strained because of Kronecker's clashes with Weierstrass, clashes that arose because of Kronecker's inability to maintain cordial relations with those whose mathematical personas and views differed from his own.

Although Kummer did not care to establish ideal theory for all number fields, one should not forget that the steps he took to do it for cyclotomic fields were revolutionary and went far beyond anything that Gauss and Dirichlet had imagined or dared to do.

Laurent Lafforgue (1966-). Laurent Lafforgue was born in France in 1966. He did his doctorate under Gérard Laumon at the University of Paris in 1994. He did his great work on the Langlands conjectures at the Orsay campus of the University of Paris. He has been a permanent member of the IHES, Bures-sur-Yvette, since 2000. He established the Langlands conjectures for $\text{GL}(n)$ for a function field of a smooth projective curve over a finite field, which included the conjectures of Ramanujan-Petersson and Deligne. He was awarded the Fields Medal for this achievement in 2002.

Joseph Louis Lagrange (1736-1813). Lagrange was the successor to Euler in the Berlin Academy when Euler left for his second and final stay in St. Petersburg. He had already achieved widespread fame with his work on the calculus of variations. Indeed, when he wrote to Euler outlining his ideas and their consequences for the problems of maxima and minima involving families of curves, Euler saw at once the overwhelming superiority of the methods of Lagrange and from then on adopted them as the preferred methods in this subject. His work on dynamical systems based on the action principle applied to a Lagrangian was a great unifying force in mechanics. The ease with which it can be generalized to higher dimensions has made it the method of choice in the classical theory of fields and indeed even in the quantum theory of fields. Modern physics without Lagrangians is unthinkable.

Lagrange was a simple man who wanted nothing but to be left alone to do his mathematics in peace. He was born in Turin, Italy, but spent the major part of his career first in Berlin and then in Paris. Euler treated him as his equal in number theory, mechanics, and the calculus of variations. Lagrange's proof that every quadratic irrational has a periodic continued fraction justified for the first time the ancient Indian solution (by the *chakravala* method) to the Pell's equation.

Robert Langlands (1936-). A Canadian by birth, Robert Langlands lives in Princeton, NJ, and is a permanent member of the Institute for Advanced Study. He did his doctoral degree at Yale under Einar Hille. His early work on semigroups of operators, natural for a student of Hille, contained no hint of the explosive and epoch-making discoveries he was due to make in a few years.

The *Langlands philosophy*, or the *Langlands program*, which he created, is one of the most remarkable mathematical programs of the second half of the 20th century. It has excited and captured the attention of a whole generation of mathematicians. Its ramifications unite Galois theory, automorphic forms, representation theory, and algebraic geometry in a remarkable manner. In its simplest form it is an audacious generalization of class field theory, conjecturing that there is a correspondence between n-dimensional representations of the Galois groups of the extensions of a given algebraic number field, say k, with the *infinite-dimensional irreducible* unitary representations of the group $GL(n, A_k)$ that occur in the space $L^2(GL(n, A_k)/GL(n, k))$, A_k being the adele ring of k. These representations are called *automorphic* because of the circumstance that when $n = 2$ they are generated mostly by automorphic forms of one kind or the other. However the program does not stop here but brings in the theory of semisimple Lie groups and algebraic geometry. It has formulations also when algebraic number fields are replaced by algebraic function fields in one variable over a finite field, namely function fields of Riemann surfaces defined over finite fields, and even when the finite fields are replaced by the complex numbers (geometric Langlands program).

He established by himself and with collaborators like Jacquet and Labesse many fundamental facts of this program which lent credence to its validity. There is no doubt that progress in number theory will ultimately depend on the successful completion of this program.

Adrien-Marie Legendre (1752-1833). Legendre's main contributions were in foundations of geometry, planetary dynamics and dynamical systems in general, calculus of variations, number theory, and elliptic functions and elliptic integrals. In number theory he attempted to prove without success the law of quadratic reciprocity, which was eventually proved by Gauss, although the modern formulation is Legendre's. His three-volume treatise on elliptic functions and integrals was a monumental enterprise and codified the theory completely. However the works of Jacobi and Abel took the subject to new levels not reached by Legendre's opus. In geometry he was able to recognize that the parallel postulate is equivalent to the fact that the sum of the three angles of a triangle is two right angles (in fact, he showed that if this is true for one triangle, then it is true for all), which brings out the basic fact that euclidean geometry is characterized by *flatness*, a fact that did not become clear till after Riemann's work. In the calculus of variations he introduced the idea of the second variation.

In spite of his many fundamental contributions, Legendre is the only one of the major mathematical figures of his time whose complete works have not been published so far.

John Edensor Littlewood (1885-1977). The inseparable mathematical companion of Hardy, Littlewood nevertheless was a major mathematician in his own right. His work on the stability aspects of nonlinear differential equations is quite well known. Still he is most known and appreciated for the great series of papers he wrote with Hardy, including the famous series in which the circle method was applied to some of the most difficult problems in analytic number theory. He wrote a delightful book, *A Mathematician's Miscellany*, which is the standard against which mathematical books that claim to entertain and edify are judged. In that book are autobiographical remarks, some details on the circumstances in which he got the idea that would be pivotal in his Tauberian theorem, and details of the Hardy-Ramanujan collaboration on their famous paper on the asymptotics of the number of partitions of a large number.

When Ramanujan wrote his now famous letter to Hardy outlining some of his discoveries, it was to Littlewood that Hardy turned to help him evaluate that letter. What happened after that is history.[*]

Srinivasa Ramanujan (1887-1920). Perhaps the single most romantic figure in the history of mathematics, Ramanujan was a self-taught genius whose formal skills and results rival those of Euler and Jacobi. His story is well known and has been told well by many, including Hardy.[**] In lectures delivered at the Institute for Advanced Study and later published as a book, Hardy gives a masterful discussion of Ramanujan's work. Born in obscurity and in a milieu which did not understand his special genius, Ramanujan nevertheless came up with extraordinary results, rediscovering for himself entire classical theories, such as the theory of elliptic functions. He filled several notebooks with his results, giving no proofs, because he modeled his early writings on the book of G. S. Carr entitled *Synopsis of Elementary Results in Pure Mathematics*, which contained only statements and no proofs. Bruce Berndt has now edited several of these notebooks and supplied proofs to most of the assertions contained in them.

Ramanujan's power of calculation, feeling for form, and extraordinary insight into properties of numbers propelled him to some of the most remarkable discoveries. He conjectured the multiplicativity of the τ-function and the Euler product for its Mellin transform (proved later by Mordell); he made the famous Ramanujan conjectures on the bounds for $\tau(p)$ for p prime; and he obtained some of the most remarkable congruence properties of the partition function, to name just a few of

---

[*]See the letter of Lord Bertrand Russell to Lady Ottoline Morrell on page 44 in *Ramanujan, Letters and Commentary*, Bruce C. Berndt and Robert A. Rankin (Eds.), History of Mathematics, Vol. 9, Amer. Math. Soc., 1995.

[**]See footnotes in the sections on Hardy and Littlewood.

his achievements. His conjectures on the bounds for $\tau(p)$ were eventually proved by Deligne as a consequence of the Weil conjectures.

There is no figure in the universe of mathematics who is even remotely like Ramanujan. He was surely one of a kind.

George Friedrich Bernhard Riemann (1826-1866). Universally regarded as the greatest mathematician since Gauss, Riemann created so many new concepts and even areas of mathematics that a large part of the mathematics of the 20th century consisted in developing the Riemannian themes. Among these are Riemannian geometry, Riemann surfaces and their moduli spaces, complex tori and theta functions, regular singular differential equations and their monodromy, trigonometric series, shock waves, rotating ellipsoids, and so on.

He was the first to treat the zeta function as a function of a complex variable, and he proved its analytic continuation and functional equation. He realized that the distribution of zeros of the zeta function was critical in the theory of primes. He is most famous for his conjecture, the *Riemann hypothesis*, that asserts that all the nontrivial zeros of the Riemann zeta function $\zeta(s)$ lie on the critical line $\Re(s) = 1/2$. Computations of millions of zeros have been performed, and all of them have been found to be on the critical line.

In his famous habilitation address in 1854, Riemann speculated on the structure of the space we live in. He advocated the revolutionary hypothesis that the metric of space depended on the matter in it. His vision was later vindicated by Einstein's general theory of relativity. Even more intriguing, Riemann suggested that in infinitely small regions, space need not look like a manifold. Recent discoveries by physicists suggest that in the Planck scale $(10^{-33}$ cm) spacetime is foamlike and is subject to quantum fluctuations, and its structure can be understood only on the basis of more radical models. It is believed that even at somewhat larger distances, spacetime is a *supermanifold*, i.e., a manifold on which one has noncommuting (actually anticommuting) coordinates as well as the usual commuting ones; the anticommuting coordinates reflect the fermionic structure of matter. Thus Riemann's vision has been pushed forward with remarkable consequences. The differential geometry of supermanifolds is flexible enough that under quantization it appears capable of supporting a unified field theory, the holy grail of modern physics.

Laurent Schwartz (1915-2003). Universally known for his creation of the theory of distributions, Laurent Schwartz was a seminal figure in the mathematical world of France till he died. Although many of the ideas underlying the theory of distributions were created by others, such as Heaviside, Mikusinski, Sobolev, Bochner, Weyl, and so on, Schwartz performed the great synthesis, which, together with his own discoveries, made the theory of distributions into a most flexible tool in linear analysis. The modern theory of partial differential equations would not have been possible without it. Distributions play a fundamental role in the theory of representations of Lie groups, as was shown by Bruhat, Gel'fand, and above all, by Harish-Chandra. In his autobiography he discusses the circumstances surrounding his discoveries in this theory.

Schwartz was a great teacher, and he had many students who later became major mathematicians in their own right: Malgrange, Lions, Bruhat, Grothendieck, and others.

Politically he was a liberal and supported causes that were quite extreme in many instances. In so doing he became controversial. His stance against the war in Vietnam made him unwelcome in political circles in the United States, but he did not change his views. He visited many developing countries, and his lectures there had great impact, especially in India.

He loved nature and liked to be a part of it. His hobby was collecting butterflies.

Teiji Takagi (1875-1960). Takagi is one of the central figures in class field theory and one of the most influential mathematicians in Japan in the modern era. After his education in Japan he went to Göttingen, but found that Hilbert was no longer interested in number theory but in spectral theory. After he returned to Japan he proved the fundamental theorems of class field theory. He generalized the Hilbert class field and constructed the class fields corresponding to arbitrary ideal class groups.

He wrote many books that were responsible for the mathematical education of a generation of Japanese mathematicians.

John Tate (1925-). Tate was a student of Artin and did his doctorate under him. His subject was harmonic analysis on the adele ring, and he defined general local and global zeta functions, which reduce to the Hecke L-functions when specialized suitably. His thesis is beautifully written and was extremely influential. He also explicitly computed the symbol in local class field theory. His work on p-adic elliptic functions and his pioneering theory of rigid analytic spaces mark him as a major contributor to modern number theory. He is currently at the University of Texas at Austin.

George Neville Watson (1886-1965). Watson was the first person to make a serious study of Ramanujan's notebooks and publish proofs of many of the results found therein. He was famous for his book on analysis with Whittaker and his monumental treatise on Bessel functions. He proved a fundamental lemma in the theory of Borel summation which has proved to be very crucial in that theory and its generalizations.

Heinrich Weber (1842-1913). Weber was a mathematician of broad interests, but his most fundamental work was in algebra and number theory. He was the first person to define what is meant by a class field and proved the fundamental analytical results concerning them. Takagi's work on class field theory was built on the foundations erected by Weber. He was deeply interested in the Kronecker *Jugendtraum* and completed the work of Kronecker; his work showed how all the abelian extensions of a quadratic imaginary field could be constructed using special values of modular functions and other explicit transcendental functions.

Weber wrote a famous three-volume work, *Lehrbuch der Algebra*, which was extremely influential. He was a great teacher and took great interest in pedagogical questions. His last academic position was in Strasbourg, where he died in 1913.

Andre Weil (1906-1998). One of the greatest mathematicians of the 20th century and a creator of modern algebraic geometry and arithmetic, Weil has had an astonishing influence on the development of mathematics in our times. He understood the deep links between geometry and number theory and proved the analogue of the Riemann hypothesis for smooth projective curves over a finite field. This is a result about the structure of the zeta function of such a curve that had been introduced by Emil Artin earlier. But to build his solution, Weil had to develop the theory of abstract algebraic varieties over any field, thus carrying algebraic geometry into entirely new frontiers. He then conjectured the corresponding result for higher-dimensional varieties. This set of conjectures, known as the *Weil conjectures*, was the driving force behind Grothendieck's theory of schemes and their cohomology. They were eventually proved by Deligne working in the Grothendieck framework. He also pioneered, along with Chevalley, the adelic approach to arithmetical problems. Finally, he inaugurated the study of holomorphic vector bundles on compact Riemann surfaces as a generalization of the abelian functions, a subject that has been intensively developed by the Indian school of geometers subsequently.

Weil was the first to study harmonic analysis on arbitrary locally compact abelian groups, establishing the complete theory of Fourier transform on such groups, including the Plancherel formula. Eventually Tate, at the suggestion of Artin, used this theory in the form of the Poisson summation formula on the adele ring of a number field to establish the analytic continuation and functional equations of all the arithmetical Dirichlet series known at that time, including those associated by Hecke to the Grössencharakters.

Weil's mathematical interests were even wider than his contributions suggest. He imposed no artificial boundaries on learning and discovery in mathematics. For years he ran a seminar at the Institute for Advanced Study under the rubric *mathematics*, as well as a seminar on current literature jointly with Princeton University. His criticism was feared, but he could be very kind also.

In his later years he occupied himself with historical writings. He brought a certain style of writing that paid far greater attention than his predecessors to the evolution of ideas. A splendid example of this is his book on the history of number theory, often referred to in this volume.

Weil's personality was many-sided. He studied Sanskrit as a student in Paris under Sylvain Lévi, and his first job was in India. During the two years he spent in India he dipped deeply into its culture and philosophy. Later on, when he had to decide whether to enlist in the war or not, he cited the *Bhagavad Gita* as the source that inspired him to refuse to serve. Eventually he had to serve a longish term in jail because of this. It was from the prison that he wrote his famous note announcing his discovery of the proof of the Riemann Hypothesis for smooth curves over a finite field. His thoughts during that period of incarceration are movingly described in his autobiography, *The Apprenticeship of a Mathematician.*[*]

Hermann Weyl (1885-1955). Arguably the greatest mathematician of the first half of the 20th century, Weyl was one of the last great universalists. His work straddled many fields of mathematics, just like that of his teacher, Hilbert.

His habilitation thesis was on the eigenfunction expansions of singular ordinary differential operators. In this seminal and profound work he created some of the most famous concepts of the subject (limit point and limit circle, Green's functions, etc.) and anticipated by two decades the work of von Neumann on the spectral resolution of unbounded self adjoint operators. Later he obtained his famous asymptotic law of the distribution of eigenvalues of a membrane. Eventually this area would bloom into a very rich and active one where the asymptotics of the

---

[*]See the review by V. S. Varadarajan, Notices of the AMS, 46 (1999), 448-456.

eigenvalues of the Laplacian acting on the exterior forms on a compact manifold (with or without boundary) and the relation of these asymptotics to the topology of the manifold would be the focus of intense study.

It was some time after these contributions that Weyl became attracted to physics at a fundamental level. First his interest went to relativity and differential geometry, where he discovered the concept of an affine connection. His theory of what are now called Weyl geometries was the first attempt at the unification of gravity and electrodynamics, which still remains today, after so many other efforts, unrivaled in simplicity and beauty. His greatest contribution in this area of physics and differential geometry is perhaps the discovery of gauge theory. His idea of electromagnetism as a U(1)-connection in a principal bundle later led directly, at the hands of Yang and Mills, to nonabelian gauge theory and the Yang-Mills equation and is the cornerstone of all theories of elementary particles today. In quantum theory he was among the first to formulate the commutation rules of Heisenberg, not just infinitesimally, but globally, in integrated form. He was the first to emphasize that the new particle predicted by Dirac's theory of antimatter must have the same mass as the electron, and he foresaw the violation of the parity principle.

In the middle of all of this, in the mid 1920's, Weyl wrote what many view as his greatest papers, namely those on the representations of compact Lie groups. By methods that were uniquely original at that time (although some beginnings may perhaps be discerned in the work of Issai Schur), he determined, among other things, all the irreducible characters of a compact Lie group and encoded them in a single formula, the *Weyl character formula*. It took almost 50 years before Harish-Chandra generalized and extended Weyl's work to all semisimple Lie groups by using deep methods of differential and global analysis.

Weyl is one of the very few mathematicians whose ideas and contributions deeply influenced physics and physicists in fundamental ways.[*]

He wrote beautifully. His books *Classical Groups*, *The Concept of a Riemann Surface*, *Theory of Groups and Quantum Mechanics*, and *Space, Time, and Matter* are great classics of the mathematics and physics of our times and have educated an uncounted number of mathematicians and physicists. Some of his lines have become legendary; here is a sample of them (all except the last are from his book *Classical Groups*; the last from his obituary of Hilbert cited in the section on Hilbert).

> *Only with the spinors do we strike that level in the theory of its representations on which, Euclid himself, flourishing ruler and compass, so deftly moves in the realm of geometrical figures. In some way Euclid's geometry must be deeply connected with the existence of the spin representation.*

> *And the determinant comes in, one does not know whence, as a* deus ex machina.

> *The gods have imposed on my writing the yoke of a foreign tongue that was not sung at my cradle.*

---

[*]See the account by C. N. Yang in *Hermann Weyl 1885-1955: Centenary Lectures*, K. Chandrasekharan (Ed.), Springer-Verlag, 1986.

The stringent precision attainable for mathematical thought has led many authors to a mode of writing which must give the reader an impression of being shut up in a brightly illuminated cell where every detail sticks out with the same dazzling clarity, but without relief. I prefer the open landscape under a clear sky with its depth of perspective, where the wealth of sharply defined nearby details gradually fades away toward the horizon.

Men like Einstein or Niels Bohr grope their way in the dark toward their conceptions of general relativity or atomic structure by another type of experience and imagination than those of the mathematician, although no doubt mathematics is an essential ingredient.

Norbert Wiener (1894-1964). A giant of modern analysis, creator of ideas, a mathematician with a compass that spanned all aspects of the sciences, both physical and social, Wiener is a towering figure in modern mathematics. He discovered what is now called Wiener measure, a probability measure on the space of continuous paths that models brownian motion, more than a decade before Kolmogorov laid the foundations of the modern theory of probability. Inspired by the remark of Jean Perrin, one of the experimental pioneers in the theory of brownian motion, he proved that the Wiener measure gives probability one to the nowhere differentiable paths. Under the influence of Feynman and Kac, the Wiener measure has become a preeminent technical tool in the study of quantum mechanics and quantum field theory. He discovered the Wiener Tauberian theorem, which subsumed all the major Tauberian theorems and led to his proof of the prime number theorem. He was the codiscoverer, along with Kolmogorov, of the prediction theory of stationary stochastic processes.

If this were all he did, Wiener would already be regarded with great reverence and admiration, but he went much further. He created the modern theory of data analysis and, more importantly, created *cybernetics*, the science of the interaction of man and machine. He foresaw the problems that would arise in the reckless automation of human tasks and was a champion of the idea of *the human use of human beings*. Perhaps he is one of the last figures in mathematics who, by the grandeur of their mathematical contributions as well as the scope of their humanistic ideas, command universal respect and admiration.*

---

*For his work see *Norbert Wiener, Collected Works*, The MIT Press, 1979. For a complete perspective of Wiener's work and its impact, the reader should consult in addition to his *Collected Works* cited above the biography of Wiener by P. Masani: *Norbert Wiener: 1894-1964*, Vita Mathematica, 5, Birkhäuser, 1990.

Sample Pages from *Opera Omnia*

THEOREMA

9. *Dico igitur huius aequationis differentialis*

$$\frac{dx}{\sqrt{(1-x^4)}} = \frac{dy}{\sqrt{(1-y^4)}}$$

aequationem integralem completam esse

$$xx + yy + ccxxyy = cc + 2xy\sqrt{(1-c^4)}.$$

DEMONSTRATIO

Posita enim hac aequatione eius differentiale erit

$$xdx + ydy + ccxy(xdy + ydx) = (xdy + ydx)\sqrt{(1-c^4)},$$

unde fit

$$dx(x + ccxyy - y\sqrt{(1-c^4)}) + dy(y + ccxxy - x\sqrt{(1-c^4)}) = 0.$$

Ex eadem vero aequatione resoluta colligitur

$$y = \frac{x\sqrt{(1-c^4)} + c\sqrt{(1-x^4)}}{1 + ccxx} \quad \text{et} \quad x = \frac{y\sqrt{(1-c^4)} - c\sqrt{(1-y^4)}}{1 + ccyy}.$$

Si enim ibi radicali $\sqrt{(1-x^4)}$ tribuitur signum $+$, hic radicali $\sqrt{(1-y^4)}$ signum $-$ tribui debet, ut posito $x = 0$ utrinque idem valor prodeat $y = c$. Erit ergo

$$x + ccxyy - y\sqrt{(1-c^4)} = -c\sqrt{(1-y^4)},$$

$$y + ccxxy - x\sqrt{(1-c^4)} = \quad c\sqrt{(1-x^4)},$$

quibus valoribus in aequatione differentiali substitutis prodit

$$-cdx\sqrt{(1-y^4)} + cdy\sqrt{(1-x^4)} = 0$$

sive

$$\frac{dx}{\sqrt{(1-x^4)}} = \frac{dy}{\sqrt{(1-y^4)}}.$$

Huius ergo aequationis differentialis integrale est

$$xx + yy + ccxxyy = cc + 2xy\sqrt{(1-c^4)},$$

et quia constantem c ab arbitrio nostro pendentem continet, erit simul integrale completum. Q. E. D.

Addition theorem for elliptic integrals, *Opera Omnia*, I-20, p. 63. See Chapter 2, Section 3.

DE SUMMIS SERIERUM RECIPROCARUM [1]

Commentatio 41 indicis Enestroemiani
Commentarii academiae scientiarum Petropolitanae 7 (1734/5), 1740, p. 123—134

1. Tantopere iam pertractatae et investigatae sunt series reciprocae potestatum numerorum naturalium, ut vix probabile videatur de iis novi quicquam inveniri posse. Quicunque enim de summis serierum meditati sunt, ii fere omnes quoque in summas huiusmodi serierum inquisiverunt neque tamen ulla methodo eas idoneo modo exprimere potuerunt. Ego etiam iam saepius, cum varias summandi methodos tradidissem, has series diligenter sum persecutus neque tamen quicquam aliud sum assecutus, nisi ut earum summam vel proxime veram definiverim vel ad quadraturas curvarum maxime transcendentium reduxerim; quorum illud in dissertatione proxime praelecta[2]), hoc vero in praecedentibus[3]) praestiti. Loquor hic autem de seriebus fractionum, quarum numeratores sunt 1, denominatores vero vel quadrata vel cubi vel aliae dignitates numerorum naturalium; cuius modi sunt

$$1 + \frac{1}{4} + \frac{1}{9} + \frac{1}{16} + \frac{1}{25} + \text{etc.},$$

item

$$1 + \frac{1}{8} + \frac{1}{27} + \frac{1}{64} + \text{etc.}$$

atque similes superiorum potestatum, quarum termini generales continentur in hac forma $\frac{1}{x^n}$.

---

1) Confer hac cum dissertatione Commentationes 61, 63, 130 huius voluminis nec non dissertationem a P. Stäckel (1862—1919) scriptam *Eine vergessene Abhandlung Leonhard Eulers über die Summen der reziproken Quadrate der natürlichen Zahlen*, quae ut introductio in Commentationem 63 modo laudatam itidem in hoc volumine invenitur.   C. B.

2) Vide Commentationem 46 huius voluminis.   C. B.

3) Vide Commentationes 20 et 25 huius voluminis.   C. B.

Beginning of the paper in which Euler obtains the value of $\zeta(2)$,
Opera Omnia, I-14, p. 73. See Chapter 3, Section 1.

équivalente à celle-là pour le cas $n = 0$; or, celle-cy nous donne ouvertement la même valeur $2l2$ que la nature des séries exige. Voilà donc une nouvelle preuve qui, étant jointe à la précédente, pourra bien tenir lieu d'une démonstration complette de notre conjecture. Cependant, on n'est que trop autorisé d'en exiger encore une démonstration directe qui renferme à la fois tous les cas possibles.

13. Notre conjecture étant donc juste pour tous les cas où n est un nombre entier positif, je m'en vai prouver à présent qu'elle est également d'accord avec la vérité, lorsqu'on prend pour n un nombre entier négatif quelconque. Or, dans ces cas la valeur de la formule $1 \cdot 2 \cdot 3 \cdots (n-1)$ devient infinie, ce qui semble troubler la démonstration que j'ai en vue; mais une observation que j'ai prouvée ailleurs[1]) levera cet obstacle. Prenant ce signe $[\lambda]$ pour marquer ce produit $1 \cdot 2 \cdot 3 \cdots \lambda$, j'ai démontré qu'il y a toujours

$$[\lambda]\,[-\lambda] = \frac{\lambda\pi}{\sin.\,\lambda\pi};$$

donc, posant $n - 1 = -m$ ou $n = -m + 1$ pour avoir cette expression

$$\frac{1 - 2^{-m} + 3^{-m} - 4^{-m} + 5^{-m} - 6^{-m} + \text{etc.}}{1 - 2^{m-1} + 3^{m-1} - 4^{m-1} + 5^{m-1} - 6^{m-1} + \text{etc.}}$$

$$= \frac{-1 \cdot 2 \cdot 3 \cdots (-m)\,(2^{-m+1} - 1)}{(2^{-m} - 1)\,\pi^{-m+1}} \cos.\,\frac{(1-m)\pi}{2},$$

où, puisque

$$1 \cdot 2 \cdot 3 \cdots (-m) = [-m] \quad \text{et} \quad [m]\,[-m] = \frac{m\pi}{\sin.\,m\pi},$$

nous aurons

$$1 \cdot 2 \cdot 3 \cdots (-m) = \frac{m\pi}{1 \cdot 2 \cdot 3 \cdots m \,\sin.\,m\pi} = \frac{\pi}{1 \cdot 2 \cdot 3 \cdots (m-1)\,\sin.\,m\pi}.$$

Ensuite on sait que

$$\cos.\,\frac{(1-m)\pi}{2} = \sin.\,\frac{m\pi}{2},$$

---

1) Voir les mémoires 19 (§ 1) et 128 (§ 10) du volume précédent. Voir aussi le mémoire 421 (suivant l'Index d'ENESTRÖM): *Evolutio formulae integralis $\int x^{f-1} dx (lx)^{\frac{m}{n}}$ integratione a valore $x = 0$ ad $x = 1$ extensa,* Novi comment. acad. sc. Petrop. 16 (1771), 1772, p. 91—139, surtout le § 43; *LEONHARDI EULERI Opera omnia,* series I, vol. 17, p. 316—357.   G. F.

Functional equation of the zeta function, *Opera Omnia*, I-15, p. 82. See Chapter 5, Section 2.

Est autem

$$\cos. \frac{2k+1}{i}\pi = 1 - \frac{(2k+1)^2}{2ii}\pi\pi,$$

unde forma factoris erit

$$\frac{4xx}{ii} + \frac{(2k+1)^2}{ii}\pi\pi$$

evanescente termino, cuius denominator est i^4. Quoniam ergo omnis factor expressionis

$$1 + \frac{xx}{1\cdot 2} + \frac{x^4}{1\cdot 2\cdot 3\cdot 4} + \text{etc.}$$

huiusmodi formam habere debet $1 + \alpha xx$, quo factor inventus ad hanc formam reducatur, dividi debet per $\frac{(2k+1)^2\pi^2}{ii}$; hinc factor formae propositae erit

$$= 1 + \frac{4xx}{(2k+1)^2\pi\pi}$$

ex eoque omnes factores infiniti invenientur, si loco $2k+1$ successive omnes numeri impares substituantur. Hanc ob rem erit

$$\frac{e^x + e^{-x}}{2} = 1 + \frac{xx}{1\cdot 2} + \frac{x^4}{1\cdot 2\cdot 3\cdot 4} + \frac{x^6}{1\cdot 2\cdot 3\cdot 4\cdot 5\cdot 6} + \text{etc.}$$

$$= \left(1 + \frac{4xx}{\pi\pi}\right)\left(1 + \frac{4xx}{9\pi\pi}\right)\left(1 + \frac{4xx}{25\pi\pi}\right)\left(1 + \frac{4xx}{49\pi\pi}\right) \text{ etc.}$$

158. Si x fiat quantitas imaginaria, formulae hae exponentiales in sinum et cosinum cuiuspiam arcus realis abeunt. Sit enim $x = z\sqrt{-1}$; erit

$$\frac{e^{z\sqrt{-1}} - e^{-z\sqrt{-1}}}{2\sqrt{-1}} = \sin. z$$

$$= z - \frac{z^3}{1\cdot 2\cdot 3} + \frac{z^5}{1\cdot 2\cdot 3\cdot 4\cdot 5} - \frac{z^7}{1\cdot 2\cdot 3\cdots 7} + \text{etc.},$$

quae adeo expressio hos habet factores numero infinitos

$$z\left(1 - \frac{zz}{\pi\pi}\right)\left(1 - \frac{zz}{4\pi\pi}\right)\left(1 - \frac{zz}{9\pi\pi}\right)\left(1 - \frac{zz}{16\pi\pi}\right)\left(1 - \frac{zz}{25\pi\pi}\right) \text{ etc.},$$

seu erit

$$\sin. z = z\left(1 - \frac{z}{\pi}\right)\left(1 + \frac{z}{\pi}\right)\left(1 - \frac{z}{2\pi}\right)\left(1 + \frac{z}{2\pi}\right)\left(1 - \frac{z}{3\pi}\right)\left(1 + \frac{z}{3\pi}\right) \text{ etc.}$$

Infinite product for $\sin z$, *Opera Omnia*, I-8, p. 168.
See Chapter 3, Section 4.

Index